Seventh Edition

Anatomy & Physiology

LABORATORY TEXTBOOK • ESSENTIALS VERSION

JASON LAPRES
Lone Star College

STANLEY E. GUNSTREAM
Pasadena City College

HAROLD J. BENSON
Pasadena City College

ARTHUR TALARO
Pasadena City College

KATHLEEN P. TALARO
Pasadena City College

ANATOMY & PHYSIOLOGY LABORATORY TEXTBOOK ESSENTIALS VERSION, SEVENTH EDITION

Published by McGraw-Hill Education, 2 Penn Plaza, New York, NY 10121. Copyright ©2021 by McGraw-Hill Education. All rights reserved. Printed in the United States of America. Previous editions ©2013, 2010, and 2007. No part of this publication may be reproduced or distributed in any form or by any means, or stored in a database or retrieval system, without the prior written consent of McGraw-Hill Education, including, but not limited to, in any network or other electronic storage or transmission, or broadcast for distance learning.

Some ancillaries, including electronic and print components, may not be available to customers outside the United States.

This book is printed on acid-free paper.

1 2 3 4 5 6 7 8 9 LMN 22 21 20

ISBN 978-0-07-809727-0
MHID 0-07-809727-4

Portfolio Manager: *Matthew Garcia*
Product Developer: *Erin DeHeck*
Marketing Manager: *Valerie Kramer*
Content Project Manager: *Jeni McAtee*
Buyer: *Laura Fuller*
Design: *David W. Hash*
Content Licensing Specialist: *Gina Oberbroeckling*
Cover Image: *©Tetra Images/Getty Images*
Compositor: *MPS Limited*

All credits appearing on page or at the end of the book are considered to be an extension of the copyright page.

The Internet addresses listed in the text were accurate at the time of publication. The inclusion of a website does not indicate an endorsement by the authors or McGraw-Hill Education, and McGraw-Hill Education does not guarantee the accuracy of the information presented at these sites.

mheducation.com/highered

Contents

Preface v
To the Student vii

Unit 1
Introduction to Anatomy

1. Introduction to Human Anatomy and Body Organization 1
2. Microscopes 18

Unit 2
Cells and Tissues

3. Cellular Anatomy and Mitosis 24
4. Membrane Transport 30
5. Histology 35

Unit 3
Protection, Support, and Movement

6. The Integumentary System 53
7. Skeletal Overview 57
8. The Axial Skeleton 62
9. The Appendicular Skeleton 74
10. Articulations and Body Movements 79
11. Muscle Organization and Contractions 85
12. Axial Muscles 91
13. Appendicular Muscles 95

Unit 4
Control and Integration

14. The Spinal Cord and Reflex Arcs 105
15. The Brain 111
16. Senses 122
17. The Endocrine System 134

Unit 5
Internal Transport and Gas Exchange

18. Blood Tests 145
19. The Heart 153
20. Blood Vessels 160
21. The Lymphoid System 169
22. The Respiratory System 173

Unit 6
Nutrient Absorption and Water Balance

23. The Digestive System 181
24. The Urinary System 192

Unit 7
Reproduction

25. The Reproductive System 201

Laboratory Reports 211
Appendices 319

PREFACE

This *Essentials Version* of the *Anatomy & Physiology Laboratory Textbook* presents the fundamentals of human anatomy and physiology in a manner that is appropriate for students in allied health programs such as practical nursing, radiologic technology, medical assisting, and dental assisting. These students usually take a one-semester course in human anatomy and physiology and need a laboratory text that provides coverage of the fundamentals without the clutter of excessive details and unneeded terminology.

Changes to the New Edition

Based on the experience and the thoughtful comments of reviewers, the following overarching changes have been made to the seventh edition:

- Updated terminology throughout the lab manual for more current presentation of material.
- Reordered 38 exercises into a new format of 25 exercises.
- Included cautionary statements to prepare students for working with potentially dangerous materials in the laboratory.
- Moved the appropriate histological photomicrographs out of the atlas at the end of the manual and into each Exercise.
- Improved the organization of histology by presenting all four tissue types together in Exercise 5.

Exercise Format

Each of the 25 exercises begins with a short list of objectives that outline the minimal learning responsibilities for the students. The exercises are basically self-directing, which minimizes the need for lengthy introductions by the instructor.

Instructors will find that the listing of required equipment and materials on the first page of each exercise facilitates laboratory preparation. The exercises use standard equipment and materials that are usually available in most biology departments.

Each exercise topic is covered at a reduced level of difficulty appropriate for the student audience and presented in a direct, concise manner to facilitate student learning. Numerous illustrations and many photomicrographs are correlated with the text to improve understanding.

The necessary key terms are in bold print to aid students in building a vocabulary of anatomical and physiological terms.

Laboratory procedures are distinct from and follow the discussion of the exercise topic. They are presented in a concise, stepwise manner to guide the student. Activities consist of (1) labeling illustrations, (2) dissections, (3) study of specimens and models, (4) physiological experiments, and (5) microscopic studies. A major dissection specimen, such as the cat or fetal pig, is not included. Instead, a rat dissection is used in a single exercise to acquaint students with the basic organization of organ systems in mammals.

Pedagogical Design

The pedagogical design calls for students to develop an understanding of each exercise topic by (1) labeling the illustrations using information presented in the text and (2) completing corresponding portions of the laboratory report. Usually, students should complete these activities *before* coming to the laboratory or proceeding with the other laboratory studies.

The *laboratory reports* are designed to guide and reinforce student learning and provide a convenient site for recording data and making diagrams of observations. They are located at the back of the book to better control the pagination and to keep the book brief. The corresponding laboratory report should be removed from the back of the book whenever students begin working on an exercise. Removing the report prevents page flipping and facilitates completing the laboratory report. Completed laboratory reports should be kept in the student's notebook. Instructors may choose to either post the answer keys from the *Instructor's Manual* so the students can check their work or collect and grade the laboratory reports. The design of the laboratory reports makes them easy to grade.

Students in the target group typically have difficulty in learning the necessary anatomical and physiological terminology, especially the correct spellings. Experience has shown that students learn key terms and their correct spellings more easily if they write the terms. Thus, the laboratory reports are structured so that students are required to write the names of key terms rather than simply provide a matching letter or number.

Microscopic study is another difficult area for these students since they have trouble locating histological structures on slides when using diagrams as a reference. This problem is largely overcome by the inclusion of full-color photomicrographs of common tissues and of selected histological subjects, which have been reorganized to be within each exercise.

Supplements

Instructor's Manual

The accompanying *Instructor's Manual* provides additional aid for the Instructor: (1) a composite list of equipment and supplies, (2) operational suggestions, and (3) answer keys for the laboratory reports.

Adopters are encouraged to contact the author regarding any problems and to submit constructive suggestions that will improve future editions.

McGraw-Hill Anatomy & Physiology Revealed®

 www.aprevealed.com

McGraw-Hill Anatomy & Physiology Revealed® is an interactive cadaver dissection tool to enhance lecture and lab, which students can use anytime, anywhere. Detailed cadaver photographs blended together with a state-of-the-art layering technique provide a uniquely interactive dissection experience. Anatomy & Physiology Revealed can now be customized to your course. Choose the specific structures that you require in your course and APR 3.0 does the rest. Once your list is generated, APR 3.0 highlights your selected structures for students. New to this edition are APR icons indicating eBook links to APR to support selected topics.

Acknowledgments

The development and production of this seventh edition has been the result of a team effort. Our dedicated and creative teammates at McGraw-Hill have contributed greatly to the finished product. We gratefully acknowledge the efforts and support of Matt Garcia, Portfolio Manager, Erin DeHeck, Product Developer, Jeni McAtee and Brent dela Cruz, Project Mangers.

The following instructors have served as critical reviewers: Stephanie R. Allen, *University of Louisiana Monroe*; Malene Arnaud, *Delgado Community College*; Barry Bates, *Atlanta Technical College*; Melody Bell, *Vernon College*; Ann M. Findley, *University of Louisiana at Monroe*; Pamela B. Jackson, *Piedmont Technical College*; Sarah Mangum, *Pitt Community College*; Sharon Hall Murff, *Grambling State University*; Susan Rohde, *Triton College*; Nicholas E. Smith, *Mountwest Community and Technical College*; and Adam C. Swolsky, *Mountwest Community and Technical College*.

To the Student

This laboratory textbook has been designed to help you master the fundamentals of human anatomy and physiology, which provide the basis of the health-related professions. An understanding of anatomical structures, physiological processes, terminology, and techniques will give you the basic knowledge that is essential for success in your chosen field.

Each exercise begins with a list of learning objectives to guide your study. The sequence of activities in each exercise is established to facilitate the development of your understanding.

The activities consist of (1) labeling illustrations, (2) dissections, (3) study of specimens and models, (4) physiological experiments, and (5) microscopic study. Follow the directions for each activity with care to enhance your success in the laboratory. Work carefully and thoughtfully. Remember, the objective is to learn the material, not just to complete the exercise.

Each exercise has a *laboratory report,* located in the back of the book, that you are to complete as you work through the exercise. Remove it from the book when you start the exercise so that you can complete it without page flipping. Completed laboratory reports may be kept in your notebook.

Safety

A list of safety guidelines is included in the inside front cover.

Some of the exercises contain a *Before You Proceed* box following the material section. Be sure to check with your instructor about using protective disposable gloves where appropriate.

Labeling Illustrations

Most of the exercises are designed so that you will learn anatomy by labeling illustrations from the information provided in the text. This process helps you to understand the relationships between anatomical features, and it facilitates learning. Correctly labeled illustrations are then used as references when examining specimens and models. Use them for study and review purposes. Usually, the illustrations should be labeled *before* coming to the laboratory session.

Dissections

The dissections will include freshly killed rats and animal parts obtained from a slaughterhouse. The purpose of dissection is to expose anatomical parts for observation and study. Strive to cut as little as possible to achieve this goal. Use the scalpel sparingly. Most dissections can be accomplished with scissors, forceps, and probe.

Experiments

Before performing any experiment, read the directions completely so that you have a good understanding of the experiment. Be sure that you have all of the equipment and materials required to complete the experiment before starting. Then, carefully follow the directions.

Microscopic Study

An examination of prepared slides enables you to visualize and understand the cytological or histological structure of specimens. Correlate your observations with the text, diagrams, and histological photomicrographs. If drawings are required, make them with care and label the structures. In this way, you will understand the microscopic structures more quickly and better prepare yourself for lab practicums.

Laboratory Reports

Complete the laboratory reports independently to maximize your understanding. The laboratory reports are purposely designed so that you must write the key terms involved in the exercise, sometimes more than once. This format is used because writing the terms enables you to learn them more easily, especially their correct spellings.

General Operations

Success in the laboratory can be increased by following a few simple guidelines.

1. Be on time to class so that you will hear the comments and directions given by your instructor.

2. Follow your instructor's directions. Take notes on any changes in the equipment, materials, or procedures.
3. Keep your work area free of clutter. Extraneous items should be located elsewhere during the laboratory session.
4. Bring your textbook to the laboratory for reference.
5. Use equipment and materials with care. Report any problems or accidents to your instructor immediately.
6. Work independently, but be cooperative and helpful in team assignments.
7. Do not eat, drink, or smoke in the laboratory.

Exercise 1

INTRODUCTION TO HUMAN ANATOMY AND BODY ORGANIZATION

Objectives

After completing this exercise, you should be able to

1. Correctly use directional and regional terms in describing the location of anatomical structures.
2. Identify the external body regions and body cavities on charts or a torso model.
3. Contrast the types of body sections used in anatomical studies.
4. Describe the organs and membranes of the body cavities.
5. List the organ systems and describe their general functions.
6. List the component organs for each organ system.
7. Locate and identify the organs in the ventral cavity of the rat.

Materials
Human torso model with removable organs
Freshly killed or preserved rat
Dissecting pan with wax bottom
Dissecting instruments and pins
Protective disposable gloves

Before You Proceed

Consult with your instructor about using protective disposable gloves when performing portions of this exercise.

Human anatomy is the study of the structures composing the human body and their interrelationships. The description of anatomical features requires the use of specific *anatomical terminology* that provides precise meaning. It is important that you learn the terminology presented in this exercise as quickly as possible because these terms will be used frequently throughout your study of the human body. In this exercise, you will be introduced to the anatomical terminology that is used to describe (1) relative positions of structures, (2) body planes and sections, (3) body regions, (4) body cavities and their membranes, and (5) body systems.

The human body is composed of a hierarchy of organizational levels. From simplest to most complex, they are chemical, cellular, tissue, organ, and organ system. See Figure 1.1. Components of each organizational level work together at an organismal level to maintain **homeostasis**, the relative constancy of the body's internal environment. Homeostatic mechanisms enable the body to maintain its own stable internal environment.

Directional Terms

The relative position of body structures is communicated through the use of **directional terms.** These terms typically occur as pairs, with the members of each pair having opposite meanings. For example, *anterior (ventral)* means toward the front of the body, and *posterior (dorsal)* means toward the back of the body. Most directional terms apply to organs of the body as well as to the body as a whole. Learn the meanings of the directional terms in Table 1.1.

When directional terms are used to describe the relative positions of body parts, it is assumed that the body is in a standard position called the **anatomical position.** In this position, the body is erect with upper limbs at the sides and with the palms of the hands and the feet facing forward. The anatomical position is shown in Figure 1.2.

Unit 1: Introduction to Anatomy

Figure 1.1 Levels of organization. (young woman) ONOKY - Fabrice LEROUGE/Brand X Pictures/Getty Images

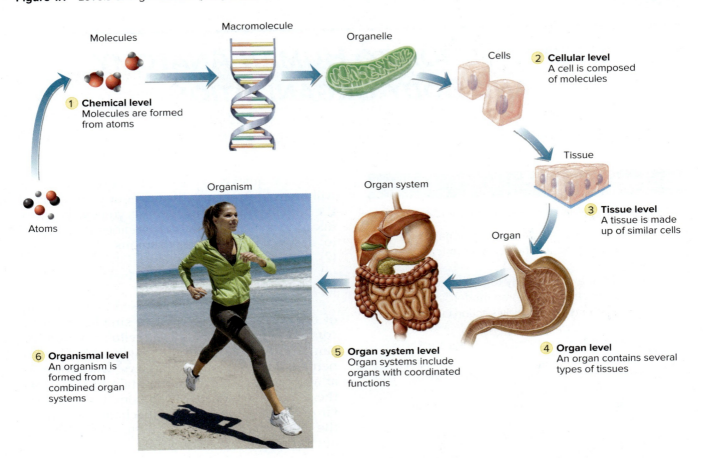

TABLE 1.1
Directional Terms

Term	Meaning
Anterior (ventral)	Toward the front or abdominal surface of the body
Posterior (dorsal)	Toward the back of the body
Superior (cephalic)	Toward the head
Inferior (caudal)	Away from the head
Medial	Toward the midline of the body
Lateral	Away from the midline of the body
Proximal	Closer to the beginning
Distal	Farther from the beginning
Superficial (external)	Toward or on the body surface
Deep (internal)	Away from the body surface
Parietal	Pertaining to the outer boundary of body cavities
Visceral	Pertaining to the internal organs

Body Planes and Sections

In studying the body or organs, you often will be observing the flat surface of a **section** that has been produced by a cut through the body or a body part. Such sections are made along specific **planes.** See Figure 1.2.

Sagittal planes are parallel to the longitudinal (long) axis of the body and divide the body into left and right portions. A **median plane** is made along the **median line** and divides the body into *equal* left and right halves.

Frontal (coronal) **planes** divide the body into front and back portions and are perpendicular to sagittal sections but parallel to the longitudinal axis.

Transverse (horizontal) **planes** divide the body into top and bottom portions and are perpendicular to the longitudinal axis of the body.

Assignment

Check your understanding of directional terms, body planes, and body surfaces by labeling Figure 1.2. Place the correct number of each plane and surface in the space in front of the labels listed. Record your responses for Figure 1.2 on Laboratory Report 1.

Introduction to Anatomy

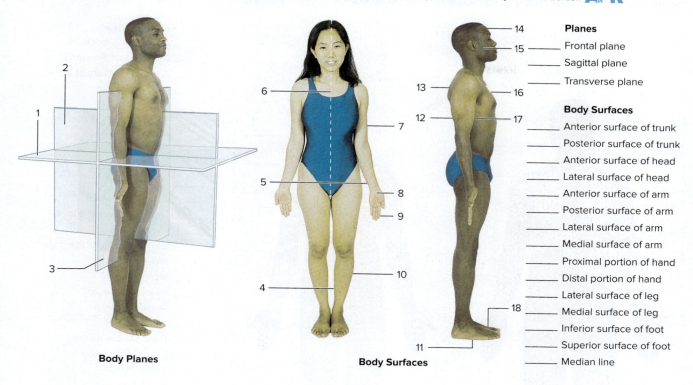

Figure 1.2 Body planes and surfaces. Note that the body is in the anatomical position. Courtesy of Harold Benson

Regional Terminology

The human body consists of an **axial portion,** the head, neck, and trunk, and an **appendicular portion,** the upper and lower limbs and their girdles. These major divisions are subdivided into smaller regions that are identified by specific anatomical terms. Except for the small regions of head and neck, these anatomical regions are shown in Figure 1.3.

Head and Neck

The **head** and **neck** contain several small regions whose names will assist you in learning bones, muscles, and other structures. **Cephalic** simply refers to the entire head. It can be subdivided into the **cranial region,** which refers to the skull, and the **facial region,** or simply the face. The facial region is further subdivided into the **buccal region** (cheeks), **nasal region** (nose), **oral region** (mouth), **orbital region** (eyes), and **otic region** (ears). The anatomical term for the next is the **cervical region.**

Trunk

The front of the trunk consists of an upper **thoracic region** (chest), which is subdivided into two **pectoral regions** separated by a central **sternal region**. The **abdominal region** is below the thoracic region. Two **inguinal** (groin) **regions** lie at the junction of the trunk with the lower limbs. The **pubic region** lies between the inguinal regions. The external reproductive organs in both sexes make up the **genital region.**

The back of the trunk is the **dorsal region.** It contains two **scapular regions,** which are located over the shoulder blades. An elongated **vertebral region** lies along the vertebral column. The small **sacral region** lies below the vertebral region and between the two **gluteal regions** (buttocks). A **lumbar region** is located on each side of the vertebral region below the ribs.

The **axillary regions** (armpits) and the **coxal regions** (hips) lie on the lateral surfaces of the trunk.

Upper Limb

The **deltoid region** (shoulder) is formed by the attachment of each upper limb to the trunk. The arm composes the **brachial region,** and the forearm composes the **antebrachial region.** The elbow forms the **cubital region,** and the front of the elbow is the **antecubital region.** The **carpal region** (wrist), the **manual region** (hand), and the **digital region** (fingers) complete the regions of each upper limb.

Introduction to Human Anatomy and Body Organization

Figure 1.3 Regions of the body. Common names are noted in parentheses. **APR**

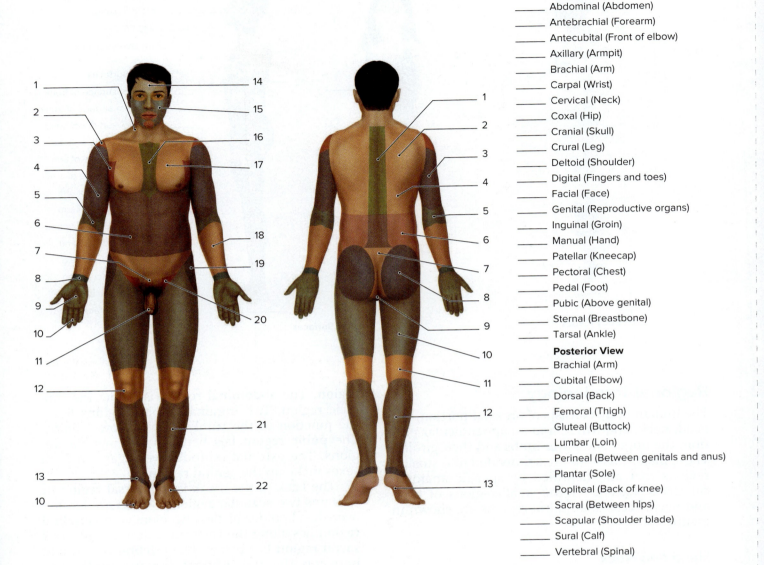

Anterior View
- Abdominal (Abdomen)
- Antebrachial (Forearm)
- Antecubital (Front of elbow)
- Axillary (Armpit)
- Brachial (Arm)
- Carpal (Wrist)
- Cervical (Neck)
- Coxal (Hip)
- Cranial (Skull)
- Crural (Leg)
- Deltoid (Shoulder)
- Digital (Fingers and toes)
- Facial (Face)
- Genital (Reproductive organs)
- Inguinal (Groin)
- Manual (Hand)
- Patellar (Kneecap)
- Pectoral (Chest)
- Pedal (Foot)
- Pubic (Above genital)
- Sternal (Breastbone)
- Tarsal (Ankle)

Posterior View
- Brachial (Arm)
- Cubital (Elbow)
- Dorsal (Back)
- Femoral (Thigh)
- Gluteal (Buttock)
- Lumbar (Loin)
- Perineal (Between genitals and anus)
- Plantar (Sole)
- Popliteal (Back of knee)
- Sacral (Between hips)
- Scapular (Shoulder blade)
- Sural (Calf)
- Vertebral (Spinal)

Lower Limb

The proximal part of the lower limb, the thigh, constitutes the **femoral region,** and the leg is the portion between the knee and ankle. The front of the leg, or shin, is the **crural region;** the back of the leg, or calf, is the **sural region.** The front of the knee is the **patellar region,** and the back of the knee is the **popliteal region.** The **tarsal region** (ankle), **pedal region** (foot), **calcaneal region** (heel), and **digital region** (toes) complete the regions of each lower limb. **Plantar region** (sole) refers specifically to the bottom of the foot.

Abdominal Divisions

Underlying internal organs may be located by dividing the abdominal surface into either nine regions or four quadrants. Anatomists prefer to divide the abdominal surface into nine regions formed by the intersecting of two vertical and two transverse planes as shown in Figure 1.4.

The **umbilical region** is the central area that includes the umbilicus (navel). On either side of the umbilical region are the left and right **flank regions.** Just above the umbilical region is the **epigastric region,** which is between the left and right **hypochondriac regions.** Below the umbilical region is the **hypogastric** (pubic) **region,** which is between the left and right **inguinal regions.**

Health-care professionals often use a simpler system to subdivide the abdominal surface into quadrants by perpendicular vertical and transverse planes that intersect at the umbilicus.

4 Introduction to Anatomy

Figure 1.4 Abdominal regions and quadrants.

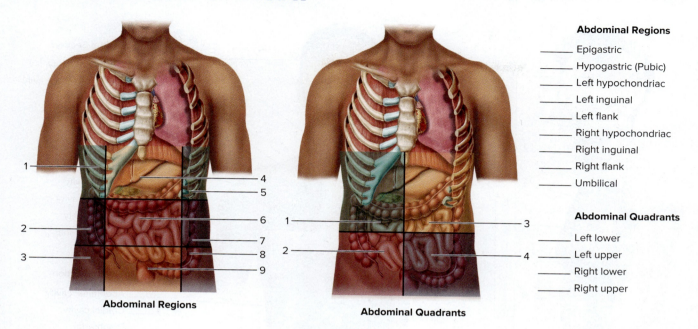

Abdominal Regions
____ Epigastric
____ Hypogastric (Pubic)
____ Left hypochondriac
____ Left inguinal
____ Left flank
____ Right hypochondriac
____ Right inguinal
____ Right flank
____ Umbilical

Abdominal Quadrants
____ Left lower
____ Left upper
____ Right lower
____ Right upper

See Figure 1.4. This division forms four quadrants: **upper right, upper left, lower right,** and **lower left.**

Assignment
Label Figures 1.3 and 1.4 and record your responses on the laboratory report.

Body Cavities

Internal organs are located within body cavities. There are two major body cavities, the dorsal aspect and the ventral cavity, as shown in Figure 1.5. These cavities are named after the body surface nearest to them.

The **dorsal aspect** consists of the **cranial cavity,** which contains the brain, and the **vertebral canal,** which contains the spinal cord.

The **ventral cavity** consists of the **thoracic cavity,** which houses the lungs, heart, and other thoracic organs, and the **abdominopelvic cavity,** which contains most of the internal organs. A thin, dome-shaped sheet of muscle, the **diaphragm,** separates the thoracic and abdominopelvic cavities.

The thoracic cavity is divided into left and right portions by the **mediastinum,** a membranous partition that contains the heart, trachea, esophagus, and thymus. The **pericardial cavity,** containing the heart, is within the mediastinum. The right and left **pleural cavities** contain the lungs.

The abdominopelvic cavity consists of (1) the **abdominal cavity,** which contains the stomach, intestines, liver, gallbladder, pancreas, spleen, and kidneys, and (2) the **pelvic cavity,** which contains the urinary bladder, sigmoid colon, and rectum. In females, it also contains the uterus, uterine tubes, and ovaries.

Membranes of Body Cavities

The body cavities are lined with membranes that provide protection and support for the internal organs. Refer to Figures 1.6 and 1.7 as you read this section.

Dorsal Aspect Membranes

The dorsal aspect is lined by three membranes that are collectively called the **meninges.** The outermost membrane is tough and fibrous; the other two membranes are more delicate. The thin innermost membrane closely envelops the brain and spinal cord. The meninges will be considered later when you study the nervous system.

Ventral Cavity Membranes

The membranes lining the ventral cavity are **serous membranes** that secrete a watery fluid generically

Introduction to Human Anatomy and Body Organization

Figure 1.5 Body cavities.

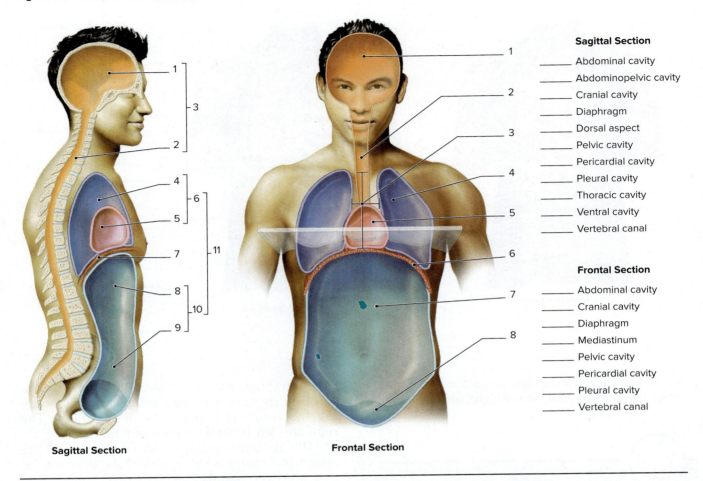

Sagittal Section

Frontal Section

Sagittal Section
- Abdominal cavity
- Abdominopelvic cavity
- Cranial cavity
- Diaphragm
- Dorsal aspect
- Pelvic cavity
- Pericardial cavity
- Pleural cavity
- Thoracic cavity
- Ventral cavity
- Vertebral canal

Frontal Section
- Abdominal cavity
- Cranial cavity
- Diaphragm
- Mediastinum
- Pelvic cavity
- Pericardial cavity
- Pleural cavity
- Vertebral canal

Figure 1.6 Thoracic membranes.

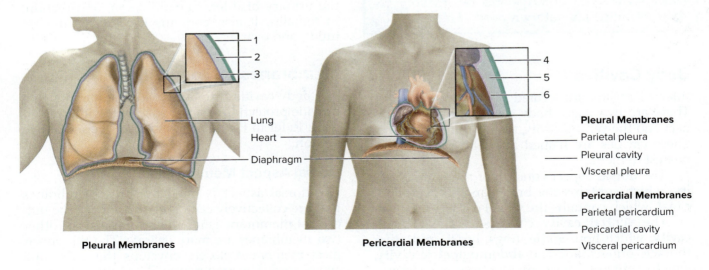

Pleural Membranes

Pericardial Membranes

Pleural Membranes
- Parietal pleura
- Pleural cavity
- Visceral pleura

Pericardial Membranes
- Parietal pericardium
- Pericardial cavity
- Visceral pericardium

called *serous fluid.* Each cavity's fluid has its own specific name that will be discussed with each membrane next. Their moist, slippery surfaces reduce friction between internal organs and the walls of the ventral cavity as the organs move.

Thoracic Cavity Membranes
The inner walls of the left and right portions of the thoracic cavity are lined by the **parietal pleurae;** the lungs are covered with the **visceral pleurae.** The parietal and visceral pleurae are

6   Introduction to Anatomy

Figure 1.7 Membranes of the abdominal cavity.

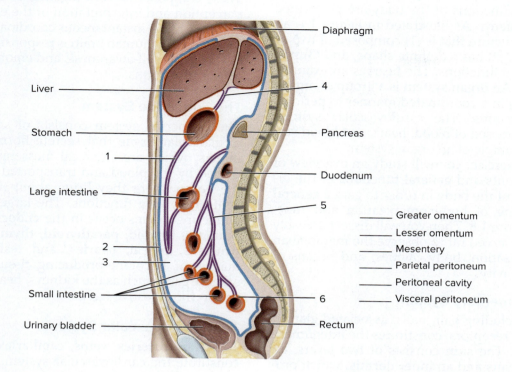

_____ Greater omentum
_____ Lesser omentum
_____ Mesentery
_____ Parietal peritoneum
_____ Peritoneal cavity
_____ Visceral peritoneum

separated only by a thin layer of **pleural fluid.** This moist space between the pleurae is known as the **pleural cavity.** The parietal pleurae are continuous with the membranes forming the mediastinum.

The heart is enclosed within the **serous pericardium.** The **visceral pericardium,** or **epicardium,** is a thin serous membrane that is tightly attached to the outer surface of the heart. The **parietal pericardium** lines the inside surface of a loosely fitting sac around the heart. The space between the visceral and parietal pericardia is the **pericardial cavity,** and it contains serous fluid, called **pericardial fluid,** that reduces friction as the heart contracts and relaxes.

Abdominal Cavity Membranes
The walls of the abdominal cavity and the surfaces of abdominal organs are covered with the **peritoneum.** The **parietal peritoneum** lines the inner abdominal wall, and the **visceral peritoneum** covers the surface of the abdominal organs. The space between the visceral and parietal peritonea is the **peritoneal cavity,** which contains **peritoneal fluid.** See Figure 1.7. The kidneys, pancreas, and parts of the large intestine are located in the **retroperitoneal region** (behind the peritoneum), so only part of their surfaces is covered with the parietal peritoneum.

Double-layered folds of the peritoneum, called **mesenteries,** extend from the cavity wall to the abdominal organs. The mesenteries support the abdominal organs, and they contain nerves and blood vessels serving the organs. A large mesenteric fold, the **greater omentum,** extends from the bottom of the stomach over the intestines and loops back up to join with the transverse colon (part of the large intestine). A smaller mesenteric fold, the **lesser omentum,** extends between the liver and the stomach.

The parietal peritoneum does not line the pelvic cavity. It only extends far enough to cover the top of the urinary bladder.

Assignment

1. Label Figures 1.5, 1.6, and 1.7 and record the labels on the laboratory report.
2. Using a human torso model, (a) locate the surface features and body cavities, (b) identify the major organs of each cavity, and (c) locate the organs within each of the nine abdominal regions and four abdominal quadrants.

Introduction to Human Anatomy and Body Organization

Organ Systems

Most of the functions of the body are performed by organ systems. As illustrated in Figure 1.1, an **organ** is a structure that is (1) composed of two or more tissues, (2) has a definite shape, and (3) performs specific functions. The heart is an example of an organ. An **organ system** is a group of organs that function in a coordinated manner to perform specific functions. The cardiovascular system, which is composed of blood, heart, and blood vessels, is an example of an organ system.

In this exercise, you will study an overview of the components and general functions of the organ systems of the body in order to gain a general understanding of body organization and function. See Figure 1.8. Then you will dissect a freshly killed or preserved rat to observe the mammalian body organization, body cavities, and organs of the ventral cavity.

The Integumentary System

The skin, including hair, nails, associated glands, and sensory receptors, constitutes the integumentary system. The **skin** consists of two layers, an outer **epidermis** and an inner **dermis,** which protect the underlying tissues from mild abrasions, excessive water loss, microorganisms, and ultraviolet radiation. Perspiration secreted by sweat glands contains water and waste materials similar to dilute urine. The evaporation of perspiration cools the body surface, thus aiding in regulating body temperature.

The Skeletal System

The skeletal system forms the framework of the body and provides support and protection for softer organs and tissues. It consists of **bones, cartilages,** and **ligaments.** In conjunction with skeletal muscles, the skeleton forms lever systems that enable movement. In addition, **red bone marrow** produces blood cells.

The Muscular System

The contraction of muscles provides the force that enables movement. **Skeletal muscles** are attached to bones by **tendons** and constitute nearly half of the body weight. Their contractions move body parts during walking, eating, and other activities.

Two other types of muscle tissue are found in the body: smooth and cardiac. *Smooth muscle* is found in the walls of hollow organs. *Cardiac muscle* is found in the walls of the heart.

The Nervous System

The nervous system is a complex, highly organized system consisting of the **brain, spinal cord, cranial** and **spinal nerves,** and **sensory receptors.** These components work together to enable rapid perception and interpretation of the environment and the almost instantaneous coordination of body functions. The human brain is responsible for intelligence, will, self-awareness, and emotions characteristic of humans.

The Endocrine System

The endocrine system consists of small masses of glandular tissue that secrete hormones. **Hormones,** which are chemical messengers, are absorbed by the blood and transported throughout the body, where they bring about the chemical control of body functions. The larger masses of endocrine tissues occur in the **endocrine glands: pituitary, thyroid, parathyroid, thymus, adrenal, pancreas, pineal, ovaries,** and **testes.** Smaller masses of hormone-producing tissues occur in other organs, such as the kidneys, heart, placenta, and in the digestive tract.

The Cardiovascular System

The **heart, arteries, veins, capillaries,** and **blood** constitute the cardiovascular system. These components work together to transport materials such as oxygen, carbon dioxide, nutrients, wastes, and hormones throughout the body. Contractions of the heart circulate the blood through the blood vessels. Blood, consisting of formed elements (cells) and plasma, is the transporting agent and also provides the primary defense against disease organisms. The spleen serves as a blood reservoir and removes worn-out red blood cells from circulation.

The Lymphoid System

The lymphoid system consists of **lymphoid tissue** and a network of **lymphatic vessels** that collect fluid from interstitial spaces (spaces between cells) and return it to large veins under the collarbones. Interstitial fluid is called **lymph** after it enters a lymphatic vessel. En route, lymph passes through **lymph nodes,** nodules of lymphoid tissue, that remove cellular debris and microorganisms. Lymphoid tissue is found in many organs, such as the spleen, thymus, tonsils, adenoids, liver, intestines, and red bone marrow. The lymphoid tissue produces immune cells and plays a central role in the defense against pathogens.

The Respiratory System

The exchange of oxygen and carbon dioxide between the atmosphere and the blood is enabled by the respiratory system. It consists of air passageways and gas-exchange organs. The passageways

Figure 1.8 The 11 organ system of the body.

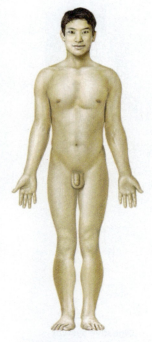

Integumentary system

Organs: skin, hair, nails, and associated glands
Functions: protects underlying tissues and helps regulate body temperature

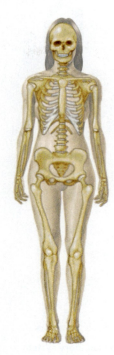

Skeletal system

Organs: bones, ligaments, and associated cartilages
Functions: supports the body, protects vital organs, stores minerals, and is the site of blood cell production

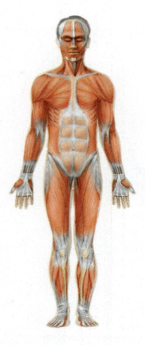

Muscular system

Organs: skeletal muscles and tendons
Functions: moves the body and body parts and produces heat

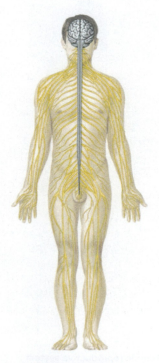

Nervous system

Organs: brain, spinal cord, nerves, and sensory receptors
Functions: rapidly coordinates body functions and enables learning and memory

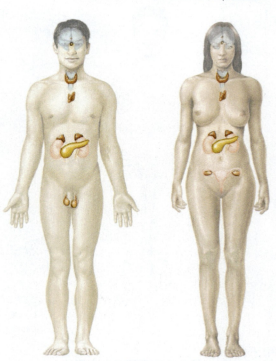

Endocrine system

Organs: hormone-producing glands, such as the pituitary and thyroid glands
Functions: secretes hormones that regulate body functions

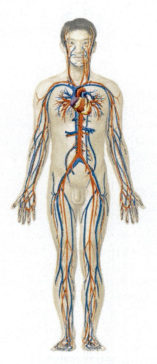

Cardiovascular system

Organs: blood, heart, arteries, veins, and capillaries
Functions: transports heat and materials to and from the body cells

Introduction to Human Anatomy and Body Organization

Figure 1.8 (*Continued*)

Lymphoid system

Organs: lymph, lymphatic vessels, and lymphoid organs and tissues
Functions: collects and cleanses interstitial fluid, and returns it to the blood; provides immunity

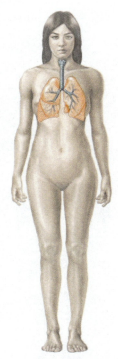

Respiratory system

Organs: nose, pharynx, larynx, trachea, bronchi, and lungs
Functions: exchanges O_2 and CO_2 between air and blood in the lungs, pH regulation, and sound production

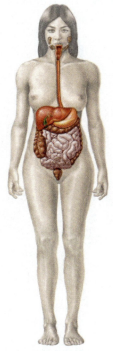

Digestive system

Organs: mouth, pharynx, esophagus, stomach, intestines, liver, pancreas, gallbladder, and associated structures
Functions: digests food and absorbs nutrients

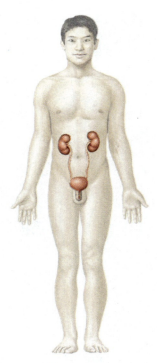

Urinary system

Organs: kidneys, ureters, urinary bladder, and urethra
Functions: regulates volume and composition of blood by forming and excreting urine

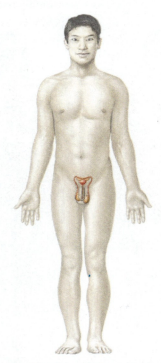

Male reproductive system

Organs: testes, epididymides, vasa deferentia, prostate, bulbo-urethral glands, seminal vesicles, and penis
Functions: produces sperm and transmits them into the female vagina during sexual intercourse

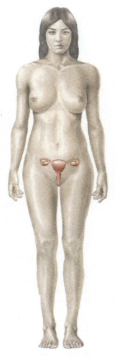

Female reproductive system

Organs: ovaries, uterine tubes, uterus, vagina, and vulva
Functions: produces oocytes, receives sperm, provides intrauterine development of offspring, and enables birth of an infant

are the **nasal cavity, pharynx, larynx, trachea,** and **bronchi.** Gas exchange occurs in the **lungs.**

The Digestive System

The digestive system converts large, nonabsorbable nutrient molecules into smaller nutrient molecules that can be absorbed into the blood. Digestion involves mechanically breaking food into smaller particles and mixing them with digestive fluids and chemically breaking large molecules into smaller molecules. The digestive tract consists of the **mouth, pharynx, esophagus, stomach, small intestine,** and **large intestine.** Accessory organs of the digestive system include the **salivary glands, gallbladder, liver,** and **pancreas.**

The Urinary System

The nitrogenous wastes of metabolism and excess water and minerals are removed from the blood and body by the urinary system. The **kidneys** remove wastes and excess materials from the blood to form urine, which is carried by two **ureters** to the **urinary bladder** for temporary storage. Subsequently, urine is voided via the **urethra.**

The Reproductive System

Continuity of the species is the function of the male and female reproductive systems. The male reproductive system consists of a pair of sperm-producing **testes** located in the **scrotum;** two tubes, the **vasa deferentia,** which carry sperm to the **urethra; accessory glands** that secrete fluids for sperm transport; and the **penis,** the male copulatory organ. The female reproductive system consists of a pair of **ovaries** that produce ova; two **uterine tubes** that carry the ova to the **uterus; accessory glands** that secrete lubricating fluids; and the **vagina,** the female copulatory organ and birth canal.

Assignment

1. Complete the laboratory report sections E, F, and G.
2. Locate the major organs of the organ systems on the human torso model.

Rat Dissection

You will work in pairs to perform this part of the exercise. Your principal objective in the dissection is to *expose the organs for study, not to simply cut up the animal.* Most cutting will be performed with scissors. The scalpel blade will be used only occasionally, but the flat blunt end of the handle will be used frequently for separating tissues.

Skinning the Ventral Surface

1. Pin the 4 feet to the bottom of the dissecting pan as illustrated in Figure 1.9. Before making any incision, examine the oral cavity. Note the large **incisors** in the front of the mouth, which are used for biting off food particles. Force the mouth open sufficiently to examine the flattened **molars** at the back of the mouth. These teeth are used for grinding food into small particles. Note that the **tongue** is attached near the throat. Lightly scrape the surface of the tongue with a scalpel to determine its texture. Most of the roof of the mouth is the bony **hard palate** and the fleshy portion behind it is the **soft palate.** The throat is called the **pharynx,** which is a component of both the digestive and respiratory systems.
2. Lift the skin in the middle of the abdomen with your forceps and make a small incision with scissors as shown in Figure 1.9. Cut the skin upward to the lower jaw, turn the pan around, and complete this incision to the anus, cutting around both sides of the genital openings. The completed incision should appear as in Figure 1.10.
3. With the handle of the scalpel, separate the skin from the musculature. You may also perform this step by gently inserting closed scissors between the skin and muscles and opening the scissors as shown in Figure 1.11. The connective tissue that lies between the skin and musculature is the **superficial fascia.**

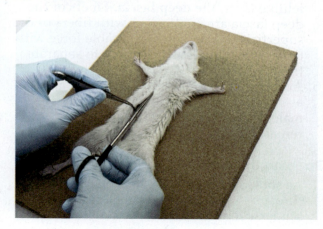

Figure 1.9 Incision is started on the median line with a pair of scissors. unoL/Shutterstock

Introduction to Human Anatomy and Body Organization

Figure 1.10 Completed incision from the lower jaw to the anus. unoL/Shutterstock

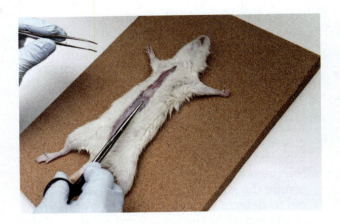

Figure 1.11 Skin is separated from the musculature using the open-scissors technique. unoL/Shutterstock

Figure 1.12 Incision of musculature is begun on the median line. unoL/Shutterstock

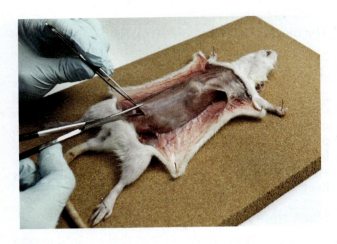

Figure 1.13 Lateral cuts at base of thoracic cage are made in both directions. unoL/Shutterstock

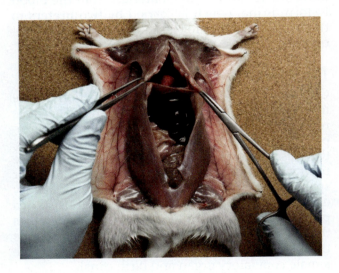

4. Skin the limbs down to the "knees" and "elbows" and pin the stretched-out skin to the wax. Examine the surfaces of the **muscles** and note that **tendons,** which consist of tough dense connective tissue, attach the muscles to the skeleton. Covering the surface of each muscle is another thin, gray feltlike layer, the **deep fascia.** Fibers of the deep fascia are continuous with fibers of the superficial fascia, so considerable force with the scalpel handle is necessary to separate the two membranes.
5. At this stage, your specimen should appear as in Figure 1.12. If your specimen is a female, the mammary glands will probably remain attached to the skin.

Opening the Abdominal Wall

1. As shown in Figure 1.12, make an incision through the abdominal wall with a pair of scissors. To make the cut, you must hold the muscle tissue with a pair of forceps. **Caution:** Avoid damaging the underlying viscera as you cut.
2. Cut upward along the median line to the thoracic cage and downward along the median line to the genitalia.
3. To completely expose the abdominal organs, make two lateral cuts near the base of the thoracic cage—one to the left and the other to the right. See Figure 1.13. The cuts should extend all the way to the pinned-back skin.
4. Fold out the flaps of the body wall and pin them to the wax as shown in Figure 1.14. The abdominal organs are now well exposed.
5. Using Figure 1.15 as a reference, identify all of the labeled viscera without moving the organs out of place. Note in particular the position and structure of the **diaphragm.**

Figure 1.14 Flaps of abdominal wall are pinned back to expose viscera. unoL/Shutterstock

Examination of the Thoracic Cavity

1. Using your scissors, cut along the left side of the thoracic cage. Cut through all of the ribs and connective tissue. Then, cut along the right side of the thoracic cage in a similar manner.
2. Grasp the tip of the sternum with forceps and cut the diaphragm away from the thoracic cage with your scissors. Now you can lift up the thoracic cage and look into the thoracic cavity.
3. With your scissors, complete the removal of the thoracic cage by cutting any remaining attachment tissue.

Figure 1.15 Viscera of a female rat.

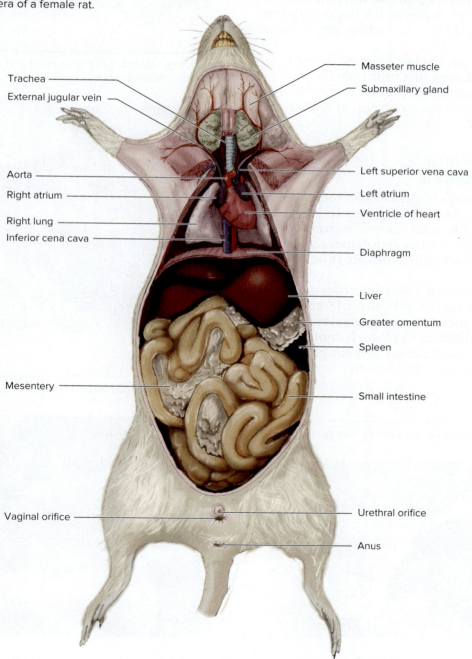

Introduction to Human Anatomy and Body Organization

4. Now, examine the structures that are exposed in the thoracic cavity. Refer to Figure 1.15 and identify all the structures that are labeled.
5. Note the pale **thymus,** which is located just above the heart. Remove this gland.
6. Carefully remove the thin **pericardial membrane** that encloses the **heart.**
7. Remove the heart by cutting through the major blood vessels attached to it. Gently sponge away pools of blood with Kimwipes or other soft tissues.
8. Locate the **trachea** in the neck region and the **larynx** (voice box) at the proximal end of the trachea. Trace the trachea down to where it divides into two **bronchi** that enter the **lungs.** Squeeze the lungs with your fingers, noting how elastic they are. Remove the lungs.
9. Probe under the trachea to locate the soft tubular **esophagus** that runs from the oral cavity to the stomach. Excise a section of the trachea to reveal the esophagus as illustrated in Figure 1.16.

Deeper Examination of Abdominal Organs

1. Lift up the lobes of the reddish brown liver and examine them. See Figure 1.17. Note that rats lack a **gallbladder.** *Carefully excise the liver* and wash out the abdominal cavity. The stomach and intestines are now clearly visible.
2. Lift out the stomach and a portion of the intestines as shown in Figure 1.18 and identify the membranous **mesentery,** which holds the intestines in place. It contains blood vessels and nerves that supply the digestive tract. If your specimen is a mature

Figure 1.16 Examination of the trachea. unoL/Shutterstock

Figure 1.17 Examination of the liver. unoL/Shutterstock

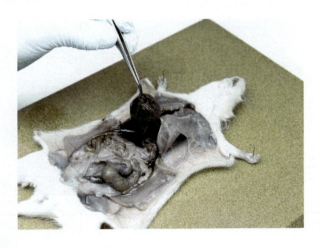

Figure 1.18 Removal of the stomach and examination of the intestines and mesentery. unoL/Shutterstock

Figure 1.19 Retroperitoneal structures. unoL/Shutterstock

healthy animal, the mesenteries will contain considerable fat.
3. Now, lift the intestines out of the abdominal cavity, cutting the mesenteries, as necessary, for a better view of the organs. Note the great length of the small intestine. Its name refers to its diameter, not its length. The first portion of the small intestine, which is connected to the stomach, is called the **duodenum.** At its distal end, the small intestine is connected to a large saclike structure, the **cecum.** The *appendix* in humans is attached to the cecum. The cecum communicates with the **large intestine.** This latter structure consists of the **ascending, transverse, descending,** and **sigmoid** segments. The last of these segments empties into the **rectum.** See Figure 1.20.
4. Try to locate the **pancreas,** which is embedded in the mesentery alongside the duodenum.

Figure 1.20 Viscera of a male rat with heart, lungs, and thymus removed.

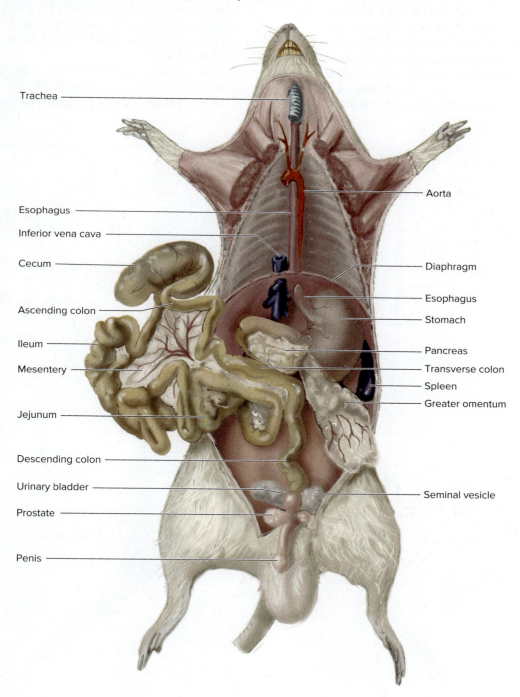

It is often difficult to see. Pancreatic enzymes enter the duodenum via the **pancreatic duct.** See if you can locate this tiny tube.

5. Locate the **spleen,** which is situated on the left side of the abdomen near the stomach. It is reddish brown and held in place with mesentery.
6. Remove remainder of the digestive tract by cutting through the esophagus next to the stomach and through the sigmoid colon. You can now see the **descending aorta** and the **inferior vena cava.** See Figures 1.19 and 1.21. The aorta carries blood to the body tissues. The inferior vena cava returns blood from below the diaphragm to the heart.
7. Peel away the peritoneum and fat from the rear wall of the abdominal cavity. *Removal of the fat will require special care to avoid damaging important structures.* The kidneys, blood vessels, and reproductive structures will now be more visible. Locate the two **kidneys** and **urinary bladder.** Trace the two **ureters,** which extend from the kidneys to the bladder. Examine the front of the kidneys and locate the **adrenal glands,** which are important components of the endocrine gland system.
8. **Female.** If your specimen is a female, compare it with Figure 1.21. Locate the two **ovaries,** which lie lateral to the kidneys. From each ovary, a **uterine tube** leads back to join the **uterus.** Note that the uterus is a Y-shaped structure joined to the **vagina.** If your specimen appears to be pregnant, open up the uterus and examine the developing embryos. Note how they are attached to the uterine wall.
9. **Male.** If your specimen is a male, compare it with Figure 1.22. The **urethra** is located in the **penis.** Apply pressure to one of the

Figure 1.21 Abdominopelvic cavity of a female rat with intestines and liver removed.

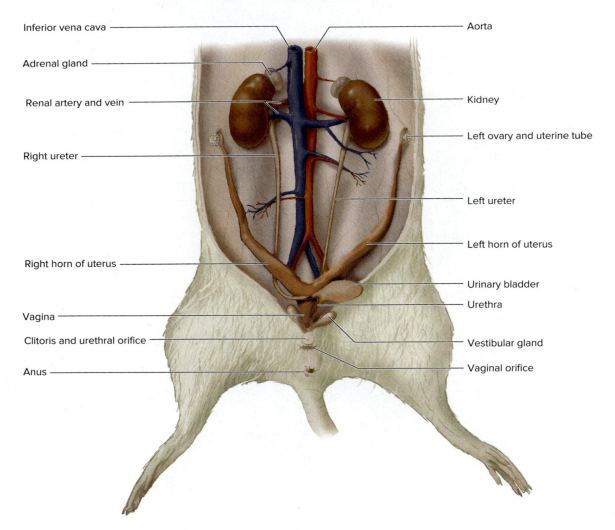

16 Introduction to Anatomy

Figure 1.22 Abdominopelvic cavity of a male rat with intestines and liver removed.

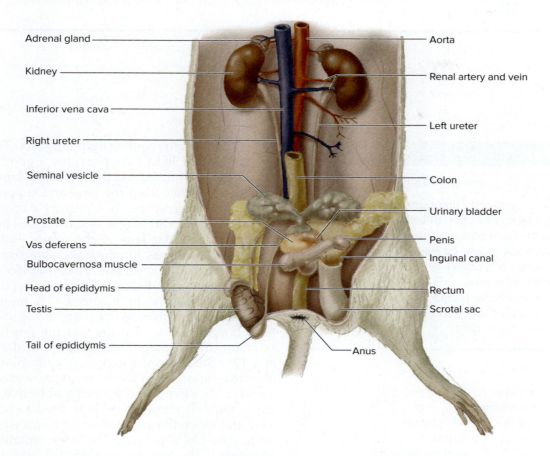

testes through the wall of the **scrotum** to see if it can be forced up into the **inguinal canal.** Carefully dissect out a testis, **epididymis,** and **vas deferens** from one side of the scrotum and, if possible, trace the vas deferens over the urinary bladder to where it penetrates the **prostate** to join the urethra.

Conclusion

In this brief dissection, you have become acquainted with the cardiovascular, respiratory, urinary, digestive, and reproductive systems. Portions of the endocrine system have also been observed. If you have done a careful and thoughtful rat dissection, you should have a good general understanding of the basic structural organization of the human body. Much that we see in rat anatomy has its human counterpart.

Cleanup Dispose of the specimen as directed by your instructor. Scrub your instruments with soap and water; rinse and dry them. Wash your hands with soap, rinse, and dry thoroughly.

Exercise 2

MICROSCOPES

Objectives

After completing this exercise, you should be able to

1. Identify the parts of a compound microscope and describe the function of each.
2. Describe and demonstrate the correct way to (a) carry a microscope, (b) clean the lenses, (c) focus with each objective, and (d) calculate the total magnification.
3. Practice microscopic techniques by viewing select prepared slides and making a wet-mount of your own cheek cells.

Materials
- Biohazard container
- Compound microscope
- Microscope slides and cover glasses
- Medicine droppers, toothpicks
- Methylene blue, 0.01%, in dropping bottles
- Prepared slides of
  - letter *e*
  - colored threads
  - human sperm
  - human blood

The effective use of a **compound microscope** is essential for the study of cells and tissues. The term *compound* indicates that multiple lenses are used to view specimens. This exercise will introduce you to the parts, care, and use of a compound microscope.

Parts of the Microscope

Figure 2.1 shows the major parts of a compound microscope. Refer to it as you study this section. Your microscope may be somewhat different from the one illustrated, but you will be able to relate the illustration and the discussion to your microscope without difficulty.

The **base** is the bottom portion of the microscope. It contains the **lamp** and the on/off switch. Ideally, the lamp should have a **light control** to vary the intensity of light. The microscope shown in Figure 2.1 has a rotatable wheel on the right side of the arm to regulate light intensity. The lowest amount of light that provides a good image should be used to extend the life of the lamp.

The **arm** rises from the base and supports the rest of the microscope. It serves as a convenient "handle" when carrying a microscope. The **body tube** has an **ocular lens** at the upper end and a **revolving nosepiece,** to which the **objective lenses** are attached, at the lower end. The nosepiece may be rotated to move an objective into viewing position.

The microscope in Figure 2.1 is monocular; however, many microscopes are binocular. The head is rotatable so that the viewing position can be changed. The magnification of the ocular is usually 10×. Student microscopes usually have three objectives attached to the revolving nosepiece. The shortest objective is the **scanning objective,** which has a magnification of 4×. It is used to scan the slide and locate the area of interest for viewing with more powerful objectives. The intermediate-length objective is the **low-power objective,** which has a magnification of 10×. The longest objective is the **high-dry objective,** which typically has a magnification of 40×. Some student microscopes have a fourth objective that is slightly longer than the high-dry objective as shown in Figure 2.1. This is the **oil immersion objective,** and it has a magnification of 100×. You probably will not use the oil immersion objective in this course. The magnifications of the oil immersion and high-dry objectives may vary slightly in different models of microscopes.

The **stage** is the platform on which a microscope slide is placed for viewing. The opening in the center of the stage is the **stage aperture.** The **mechanical stage** is the device on the stage surface that holds the slide and enables precise movement of the slide. It is operated by the **mechanical stage control knobs.**

Below the stage is the **condenser,** which concentrates the light on the microscope slide. It may be raised or lowered by the **condenser control**

Unit 1: Introduction to Anatomy

Figure 2.1 A compound light microscope. Kovalchuk Oleksandr/Shutterstock

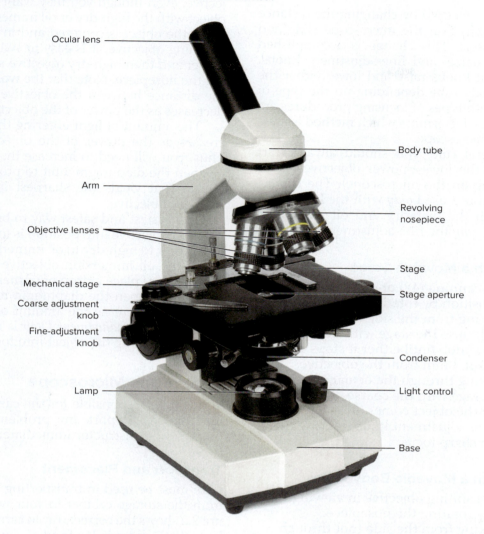

knob located under the stage. Usually, it should be raised to its highest position.

An **iris diaphragm** within the condenser regulates the amount of light that reaches the slide. On the microscope shown in Figure 2.1, the diaphragm is adjusted by turning a ring. Some microscopes have a diaphragm control lever.

Two focusing knobs are used to bring an object into clear focus. The **coarse-adjustment knob** has a larger diameter and is used to bring objects into rough focus. The **fine-adjustment knob** has a smaller diameter and is used to bring objects that are in rough focus into sharp focus.

Magnification

The microscope lens system magnifies specimens so that very small structures can be distinguished. The **total magnification** is determined by the power of the ocular and objective being used. It is calculated by multiplying the power of the ocular by the power of the objective. For example, a 10× ocular and a 40× high-dry objective yield a total magnification of 400×.

Resolution

Magnification without resolution is of little value. **Resolution** is the ability to distinguish tiny adjacent objects as two distinct objects. **Resolving power** is a function of the wavelength of light and the design of the microscope lenses. The shortest wavelengths (blue) of visible light provide maximum resolution. This is why microscopes have a blue light filter over the lamp. Use of the oil immersion objective is required for maximum resolution, and on the best light microscopes, it will enable the distinction of microscopic objects that are 0.2 μm apart. If they are closer together, they will be seen as one object because of a fusion of the images.

Microscopes   19

Focusing

A microscope is focused by changing the distance between the object on the microscope slide and the objective lens. This change is accomplished by using the coarse- and fine-adjustment knobs. The adjustment knobs raise and lower either the stage *or* the body tube depending on the type of microscope. Both types of focusing procedures are described below. Determine which method is to be used for your microscope.

As a general rule, you should always start focusing with the lowest-power objective (4× or 10×, depending on the microscope). The coarse-adjustment knob is used *only* with the low-power objective. With the high-dry and oil immersion objectives, use *only* the fine-adjustment knob.

Focusing with a Movable Stage

1. Place the scanning (4×) objective into viewing position by rotating the nosepiece.
2. While looking from the side (not through the ocular), raise the stage with the coarse-adjustment knob until either it stops or the slide is about 3 mm from the objective.
3. While looking through the ocular, slowly lower the stage with the coarse-adjustment knob until the object comes into view.
4. Use the fine-adjustment knob to bring the object into sharp focus.

Focusing with a Movable Body Tube

1. Place the scanning objective in viewing position by rotating the nosepiece.
2. While looking from the side (not through the ocular), lower the body tube with the coarse-adjustment knob until either it stops or the slide is about 3 mm from the objective.
3. While looking through the ocular, slowly raise the body tube with the coarse-adjustment knob until the object comes into view.
4. Use the fine-adjustment knob to bring the object into sharp focus.

Switching Objectives

Modern microscopes are usually **parcentric** and **parfocal**. This means that when an object is centered in the field and in sharp focus with one objective, it will be centered and in focus when another objective is rotated into the viewing position. However, it may be necessary to make slight adjustments to re-center the object with the mechanical stage or bring it into sharper focus with the fine-adjustment knob.

Start your observations with the scanning objective, even though you may want to observe the object with the high-dry or oil immersion objective. Once the object is centered and in focus with the scanning objective, it is easy to switch to the low-power and then high-dry objective simply by rotating the nosepiece. Note that the **working distance,** the distance between the objective and the slide, decreases as the power of the objective increases.

The amount of light entering the objective decreases as the power of the objective increases. Thus, you will need to increase the light intensity or open the diaphragm a bit to provide a light intensity that yields the sharpest image with the high-dry objective.

The easiest and safest way to bring the oil immersion objective into position is to progress from low-power to high-dry to oil immersion. Before rotating the oil immersion objective into the viewing position, place a drop of immersion oil on the slide. Also, open the diaphragm to its maximum aperture to increase the amount of light. A *slight turn of the fine-adjustment knob* is all that will be required to bring the object into focus.

Care of the Microscope

You will be responsible for the care of laboratory microscopes. Report any problems or malfunctions to your instructor immediately.

Transport and Placement

Care must be used in transporting the microscope from the storage cabinet to your workstation. Figure 2.2 shows the correct way to carry a microscope. Note that it is carried in front of you, with one hand supporting the base while the other grasps the arm. *Never* carry it with one hand at your side because a microscope in this position is apt to collide with furniture in the lab.

Most modern microscopes have an inclined body tube and a rotatable head that is secured by a set screw. The position can be changed so that either the microscope arm or the mechanical stage is facing the student. In the latter position, the stage is more easily accessible. Care must be taken when changing the position of the microscope. Use the position recommended by your instructor.

Lens Care

Develop the habit of cleaning the lenses with **lens paper** before using the microscope. *Use only lens paper for cleaning the lenses.* If liquid gets on the objectives or stage of the microscope, wipe it off immediately with lens paper. If simply wiping the lenses with lens paper doesn't get them clean,

Figure 2.2 The microscope should be held firmly with both hands while being carried.

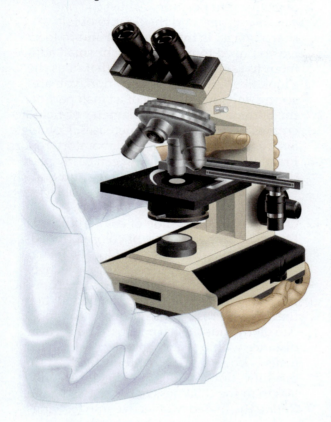

Figure 2.3 When oculars are removed for cleaning, cover the ocular opening with lens tissue. A blast from an air syringe or gas canister removes dust and lint.

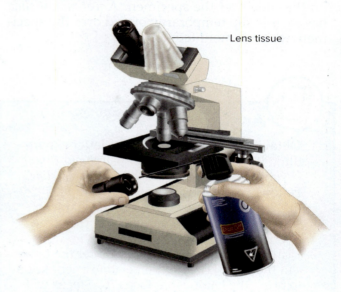

— Lens tissue

it may be necessary to inform your instructor so they may have the microscope properly serviced. *Never try to disassemble any part of the microscope.*

The best way to determine if the ocular is clean is to rotate it between your thumb and forefinger while looking through the microscope. A rotating pattern is evidence of a dirty lens. If cleaning the top lens fails to remove the debris, *inform your instructor. If the ocular is removed to clean the lower lens, it is imperative that a piece of lens paper be placed over the open end of the body tube as shown in Figure 2.3.*

An image that appears cloudy or blurred indicates a dirty objective. If cleaning with lens paper moistened with water fails to clear the image, *consult your instructor* about getting the microscope professionally serviced.

Routinely wipe off the upper surface of the condenser lens with lens paper to remove any accumulated dust.

Cleanup and Storage

When you have finished using the microscope, complete the following steps before returning it to the cabinet.

1. Remove the slide from the stage.
2. Clean the ocular, objective lenses, and the stage.
3. Rotate the nosepiece to place the lowest-power objective in the viewing position.
4. Depending on the type of microscope, lower the body tube to its lowest position *or* raise the stage to its highest position.
5. Raise the condenser to its highest position.
6. Adjust the mechanical stage so that it projects minimally from each side of the stage.
7. Place the dustcover over the microscope.

Assignment

Complete Sections A through C of the laboratory report.

Using the Microscope

Obtain the microscope and carry it to your station in the manner just described. Locate on your microscope the parts labeled in Figure 2.1. Manipulate the control knobs and levers to see how they operate. Set up your microscope for viewing as directed by your instructor.

Plug in the cord, turn on the lamp, and look through the ocular. The circle of light that you see is called the **field of view** or just the **field**. If your microscope is binocular, adjust the oculars to match the distance between your eyes.

Clean the lenses with lens paper. Raise the condenser to its highest position. Once you feel comfortable with the microscope controls, proceed with the assignment below.

In this course, you will examine specimens on prepared slides and on liquid-mount slides. A *prepared slide* has a cover slip permanently mounted on the slide over the specimen. A *wet-mount slide* has a cover slip temporarily placed over the specimen, which is mounted in liquid.

Assignment

1. Obtain a prepared slide of the letter *e*, handling it by the edges. *Never place your thumb or finger on the cover glass of a prepared slide.* Following the procedures shown in Figure 2.4, place the slide in the mechanical stage and center the letter *e* over the stage aperture.
2. Using the scanning objective, center the letter *e* in the field. Use the course-adjustment knob to bring the letter *e* into focus. The focus will not be great with just the course-adjustment knob, but that is okay as the scanning objective is only used to find your specimen.
3. Rotate the low-power objective (10×) into position. Then, use the focusing technique described earlier for your microscope to focus on the letter *e*. Adjust the light intensity to the lowest level that provides a sharp image. Practice moving the slide with the mechanical stage until you can center the *e* easily. What is different about the orientation of the *e* when viewed with the naked eye and through the microscope?
4. Center the *e* in the field. While viewing through the ocular, increase and decrease the light by adjusting the iris diaphragm or the light intensity dial, noting how the sharpness of the image changes. Make drawings of your observations in Section D, number 1 of the laboratory report.
5. With the *e* centered, rotate the high-dry objective into position. Adjust the *fine-adjustment knob* and the light to obtain a sharp image. Is the *e* centered and in focus? Do you need to make a large or small adjustment to see a sharp image? What does the *e* look like? At this magnification, you are seeing only ink blotches of a small part of the letter. Note how the **diameter of field** decreases as magnification increases. Complete Section D, number 1 of the laboratory report.
6. Practice steps 1–5 until you can easily place the slide in the mechanical stage and quickly focus on the *e* with the low-power objective and then with the high-dry objective.
7. Obtain a prepared slide of colored threads. Place the prepared slide of colored threads on the mechanical stage and locate, focus, and center it using the scanning objective.
8. Rotate to the low-power objective to examine the color threads. Focus on the point where the threads cross. Which thread is on top? Are all threads in sharp focus? The vertical distance within which an object is in sharp focus is known as the **depth of field.** At this magnification, the depth of field is slightly less than the thickness of the crossed hairs.
9. Rotate the high-dry objective into viewing position and examine the threads. Are all threads in sharp focus? At this magnification only a portion of one thread is in sharp focus at a time.
10. Carefully focus through the thickness of the thread. In a surface view, you will see the edges of the thread will be out of focus. When you focus through the midsection of the thread the central core and edges of the thread will be in sharp focus. The depth of field is less than the thickness of a thread. Complete Section D, number 2 of the laboratory report.

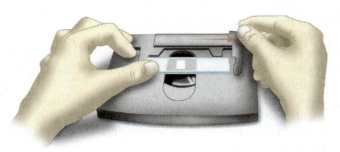

Figure 2.4 The slide must be properly positioned as the retainer lever is moved to the right or left.

Human Sperm

Examine a prepared slide of human sperm set up under the oil immersion objective of a demonstration microscope. Note their small size. The head consists primarily of the sperm nucleus. Movements of the flagellum enable forward movement by a sperm.

Human Blood Cells

Sharpen your microscopy skills by examining a prepared slide of human blood with the low-power and high-dry objectives. The blood cells have been

Figure 2.5 Making a wet-mount slide of human hairs.

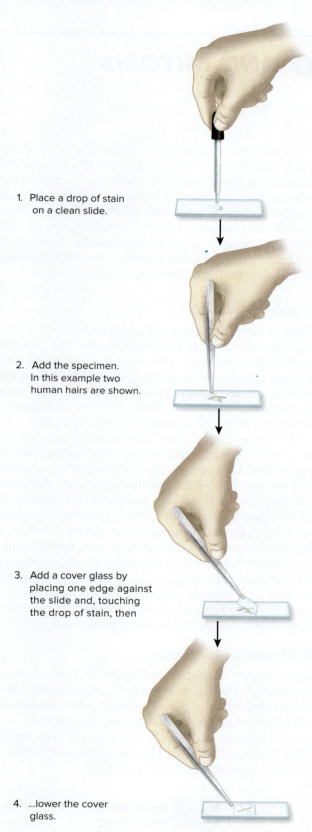

1. Place a drop of stain on a clean slide.
2. Add the specimen. In this example two human hairs are shown.
3. Add a cover glass by placing one edge against the slide and, touching the drop of stain, then
4. ...lower the cover glass.

stained to enable recognition of cell types. Most of the cells that you see are red blood cells. They are stained pink and lack a nucleus. There are five types of white blood cells. A white blood cell is easy to recognize because its nucleus is stained purple. Some white blood cells have cytoplasmic granules stained lavender, red, or blue, but they are difficult to see at 400×.

Locate a white blood cell with the low-power objective, then switch to the high-dry objective for your observations. Locate as many different types of blood cells as you can.

Epithelial Cells

The epithelial cells lining the inside of your cheek are easily obtained for study. The cells that you will remove are dead and are continuously sloughed off in healthy individuals. See Figure 2.5, which shows how to make a wet mount slide using human hairs. The procedure is the same no matter what the specimen is.

1. Place a drop of methylene blue in the center of a clean slide.
2. *Gently* scrape the inside of your cheek with a clean toothpick, then swirl the end in the drop of methylene blue solution to dislodge the cells. *Immediately place the toothpick in the biohazard container. Do not place it on the tabletop.*
3. Add a cover glass and locate the cells with the scanning objective.
4. Center the cells and rotate to the low-power objective. Focus and center the specimen. Now move to the high-dry objective into position for careful examination. Locate the nucleus, cytoplasm, and plasma membrane.
5. *When you are finished examining your slide, place the slide with the cover glass in the biohazard container.*

Assignment

Complete Section E of the laboratory report.

Exercise 3

CELLULAR ANATOMY AND MITOSIS

Objectives

After completing this exercise, you should be able to

1. Identify the parts of a cell on diagrams or models.
2. Describe the basic structure of cellular organelles and their functions.
3. Contrast interphase, mitosis, and cytokinesis.
4. List the stages of mitosis and describe the characteristics of each stage.
5. Diagram and label cells in interphase and in stages of mitosis.
6. Identify the stages of mitosis when viewed microscopically.

Materials
  Model of a generalized cell
  Model(s) depicting the phases of mitosis
  Prepared slides of
    whitefish mitosis
    onion root tip (alternative)

The **cell** is the structural and functional unit of all organisms, including humans. The processes of life take place within cells. An understanding of cell anatomy is essential to understanding the physiological processes that occur within cells.

All cells have many anatomical and functional characteristics in common, although differences in size, shape, and internal composition may result from specialization for particular functions. The human body is composed of trillions of cells. Most of these cells are specialized to perform specific functions. In this exercise, you will study the anatomical features common to human cells and how cells divide in growth and repair via mitosis.

Mitosis is the type of cell division by which a parent cell divides to form two daughter cells that have the same genetic composition as the parent cell. All new body cells are formed this way, and billions of new cells are formed each day. These new cells enable either growth or replacement of damaged and worn-out cells.

Hereditary information (directions controlling cellular functions) is encoded in the DNA molecule forming the core of each chromosome. Human body cells contain 23 pairs of chromosomes—a total of 46 chromosomes. Each pair of chromosomes contains its own unique encoded information. In cell division, it is vital that each chromosome is precisely replicated and that the replicated chromosomes are equally distributed to the new daughter cells.

Basic Structure

When viewed with your compound microscope, three major components of a cell are visible. The **plasma membrane** forms the outer boundary of a cell, the **nucleus** appears as a dense spherical body within the cell, and the **cytoplasm** is the relatively clear area between the nucleus and the plasma membrane. However, cell structure is much more complex.

Electron microscopy has shown the cell to be a complex organization of components called **organelles.** Organelles are subcellular structures that are specialized for specific functions. Because specific cellular functions are compartmentalized within specific organelles, many cellular functions can take place at the same time. Figure 3.1 shows the ultrastructure of a pancreatic cell, a rather unspecialized cell, as it might appear when viewed with an electron microscope. It is not possible to see the structure of organelles with your microscope. As you read the discussion of cell anatomy that follows, locate the structures in Figure 3.1.

The Plasma Membrane

Cell contents are separated from the surrounding environment by the **plasma** (cell) **membrane.** Like all membranes, the plasma membrane consists of two layers of phospholipid molecules (label 6), arranged back to back, in which protein molecules (label 5) are embedded. The membrane is flexible, with the consistency of a liquid film.

Unit 2: Cells and Tissues

Figure 3.1 The structure of a cell.

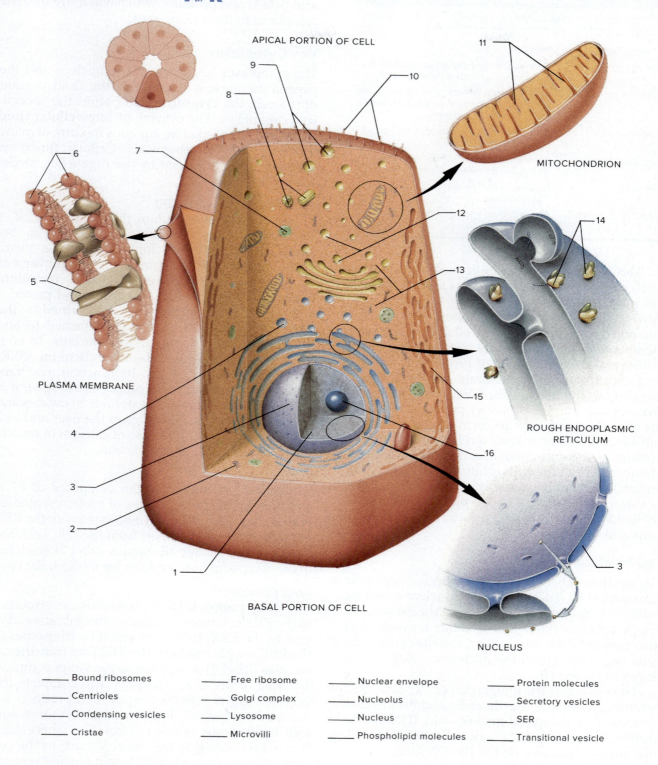

_____ Bound ribosomes _____ Free ribosome _____ Nuclear envelope _____ Protein molecules
_____ Centrioles _____ Golgi complex _____ Nucleolus _____ Secretory vesicles
_____ Condensing vesicles _____ Lysosome _____ Nucleus _____ SER
_____ Cristae _____ Microvilli _____ Phospholipid molecules _____ Transitional vesicle

Cellular Anatomy and Mitosis

TABLE 3.1
Proteins of the Plasma Membrane and Their Functions

Protein Type	Function
Receptors	Receive chemical signals from other cells; enable communication between cells.
Channel proteins	Allow passage of water and solutes; some are always open; others open and close under different conditions.
Carrier proteins	Actively transport substances into, or out of, the cell.
Identity markers	Serve as an identity tag so other cells can distinguish between self and foreign cells.
Enzymes	Catalyze reactions that break down specific molecules.
Attachment proteins	Attach cells to each other or to extracellular material.

The **selective permeability** of the plasma membrane allows some molecules to pass through the membrane while preventing the passage of others. In this way, the movement of materials into and out of the cell is controlled by the plasma membrane in both active and passive transport.

Several different types of proteins are associated with the plasma membrane where they perform important and specific functions. Some (e.g., channel and carrier proteins) penetrate through the plasma membrane to form passageways for substances to enter or leave a cell. These transport proteins will be discussed further in Exercise 4. Membrane proteins with other functions (e.g., receptors) are found primarily on the intracellular or extracellular surface of the plasma membrane. Table 3.1 lists the major types of membrane proteins and their functions.

The Nucleus

The **nucleus** (label 1) is a large, spherical organelle surrounded by a double-layered **nuclear envelope** (label 3). Unlike other membranes, the nuclear envelope contains pores that facilitate movement of materials between the nucleus and the cytoplasm. Note that a portion of the nuclear envelope is enlarged in Figure 3.1.

The nucleus is the integration center of the cell because it contains the threadlike **chromosomes** composed of **deoxyribonucleic acid (DNA)** and protein. The genetic information that controls cellular functions is encoded in the DNA molecules. In nondividing cells, the uncoiled and dispersed chromosomes appear as **chromatin** (not shown). During cell division, they coil tightly to form the familiar rodlike shape of chromosomes.

A single spherical **nucleolus** (label 16) is shown within the nucleus in Figure 3.1, although some types of cells may contain two or more nucleoli. Nucleoli are the assembly sites for **ribonucleic acid (RNA)** and proteins, which will move into the cytoplasm to form ribosomes.

The Cytoplasm

The **cytoplasm** lies between the nucleus and the plasma membrane. It contains the fluid portion of the cell, the **cytosol**, which bathes the specialized organelles. The cytosol, or **intracellular fluid (ICF)**, is composed of an aqueous mixture of many different kinds of chemicals. Cellular functions are primarily performed by the organelles, as described below.

Endoplasmic Reticulum (ER)
The endoplasmic reticulum (ER) is a network of membranes that permeates the cytoplasm between the nuclear envelope and plasma membrane and provides channels for the movement of molecules. **Rough endoplasmic reticulum (RER)** has ribosomes on its surface and provides storage and transport for proteins formed by the ribosomes. The RER is directly attached to and therefore usually surrounds the nucleus. In contrast, the **smooth endoplasmic reticulum (SER)** (label 15) lacks ribosomes. Its function may vary in different types of body cells. For example, it is involved in the synthesis of steroids in ovaries and testes, detoxification of drugs in the liver and kidneys, and storage and release of calcium in muscle cells to trigger muscle contraction.

Ribosomes
Ribosomes are tiny organelles composed of RNA and protein and serve as sites of protein synthesis. **Bound ribosomes** (label 14) associated with the RER produce proteins for export from the cell. The clusters or chains of **free ribosomes** (label 2) scattered in the cytoplasm form proteins for intracellular use.

Golgi Complex
The **Golgi complex** (label 13) is similar in structure to the ER. It consists of a stack of membranes adjacent to the RER. Proteins formed by ribosomes of the RER are packaged by the RER into **transitional vesicles** (label 4) that move to the Golgi complex. The role of the Golgi complex is to complete the processing and repackaging of such products for export from the cell. It places products for export into **condensing vesicles** (label 12), where final processing occurs as the vesicles move to the cell surface. Near the cell surface, condensing vesicles become **secretory vesicles** (label 9) that release their products outside the cell by *exocytosis*. The reverse process, endocytosis, moves substances into the cell within vesicles formed by the plasma membrane. The term **vesicle** refers to any small membranous sac of material within a cell.

Mitochondria

Mitochondria are ovoid or elongated organelles that have an outer membrane enclosing an inner membrane that has numerous partition-like folds called **cristae** (label 11). The folds increase the surface area of the inner membrane to facilitate chemical reactions. Mitochondria can move about in the cytoplasm and can replicate themselves. They contain small amounts of RNA and DNA. They are sometimes called the "powerhouses" of the cell because they are sites of ATP synthesis by aerobic respiration.

Lysosomes

Lysosomes are tiny oval sacs (label 7) that contain digestive enzymes and are formed by the Golgi complex. They release their enzymes into vesicles within the cytoplasm, to digest engulfed foreign particles or worn-out organelles. In fatally damaged cells, the lysosome membrane ruptures and the released enzymes digest the entire cell.

Cytoskeletal Elements

Very thin protein fibers extend throughout the cytoplasm, providing support, contractility, and the movement of organelles within the cell. Two types of protein fibers compose these cytoskeletal elements: microtubules and microfilaments. They are not shown in Figure 3.1.

The cylindrical **microtubules** are the larger of the two. They serve as pathways for the movement of vesicles and other organelles within the cytoplasm and provide support for the cell. The thinner **microfilaments** are solid rather than tubular, and they are responsible for cell movements because of their contractility (shortening). Microfilaments are best developed in muscle cells.

Centrioles

Cells capable of cell division contain a pair of **centrioles** (label 8) located close together and at right angles to each other. Together they constitute the **centrosome.** Each centriole is a short hollow tube whose wall is formed by nine microtubules. Centrioles are centers of microtubule production, especially those that form the astral fibers and spindle fibers during cell division. These fibers are labeled in Figure 3.3. In cells viewed with a light microscope, centrioles appear as a pair of dots near the nucleus.

Microvilli

Microvilli are tiny cytoplasmic projections on the free surfaces of certain cells. They greatly increase the surface areas of the cells. Microvilli (label 10) constitute the brush border on cells involved in the absorption of substances.

Cilia and Flagella

Both cilia and flagella (not shown in Figure 3.1) are composed of microtubules derived from centrioles or their products. **Cilia** are numerous, short, hairlike projections on the surface of certain cells that are used to move particles along the cell surface. They occur on cells lining the uterine tubes and respiratory passages. Sperms are the only human cells with a **flagellum.** Each sperm has a single flagellum whose whiplike movements propel the sperm.

Assignment

1. Label Figure 3.1 and record the labels in the laboratory report.
2. Locate the parts of a cell on the cell model.
3. Complete Section B of the laboratory report.

The Cell Cycle

The life cycle of a cell consists of three distinct stages: interphase, mitosis, and cytokinesis.

Interphase

Interphase is the stage of the life cycle when a cell is carrying out its usual metabolic functions and is not undergoing cell division. It is the stage between the end of one cell division and the start of the next cell division. During interphase, chromosomes are long, thin strands, which are loosely packed in the nucleus. When viewed microscopically, they appear as tiny dots and lines called **chromatin.**

As shown in Figure 3.2, interphase consists of three periods: G_1, S, and G_2. G_1 (gap 1) is a period

Figure 3.2 The cell cycle.

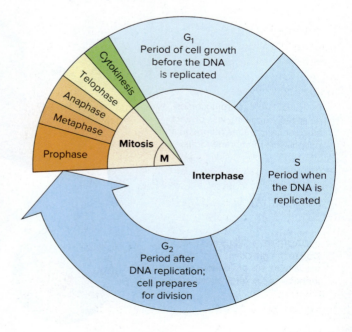

Cellular Anatomy and Mitosis

Figure 3.3 Mitosis. **A&PR** (a–f) Ed Reschke

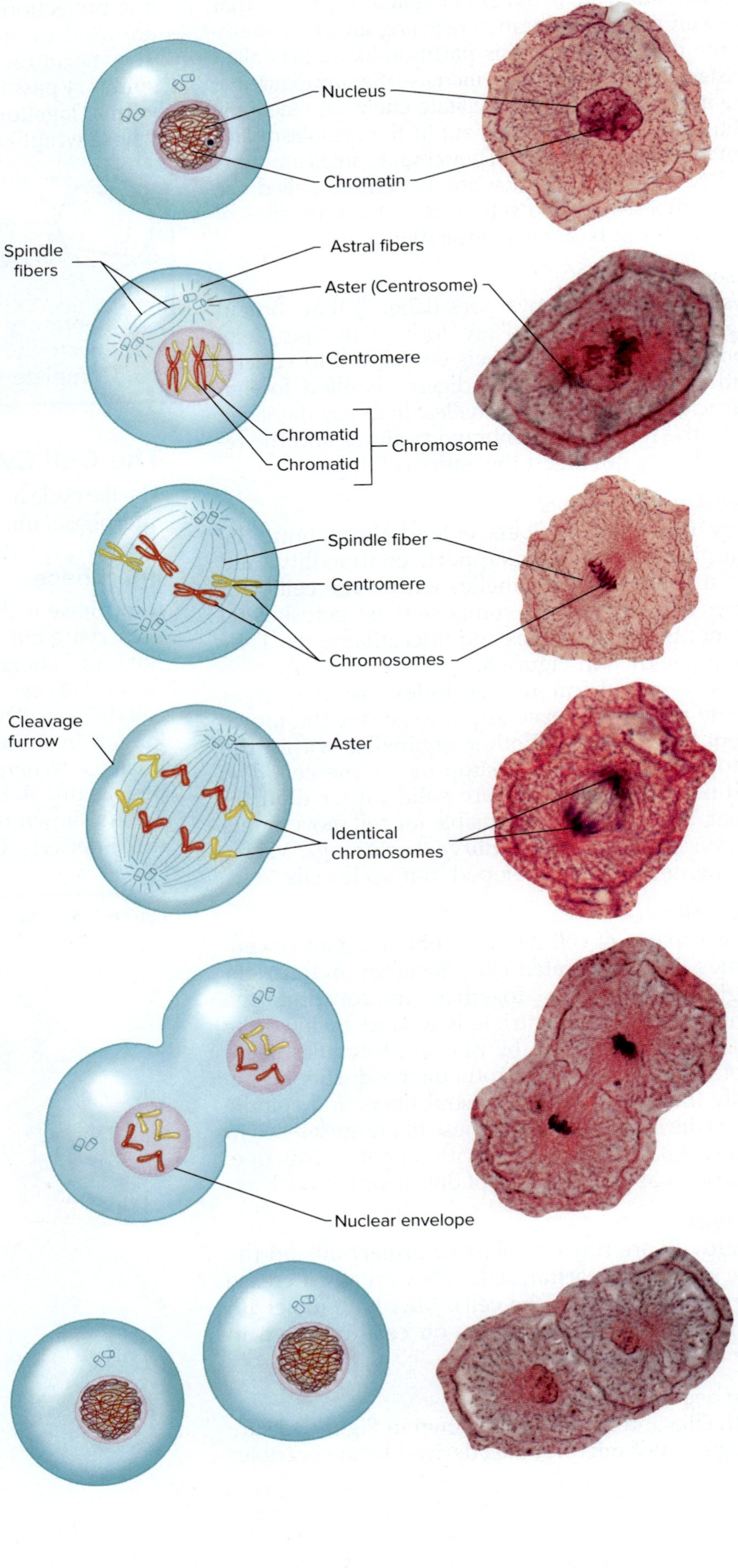

(a) **INTERPHASE** The nucleus is distinct with an intact nuclear envelope and chromatin.
- Nucleus
- Chromatin

(b) **PROPHASE** The nuclear envelope disintegrates, and the replicated chromosomes become visible as they coil to appear as rodlike structures.
- Spindle fibers
- Astral fibers
- Aster (Centrosome)
- Centromere
- Chromatid
- Chromatid
- Chromosome

(c) **METAPHASE** Chromosomes become arranged in equatorial plane of cell. Note relationship of centromeres to spindle fibers.
- Spindle fiber
- Centromere
- Chromosomes

(d) **ANAPHASE** Chromatids of each chromosome are separated and move along spindle fibers to opposite poles of cell.
- Cleavage furrow
- Aster
- Identical chromosomes

(e) **TELOPHASE** Chromosomes lengthen, becoming less distinct. Nuclear membranes reappear, and new plasma membrane forms.
- Nuclear envelope

(f) **DAUGHTER CELLS (INTERPHASE)** Due to the preciseness of mitosis these two cells are genetically identical to the original cell.

28   Cells and Tissues

of growth in a newly formed cell. Cells that will not divide again remain in this stage, now called G_0, and carry on their usual functions.

In the S (synthesis) stage, chromosome replication occurs. The precise replication of the DNA molecules is the key component of the replication process.

In G_2, the cell makes final preparations, including completion of centriole replication, for the impending cell division.

Mitosis

Mitosis is the orderly separation and equal distribution of the replicated chromosomes to form two daughter nuclei with identical chromosomes and, therefore, the same genetic composition. Mitosis accounts for only 5% to 10% of the cell cycle. Once it begins, it is a continuous process, but it is subdivided into four phases for ease of understanding: prophase, metaphase, anaphase, and telophase. Examine Figure 3.3 as you study these phases.

Prophase

During **prophase,** the nuclear envelope disintegrates. The replicated chromosomes coil tightly and become visible as rodlike structures. Each replicated chromosome consists of two **sister chromatids** joined by the paired **centromeres.** The two centrosomes move to opposite ends of the cell. By the end of prophase, each pair of centrioles in each centrosome has formed **spindle fibers** (microtubules), which extend to the equator of the cell, and **astral fibers** (microtubules), which radiate to the plasma membrane. Each centrosome forms one pole (end) of the spindle and is called an **aster** because of the radiating astral fibers. The replicated chromosomes begin their migration to the equator of the spindle.

Metaphase

During the brief **metaphase,** the replicated chromosomes line up on the equatorial plane of the spindle. The centromeres of sister chromatids are attached to spindle fibers extending to opposite poles of the spindle.

Anaphase

In **anaphase,** the paired centromeres separate, separating the sister chromatids, which migrate along spindle fibers to opposite poles of the cell. They appear to be pulled to the poles by the spindle fibers. The separated sister chromatids are now called **daughter chromosomes.**

Telophase

During **telophase,** the daughter chromosomes extend and become less distinct. A new nuclear envelope forms around each set of chromosomes, forming the daughter nuclei. The spindle fibers gradually disappear.

Cytokinesis

Cytokinesis is the division of the parent cell into two daughter cells. It begins during late anaphase or early telophase with the formation of a **cleavage furrow,** which forms as the plasma membrane constricts at the equator of the cell. The cleavage furrow gradually constricts and finally separates the parent cell into two daughter cells. The daughter cells then enter interphase of the cell cycle.

Assignment
1. Complete Sections C and D of the laboratory report.
2. Study anatomical models of mitosis and locate the structures described in this lab.

Microscopic Study

In this section, you will examine mitosis in cells of whitefish blastula, an early embryonic stage. After initially finding the blastula using the scanning objective switch to the low-power objective to locate cells in interphase and the stages of mitosis. Then switch to the high-dry objective for careful examination. Your study will be simplified by studying mitotic stages as you encounter them on your slides rather than by trying to locate them as they occur in the mitotic sequence. There are several clusters of cells on your slides, and you may need to examine all of them, or even additional slides, to locate cells in all of the mitotic stages. This portion of the lab can alternately be done with the tips of onions roots. Although these are plant cells and have distinct differences from animal cells, the phase of mitosis can still be explored relatively accurately.

Assignment
Examine a prepared slide of whitefish mitosis (or onion root tip mitosis) and locate cells in interphase and in each mitotic stage. Whitefish cells contain many chromosomes, but you can identify the various stages by the chromosome distribution. Diagram and label these cells in Section E of the laboratory report.

Exercise 4

Membrane Transport

Objectives

After completing this exercise, you should be able to

1. Explain the cause of Brownian movement, diffusion, and osmosis.
2. Describe the effect of molecular weight and temperature on the rate of diffusion.
3. Describe the relationship between osmotic equilibrium and the integrity of plasma membranes.
4. Explain carrier-mediated active transport.

Materials
- Compound microscopes
- Hot plates
- Beakers, 250 ml
- Biohazard container
- Depression slides and cover glasses
- Forceps
- Mechanical pipetting devices
- Petri dishes of agar-agar
- Serological pipettes
- Serological test tubes and test tube racks
- Syringes and needles
- Thermometers, Celsius
- Thistle-tube osmometers
- Fresh blood, mammalian
- Glucose solutions, 2.0%, 5.0%
- Ice
- Methylene blue granules
- NaCl solutions, 0.3%, 0.9%, and 2.0%
- Potassium permanganate granules
- Sucrose solutions, 5% and 15%, for osmometers
- Small paper sheets that say ATP on them
- Small paper sheets that say Na^+ and K^+ on them

Before You Proceed

Consult with your instructor about using protective disposable gloves when performing portions of this exercise.

Materials are constantly exchanged between body cells and the interstitial fluid (tissue fluid) that bathes them. These substances, including organic nutrients, inorganic ions, or water, pass into and out of cells through the plasma membrane. Some molecules, usually small ones, move passively through the membrane by diffusion along a concentration gradient. Others move either with or against a concentration gradient by active transport. In either case, the selective permeability of the plasma membrane plays an important role.

Brownian Movement

Molecules in a liquid or gaseous state are in constant, random motion. Each molecule moves in a straight line until it bumps into another molecule, and then it bounces off in another straight-line path. This molecular motion cannot be observed directly, but it may be observed indirectly by noting the random motion of microscopic particles suspended in water. The random bombardment of these particles by water molecules causes their random, vibratory movement, which is called **Brownian movement** after Robert Brown, a Scottish botanist who first described it in 1827.

Diffusion APR

The net movement of the same kind of molecules from an area of their higher concentration to an area of their lower concentration is called **diffusion.** Molecules move away from the area of higher concentration because the area of lower concentration has fewer molecules to obstruct straight-line movement of the molecules. For example, if a water-soluble substance such as a lump of sugar is placed in water, it will gradually diffuse throughout the liquid until it is equally distributed.

Diffusion occurs in liquids and gases and is an important means of distributing materials within

Unit 2: Cells and Tissues

cells and of moving some materials into and out of cells. It is a passive process that results from the random motion of molecules and does not require an expenditure of energy by the cells. However, the rate of diffusion is affected by both temperature and molecular size.

Osmosis APR

Water is the most abundant substance in living cells. It is the **solvent** of living systems—the liquid medium in which the chemical reactions of life occur. Organic and inorganic substances constitute the **solutes.** Water molecules are small and move freely through plasma membranes. Like other molecules, water molecules move from an area of their higher concentration to an area of their lower concentration. The movement of water through a semipermeable, or selectively permeable, membrane is called **osmosis.** Note that the phase "diffusion of water" was not used to describe osmosis. Osmosis and water diffusion are not the same phenomena. This is explained next.

The direction of a net movement of water across a membrane is dependent upon the concentration of water and thus the concentration of solutes on each side of the membrane. If pure water is separated from a 10% sucrose solution by a semipermeable membrane that sucrose can not cross, there will be a net movement of water into the sucrose solution. The force required to prevent this movement is called **osmotic pressure,** and its value is assigned to the sucrose solution. The greater the concentration of impermeable solute particles (molecules or ions), the greater is the osmotic pressure of the solution. Thus, the degree of ionization of solute molecules markedly affects the osmotic pressure of a solution.

If two solutions with the same osmotic pressure are separated by a semipermeable membrane, there will be no net movement of water from one solution into the other. Such solutions are said to be **isotonic,** and they are at **osmotic equilibrium.** The integrity of the plasma membrane is dependent upon osmotic equilibrium between the cells and the interstitial fluid.

Now consider hypothetical sucrose solutions A and B that are separated by a semipermeable membrane that sucrose can not cross. Solution A has a lower solute concentration but a greater water concentration than B. Thus, A has a lower osmotic pressure than B, and the net movement of water will be from A to B. Solution A is said to be **hypotonic** to B, because it has a lower solute concentration and a lower osmotic pressure. Solution B is **hypertonic** to A, because it has a higher solute concentration and a higher osmotic pressure. Whenever aqueous solutions with unequal osmotic pressures are separated by a semipermeable membrane, the net movement of water is always from the hypotonic solution into the hypertonic solution.

Assignment
Complete Section A of the laboratory report.

Carrier-Mediated Active Transport APR

In **carrier-mediated active transport,** carrier proteins use ATP to move substances across the plasma membrane against (opposite to) their concentration gradient, meaning from an area of low concentration to an area of high concentration. The most important of these is the *sodium-potassium pump* (Na^+/K^+ pump), a type of carrier protein, that establishes and maintains sodium and potassium gradients across the plasma membrane. The Na^+/K^+ pump uses the energy from an ATP molecule to move three sodium ions out of the cell and two potassium ions into the cell. The continuous action of the Na^+/K^+ pumps in the plasma membrane creates a sodium gradient with the highest concentration on the outside of the cell and a potassium gradient with the highest concentration on the inside of the cell. These sodium and potassium gradients are very important to the overall functioning of the human body.

Demonstrations and Experiments

The demonstrations and experiments in this section will aid your understanding of diffusion and osmosis. Record your observations and results on the laboratory report.

Diffusion and Temperature

1. Fill three 250-ml beakers about two-thirds full with tap water at different temperatures.

Beaker A	Ice water
Beaker B	Room temperature
Beaker C	60°C to 70°C

2. Measure and record the temperature of the water in each beaker.
3. Place the beakers on your table where they will not be disturbed and can remain motionless.
4. Drop a small granule of potassium permanganate into each beaker without disturbing the water.

Membrane Transport 31

Figure 4.1 Determining the effect of molecular weight on the rate of diffusion.

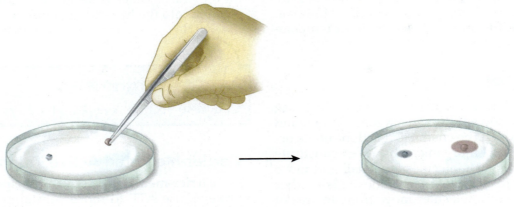

Equal-sized granules of methylene blue and potassium permanganate placed on agar medium.

Diffusion distances measured after 1 hour.

5. Observe the beakers at 5-minute intervals and record the time required for the molecules of potassium permanganate to diffuse throughout the water in each beaker.

Diffusion and Molecular Weight

1. Obtain a Petri dish containing about 12 ml of 1.5% agar-agar. The agar gel is 98.5% water, and water-soluble molecules readily diffuse through it.
2. Use forceps to place approximately equal-sized granules of potassium permanganate and methylene blue on the agar about 5 cm apart as shown in Figure 4.1. Set the dish aside where it will not be disturbed.
3. After 1 hour, or as much time as your lab allows, measure and record the diameter (in mm) of the colored area around each granule to determine any difference in the rate of diffusion. The molecular weight of potassium permanganate is 158; the molecular weight of methylene blue is 320.

Osmosis

1. Examine the two thistle-tube osmometers, like the one shown in Figure 4.2, that were set up at the start of the laboratory session. Each osmometer was prepared by securing a semipermeable membrane over the bulb of a thistle-tube and filling the bulb with a sucrose solution. One tube received a 5% solution, and the other received a 15% solution. Then, the bulb of each thistle-tube was immersed in a beaker of distilled water as shown in Figure 4.2.
2. Measure and record the height (mm) of the water-sucrose column in osmometers A and B at the start of the experiment and 30 minutes later.

Osmosis and Plasma Membrane Integrity

In this experiment, you will determine the effect of the osmotic pressure of five solutions on red blood cells. Figure 4.3 shows the protocol for the experiment. Recall that osmotic equilibrium between cells and the fluid bathing them is required for the healthy functioning of cells. Figure 4.4 shows the effect of hypertonic, isotonic, and hypotonic

Figure 4.2 Osmosis setup.

32 Cells and Tissues

Figure 4.3 Routine for making cell suspensions.

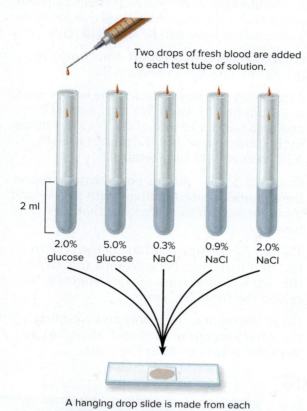

next. Your instructor will place *all items that come in contact with blood in a biohazard container immediately after use.*

1. Label five clean serological test tubes 1 to 5 and place them in a test tube rack.

2. Use a 5-ml pipette to place 2 ml of solution in each tube:

   Tube 1—2.0% glucose
   Tube 2—5.0% glucose
   Tube 3—0.3% NaCl
   Tube 4—0.9% NaCl
   Tube 5—2.0% NaCl

   Use the mechanical pipetting device as shown in Figure 4.5 to dispense the solutions. Rinse the pipette with distilled water after the delivery of each solution.

3. Use a syringe to add two drops of blood to each tube. Shake the tubes from side to side to mix thoroughly and return them to the test tube rack. The test tube rack and test tubes are on the demonstration table for your observation. *Do not handle the tubes.*

4. After 5 minutes, examine the relative transparency of the liquid in the tubes by holding a page of this text behind the test tube rack and viewing the text through the tubes. Interpret the clarity of the solutions as follows:

   Hypotonic solution: transparent due to hemolysis
   Isotonic solution: cloudy due to the intact red blood cells
   Hypertonic solution: clarity intermediate between transparent and cloudy due to the crenation of the cells

5. Examine the demonstration microscope setups using hanging drop slides and oil immersion objectives, which show the effect of these solutions on the shape of red blood cells.

solutions on red blood cells. Hypertonic solutions cause water to leave the cells, and they shrivel (a process called crenation). Isotonic solutions cause no net movement of water. Hypotonic solutions cause water to enter the cells, and they ultimately burst (lyse), because of the excessive internal pressure that is generated. The rupture of red blood cells is called **hemolysis.**

Your instructor has prepared a demonstration of osmotic pressure on red blood cells as described

Figure 4.4 The effects of hypertonic, isotonic, and hypotonic solutions on red blood cells.

Hypertonic
(Crenation)

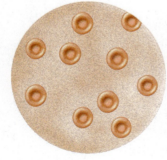
Isotonic
(Osmotic Equilibrium)

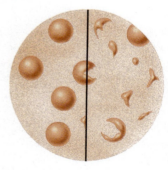
Hypotonic
(Hemolysis)

Membrane Transport

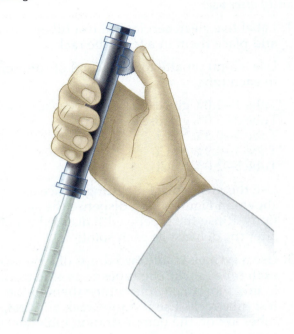

Figure 4.5 Use a mechanical pipetting device for all pipetting.

Carrier-Mediated Active Transport

For this activity you need the entire class to participate. All students should align themselves in three rows. The middle row represents the Na$^+$/K$^+$ pumps in the plasma membrane. One outer row will be designated the cytosol and needs to have three Na$^+$ sheets of paper and most, but not all students, should also have one ATP sheet of paper. The other row is the interstitial fluid outside of the cell and needs to have two K$^+$ sheets of paper.

1. Go to the laboratory report and complete the Before column in Section F.
2. After everyone has recorded the before data, any student with three Na$^+$ and one ATP needs to hand it to one of the Na$^+$/K$^+$ pumps in the middle row. Na$^+$/K$^+$ pumps use the ATP as energy to turn around.
3. Once the Na$^+$/K$^+$ pumps have turned around they should locate a student with two K$^+$ and trade them with the three Na$^+$ that they now have.
4. The Na$^+$/K$^+$ pumps need to rip the ATP in half and then turn back around and give the two K$^+$ to the student they originally received the three Na$^+$ from.
5. Go to the laboratory report and complete the After column in Section F as well as answer the questions after the table.

Assignment

Complete the laboratory report.

Exercise 5

HISTOLOGY

Objectives

After completing this exercise, you should be able to

1. List the types of epithelial and connective tissues and, for each, describe its structure, functions, and locations.
2. Recognize each type of epithelial and connective tissue when viewed with a microscope.
3. Describe the characteristics of smooth, skeletal, and cardiac muscle tissue and identify each type when viewed microscopically.
4. Describe the characteristics of nervous tissue.
5. Describe the characteristics of motor neurons, sensory neurons, and interneurons.
6. Identify nervous tissue and neurons when viewed microscopically.

Materials

Prepared slides of epithelial tissues:
   simple squamous
   simple cuboidal
   simple ciliated columnar
   simple nonciliated columnar
   stratified squamous
   stratified cuboidal
   stratified columnar
   transitional
   pseudostratified ciliated columnar
   pseudostratified nonciliated columnar
Prepared slides of connective tissues:
   areolar
   adipose tissue
   reticular
   dense regular
   dense irregular
   elastic
   hyaline cartilage
   elastic cartilage
   fibrocartilage
   compact bone l.s., x.s.
Prepared slides of
   cardiac muscle tissue
   skeletal muscle tissue, teased and unteased
   smooth muscle tissue, teased and unteased
   giant multipolar neurons
Models of bone tissue and the neuron

Although body cells share many common structures and functions, they may differ considerably in size, shape, and structure in accordance with their specialized functions. Cells of a similar type usually occur in groups. An aggregation of cells that are similar in structure and function is a **tissue.** The scientific study of tissues is **histology.**

There are four basic types of tissues in the body: epithelial, connective, muscle, and nervous tissues. Each basic tissue will be introduced before an in-depth discussion of the subtypes in each category. We will begin with epithelial tissues.

Epithelial Tissues

Epithelial tissues cover surfaces of the body and internal organs, line cavities, and form glands. They may be protective, absorptive, secretory, or excretory in function. They are characterized by (1) closely packed cells without intercellular substances or blood vessels and (2) the presence of a noncellular **basement membrane** that attaches the tissue to underlying connective tissue. Most epithelial tissues have a layer of epithelial cells with a **free surface** (i.e., cells with one surface that is not in contact with other cells). Epithelial cells have a high rate of cell division to replace cells that are worn out or sloughed off by typical processes.

Epithelial tissues may be composed of one or more layers of cells. The number of cell layers and the shape of the cells provide the basis for classifying and naming epithelial tissues. See Figures 5.1 and 5.2. Figure 5.3 depicts the morphological classification of epithelial tissues.

Unit 2: Cells and Tissues

Figure 5.1 Shapes of epithelial cells.

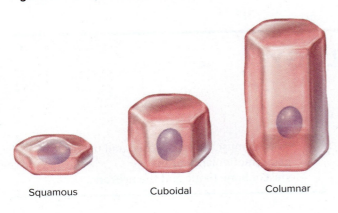

Squamous   Cuboidal   Columnar

Figure 5.2 Epithelial classifications based on the number of cell layers.

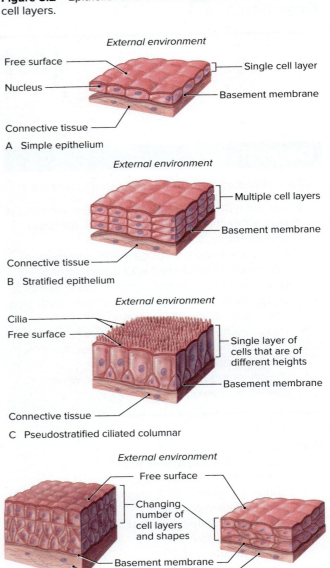

A Simple epithelium

B Stratified epithelium

C Pseudostratified ciliated columnar

D Transitional epithelium

Simple Epithelia

Simple epithelial tissues consist of a single layer of cells.

Simple Squamous Epithelium
Simple squamous cells are thin and flat and appear as irregular polygons in a surface view. The peritoneum, pleurae, pulmonary alveoli (air sacs of the lungs), and endothelium (lining of vessels) are formed of simple squamous epithelium. This epithelium is involved in filtering fluids and in the exchange of materials. Figure 5.4 illustrates the parietal pleura and pulmonary alveoli.

Simple Cuboidal Epithelium
The cells of simple cuboidal epithelium are block-like in cross section and hexagonal in a surface view. Secretion and absorption are its chief functions. Simple cuboidal epithelium occurs in renal (kidney) tubules and in both endocrine and exocrine glands. For example, it is found in the thyroid gland, pancreas, ovaries, sweat glands, and salivary glands. Figure 5.5 shows simple cuboidal epithelium in the renal tubules.

Figure 5.3 A morphological classification of epithelial types.

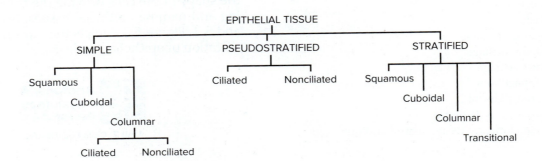

Figure 5.4 Simple squamous epithelium (250×).
(Bottom) Ed Reschke

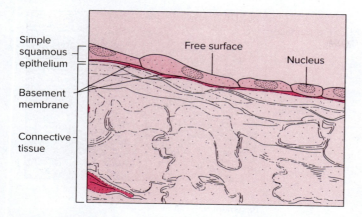
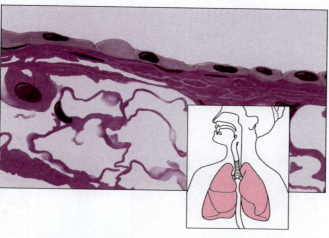

Figure 5.5 Simple cuboidal epithelium (250×).
(Bottom) Ed Reschke

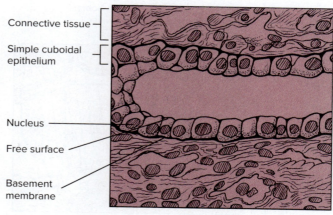
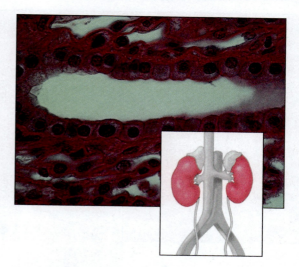

Simple Columnar Epithelium

The cells of simple columnar epithelium are elongated and appear polygonal on their free surfaces. Scattered among the columnar cells are special **goblet cells** that secrete **mucus,** which flows over the surfaces of adjacent cells and protects the tissue.

There are two types of simple columnar epithelium. **Ciliated simple columnar epithelium** lines the uterine tubes and the small bronchi of the lungs. **Nonciliated simple columnar epithelium** lines the digestive tract from stomach to anus. These intestinal cells have **microvilli** on their free surface, which greatly increase the absorptive surface area of the cells. Nonciliated simple columnar epithelium is shown in Figure 5.6.

Stratified Epithelia

Stratified epithelial tissues consist of two or more layers of cells. Four types of stratified epithelia are shown in Figures 5.7 to 5.10. The tissues are classified according to the shape of the surface cells.

Stratified Squamous Epithelium

The superficial cells of stratified squamous epithelium are distinctly squamous in shape, but the deepest cells are cuboidal or columnar. Cells change in shape as they migrate to the surface of the tissue. Protection is the chief function of this tissue, found in regions where abrasions or friction occur, such as the outer layer of the skin, oral cavity, vagina, cornea of the eye, and esophagus. Figure 5.7 depicts the stratified squamous epithelium that lines the esophagus.

Stratified Cuboidal Epithelium

The shape of the surface cells is cubelike or round. Stratified cuboidal epithelium is found in ducts of sweat glands, in sperm-forming tubules of the testes, and in egg-producing follicles of the ovaries. Secretion is a major function. Figure 5.8 shows the stratified cuboidal epithelium lining a duct of a sweat gland.

Figure 5.6 Nonciliated simple columnar epithelium (400×). (Bottom) Ed Reschke/Photolibrary/Getty Images

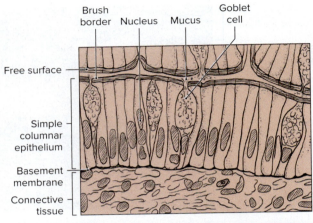
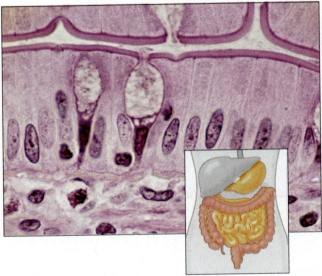

Figure 5.7 Stratified squamous epithelium (70×). (Bottom) Biophoto Associates/Science Source

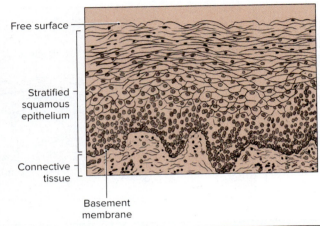
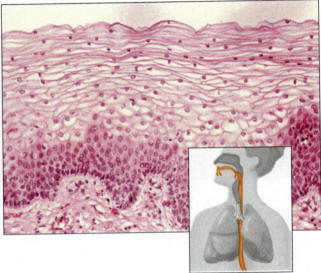

Figure 5.8 Stratified cuboidal epithelium (500×). (Right) Al Telser/McGraw-Hill Education

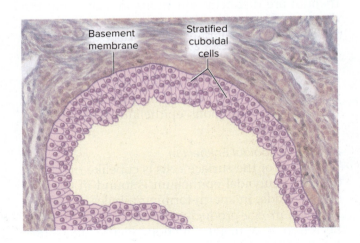
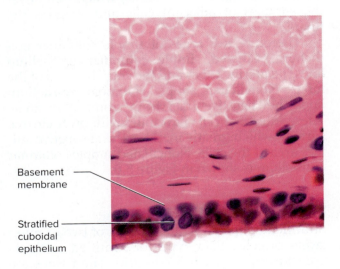

38 Cells and Tissues

Figure 5.9 Stratified columnar epithelium (500×).
(Bottom) Al Telser/McGraw-Hill Education

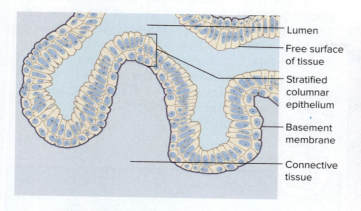

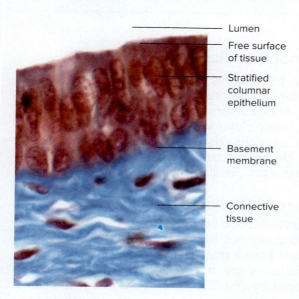

Figure 5.10 Transitional epithelium (250×). (Top) Ed Reschke/Photolibrary/Getty Images

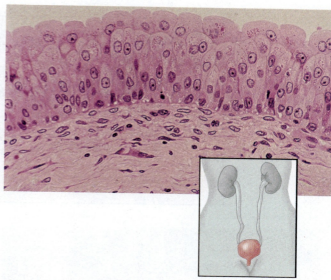

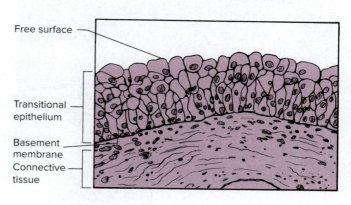

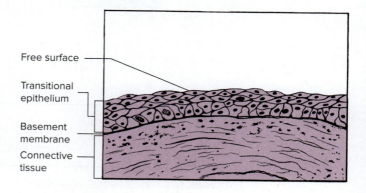

Stratified Columnar Epithelium
The columnar cells at the surface of stratified columnar epithelium are formed from the deeper rounded cells as they migrate upward. This tissue occurs in only a few areas of the body, such as the conjunctiva, parts of the pharynx and epiglottis, the anal canal, and part of the male urethra, which is shown in Figure 5.9. Protection and secretion are its primary functions.

Transitional Epithelium
Transitional epithelium occurs in organs of the body where stretching of the tissue may occur, such as the urinary bladder and ureters. Elasticity is a key characteristic. The deepest cells are columnar and loosely held together, whereas the surface cells may be dome-shaped and cuboidal or squamous, depending upon the degree of stretching. Figure 5.10 shows the transitional epithelium lining the urinary bladder and a drawing of what the tissue looks like when stretched.

Pseudostratified Epithelia

Pseudostratified epithelial tissues consist of a single layer of columnar cells but appear to be composed of more than one layer (Figure 5.11). This

Histology 39

Figure 5.11 Ciliated pseudostratified columnar epithelium (500×). (Bottom) Ed Reschke/Photolibrary/Getty Images

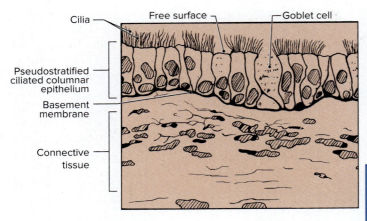

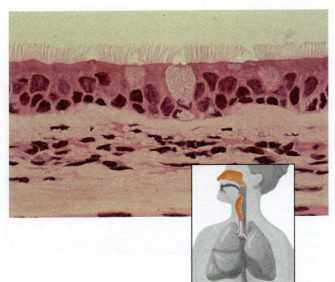

illusion occurs because (1) each cell is in contact with the basement membrane, but not all cells reach the free surface of the tissue; and (2) the nuclei occur at different levels in the cells. Ciliated pseudostratified columnar epithelium lines the nasal cavity, trachea, main (primary) bronchi, and auditory tubes. Nonciliated pseudostratified columnar epithelium is found in parts of the male urethra and in large ducts of the parotid salivary gland. Mucus-secreting goblet cells occur in both ciliated and nonciliated types.

Assignment
Complete Section A of the laboratory report.

Connective Tissues

Connective tissues provide support and attachment for various organs and fill spaces in the body. They are characterized by an abundance of a nonliving intercellular substance, the **matrix,** and few cells. Connective tissue may be classified in several ways. The system used here is based on the nature of the matrix. See Figure 5.12.

Connective Tissue Proper

Connective tissue proper consists of connective tissues that are not highly specialized. They are pliable and have a matrix that consists of a semifluid or gelatinous **ground substance** and protein fibers produced by **fibroblasts.** Three types of fibers are produced by fibroblasts. The strong, nonelastic, **collagenous fibers** provide great tensile strength. **Elastic fibers,** formed of the protein elastin, provide elasticity. They are often branched and not as

Figure 5.12 Types of connective tissue.

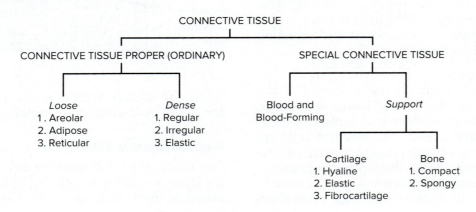

40   Cells and Tissues

strong as collagenous fibers. The thin, branching **reticular fibers** form a network of very fine collagenous threads.

Loose Connective Tissue

Loose connective tissues help to bind together other tissues and form the basic supporting framework for organs. Their matrix consists of a semifluid or jellylike ground substance in which protein fibers and fibroblasts are embedded. The word "loose" describes how the protein fibers are widely spaced and intertwined between the fibroblasts. There are three types of loose connective tissue: areolar connective tissue, adipose tissue, and reticular tissue.

Areolar Connective Tissue Areolar connective tissue is widespread in the body. Fibroblasts are the most common cells, although mast cells and macrophages are present. See Figure 5.13. Areolar connective tissue provides a flexible supporting framework within and between organs, contains blood vessels that nourish surrounding tissues, and is a site of immune reactions. It is found beneath epithelial tissues, around and within muscles and nerves, and in serous membranes.

Adipose Tissue Small groups of **adipocytes** (fat cells) are scattered throughout the body. Accumulations of fat cells form adipose tissue that serves as a protective cushion and insulation for the body. The vacuoles of adipocytes are filled with lipids and usually compress the cytoplasm and nucleus to the edge of the cell. The lipids stored in adipocytes provide an important energy reserve for the body. See Figure 5.14.

Figure 5.13 Areolar connective tissue (250x). (Bottom) Ed Reschke/Photolibrary/Getty Images

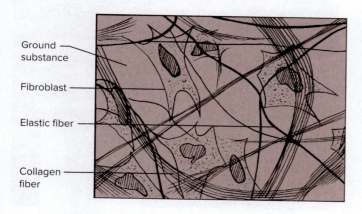

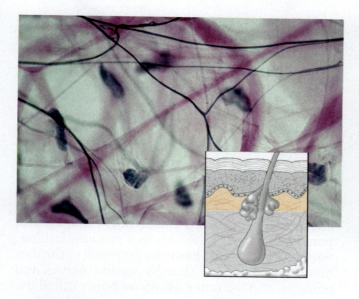

Figure 5.14 Adipose tissue (250x). (Bottom) Ed Reschke

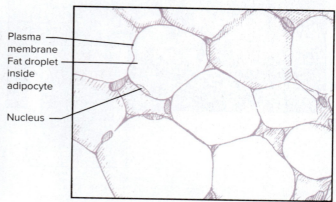

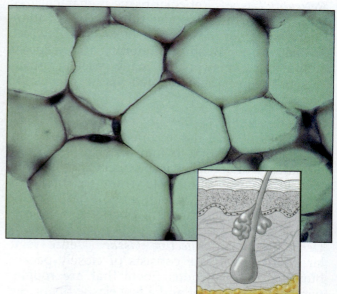

Histology

Figure 5.15 Reticular tissue (400×). (Bottom) Courtesy of Beth Ann Kersten **A&PR**

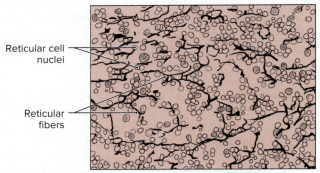

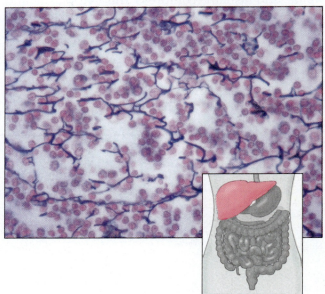

Figure 5.16 Dense regular connective tissue (100×). (Bottom) Christine Eckel/McGraw-Hill Education **A&PR**

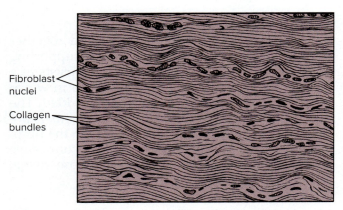

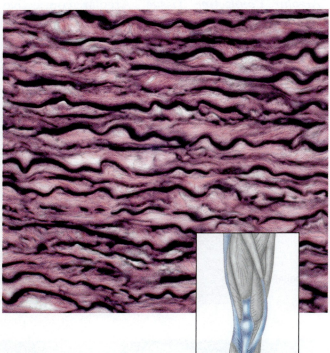

Reticular Tissue Reticular tissue is characterized by a network of fine reticular fibers plus an abundance of leukocytes (white blood cells), which are active in providing immunity. It provides a supporting framework for many vascular organs such as the liver, hematopoietic tissue, and lymph nodes. See Figure 5.15.

Dense Connective Tissue
Dense connective tissue differs from loose connective tissue in that its matrix consists of a predominance of densely packed fibers and relatively few cells, mostly fibroblasts. There are three subcategories of dense connective tissue based on the type and arrangement of the fibers.

Dense Regular Connective Tissue Dense regular connective tissue consists of closely packed bundles of collagenous fibers that are roughly parallel to each other and a few elastic fibers. See Figure 5.16. The few cells present are nearly all fibroblasts. There are relatively few blood vessels present. This tissue is noted for its tensile strength. It composes tendons, which attach muscles to bones, and ligaments, which join bones together, and it forms the deep fascia.

Dense Irregular Connective Tissue Dense irregular connective tissue consists of bundles of collagenous fibers and a few elastic fibers that are not parallel but are interwoven in a random fashion. See Figure 5.17. This tissue forms much of the dermis of the skin, the sheaths around nerves and tendons, and the outer sheath on bones called the periosteum.

Figure 5.17 Dense irregular connective tissue (400×).
(Bottom) Courtesy of Beth Ann Kersten

Figure 5.18 Elastic connective tissue (400×).
(Bottom) Courtesy of Beth Ann Kersten

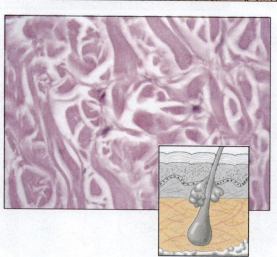

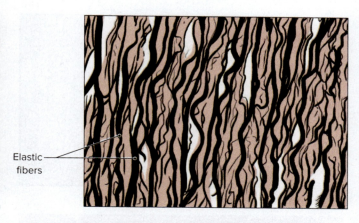

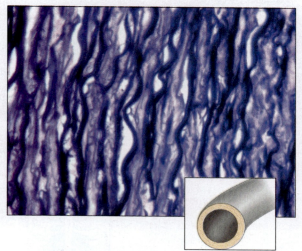

Elastic Connective Tissue An abundance of elastic fibers in the matrix distinguishes **elastic connective tissue.** Collagen fibers are also present, and fibroblasts are scattered between the fibers. Elastic connective tissue occurs where extensibility and elasticity are advantageous, such as in the lungs, air passages, vocal folds, and arterial walls. For example, elastic connective tissue enables the expansion of the lungs as air is inhaled and the recoil of the lungs as air is exhaled. See Figure 5.18.

Special Connective Tissue

Special connective tissue is subdivided into blood and hematopoietic (blood-forming) tissue and the specialized support tissues, cartilage tissue and bone tissue.

Blood is an atypical fluid connective tissue consisting of formed elements (blood cells and platelets) transported in a liquid matrix, the plasma. Blood is formed by the hematopoietic tissue that makes up the red bone marrow. Blood will be discussed in Exercise 18. The red bone marrow is labeled in the spongy bone in Figure 5.23.

Cartilage tissue and bone tissue possess (1) a matrix that is solid and strong; (2) cells located in **lacunae,** tiny cavities in the matrix; and (3) an external membrane capable of producing new tissue.

Cartilage Tissue
All three types of cartilage tissue are surrounded by a dense irregular connective tissue membrane, the **perichondrium,** and have cartilage cells, **chondrocytes,** in the lacunae of the matrix. Unlike other connective tissue, cartilage tissue lacks blood vessels, so its cells derive nourishment by diffusion from capillaries in the perichondrium.

Hyaline Cartilage Hyaline cartilage is the most common type of cartilage tissue. It has a white, glassy appearance due to the absence of visible

Histology 43

Figure 5.19 Hyaline cartilage (250×). (Bottom) Al Telser/McGraw-Hill Education

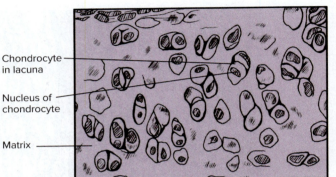
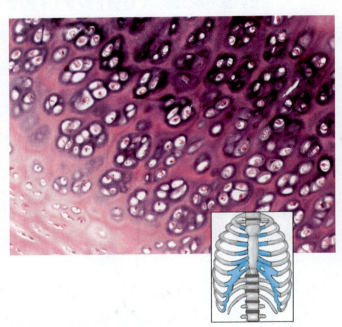

Figure 5.20 Elastic cartilage (100×). (Bottom) Al Telser/McGraw-Hill Education

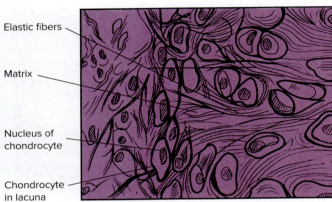
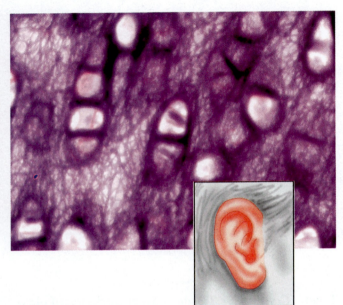

fibers in the matrix. Hyaline cartilage composes much of the fetal skeleton before ossification (bone formation) and covers the ends of long bones in adults. It also forms the cartilaginous portion of the nose and the cartilaginous supports of the respiratory passages and thoracic cage. See Figure 5.19.

Elastic Cartilage The presence of an abundance of elastic fibers makes elastic cartilage more flexible than hyaline cartilage. It is located in such places as the external ear and the epiglottis. See Figure 5.20.

Fibrocartilage Fibrocartilage is a very tough cartilage tissue that contains densely packed bundles of collagenous fibers that separate short rows of chondrocytes. It forms the intervertebral discs of the vertebral column, and parts of the knee joint and pelvic girdle, where it serves as a cushioning shock absorber. See Figure 5.21.

Bone Tissue
Bone tissue, the most rigid connective tissue, is hard, strong, and lightweight. It protects vital organs, provides an internal support for the body, and serves as a site for the attachment of muscles. As discussed earlier, bones contain the **red bone marrow** that produces the formed elements of blood.

In embryonic and fetal development, bones are preformed in either hyaline cartilage or dense connective tissue. These tissues provide a framework for the subsequent deposition of calcium salts, which form the bone matrix. Like other

Figure 5.21 Fibrocartilage (250×). (Right) Courtesy of Beth Ann Kersten A&PR

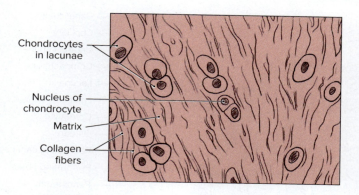
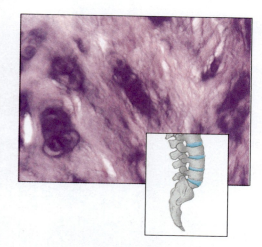

connective tissues, bone tissue is continually modified and reconstructed.

There are two basic types of bone tissue, as shown in Figure 5.22. Solid, dense **compact bone** (label 1) forms the outer portion of the bone, and porous **spongy** (cancellous) **bone** (label 2) forms the inner framework in the ends of long bones and the inside of all other bones.

Compact Bone The structural and functional unit of compact bone is the **osteon** (Haversian system), label 10 in Figure 5.22. Each osteon is an elongated hollow cylinder formed of concentric rings of bone matrix called **lamellae** (label 11). Running through the center of each osteon is a **central** (Haversian) **canal** (label 9) that contains blood vessels and nerves serving bone cells in the osteon.

The lamellae are secreted by **osteoblasts** that have numerous extensions radiating from the cell. Once osteoblasts are trapped in lacunae (label 5) by their own depositions, they are called **osteocytes**. The radiating microcanals between lacunae are the **canaliculi** (label 12). They contain the cellular extensions of the osteocytes and serve as passageways for materials moving between blood vessels in the central canals and the lacunae.

Adjacent osteons are fused together at their outer lamellae and by bone deposits between them to yield a solid, hard matrix of calcium salts. Central canals of adjacent osteons are connected by **perforating canals** (label 4) that penetrate osteons at right angles to the osteon axis. These canals carry blood vessels and nerves from the periosteum to the central canals.

The outer surface of compact bone is covered by the **periosteum** (label 8). It consists of (1) an outer, fibrous membrane and (2) an inner layer that is a source of new bone-forming cells and capillaries. The periosteum is firmly attached to the compact bone by **perforating fibers** (label 7) that penetrate the outer lamellae. See Figure 5.23.

Spongy Bone Spongy bone tissue is beneath and continuous with the compact bone. There is no distinct line of demarcation between the two tissues. The distinctive feature is the presence of numerous branching bony plates, called **trabeculae** (label 3), with interconnecting spaces between them. Osteons are absent. Instead, osteocytes are scattered through the trabeculae and are nourished by diffusion from blood vessels that meander through the spaces. The spaces are filled with red bone marrow and serve to decrease the weight of the bone. See Figure 5.23.

Assignment

1. Label Figure 5.22.
2. Compare Figure 5.22 with an anatomical model of bone tissue.
3. Complete Sections B, C, and D of the laboratory report.

Muscle and nervous tissues exhibit **excitability**, the ability to transmit electrochemical impulses (*graded potentials* and *action potentials*). The adaptive use of excitability results in (1) the

Figure 5.22 Bone tissue.

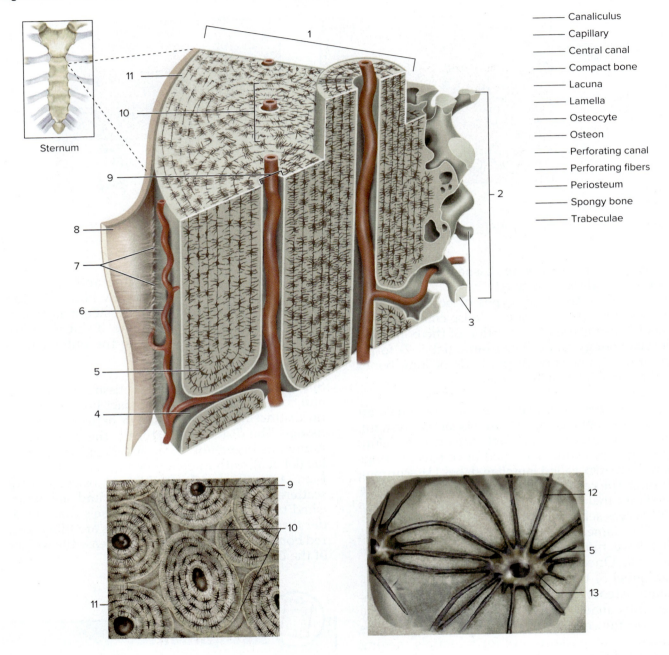

Canaliculus
Capillary
Central canal
Compact bone
Lacuna
Lamella
Osteocyte
Osteon
Perforating canal
Perforating fibers
Periosteum
Spongy bone
Trabeculae

contraction of muscle cells to enable movement and (2) the transmission of action potentials by neurons to coordinate body functions.

Muscle Tissues

Contractility, the ability to shorten or contract, is a major characteristic of muscle tissue. During contraction the elongated muscle cells shorten and exert a pull at their attached ends that causes movement of body parts. Muscle tissues may be classified according to location, structure, and function.

Skeletal Muscle

Skeletal muscle tissue occurs in muscles attached to bones, skin, and other muscles. The

46   Cells and Tissues

Figure 5.23 Bone tissue. (A) and (B) Compact bone (160×), (C) spongy bone (100×). (B) Victor B. Eichler, Ph.D.; (C) Courtesy of Beth Ann Kersten

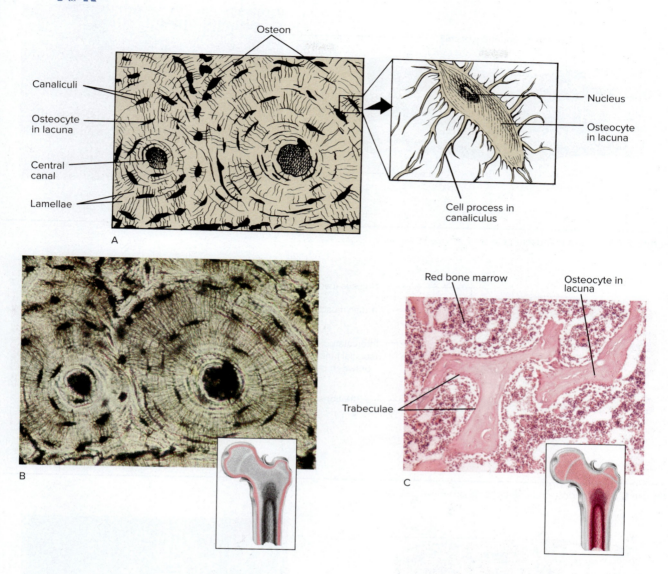

actions of skeletal muscles are consciously controlled. Thus, this tissue is classified functionally as **voluntary**.

Skeletal muscle cells are called **muscle fibers** because they are elongate and cylindrical. Each fiber is distinguished by numerous peripherally located nuclei and alternating dark and light bands called **striations** in the **sarcoplasm** (cytoplasm). The plasma membrane of the skeletal muscle fiber is known as the **sarcolemma**. See Figure 5.24.

Cardiac Muscle

The muscular walls of the heart contain **cardiac muscle tissue** that contracts rhythmically without conscious control or neural stimulation. Thus, it is classified functionally as **involuntary**.

Cardiac muscle tissue is composed of cells joined end to end and separated from each other by **intercalated discs.** The interconnected, branching cells form a complex network. Each cell has a single nucleus, and striations are present in the sarcoplasm. See Figure 5.25.

Smooth Muscle

Smooth muscle tissue occurs in the walls of hollow internal organs and blood vessels. Individual cells are spindle-shaped, have a single centrally located nucleus, and lack striations in the sarcoplasm. Like cardiac muscle tissue, smooth muscle

Histology 47

Figure 5.24 Skeletal muscle fibers. (Right) Victor P. Eroschenko

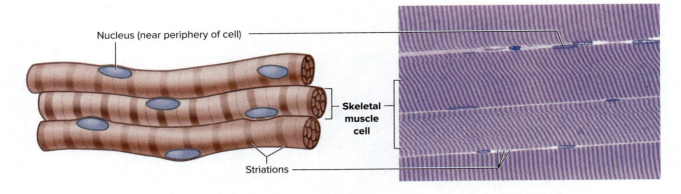

Figure 5.25 Cardiac muscle cells. (Right) Victor P. Eroschenko

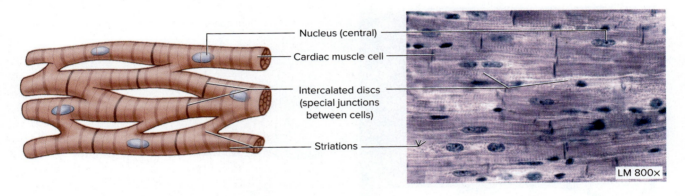

Figure 5.26 Smooth muscle cells. (Right) Victor P. Eroschenko

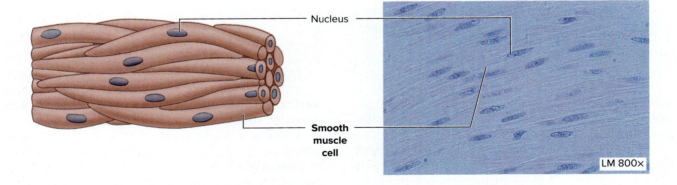

tissue is not under conscious control, it is **involuntary** in function. See Figure 5.26.

Nervous Tissue

Nervous tissue occurs in the central nervous system (brain and spinal cord) and in peripheral nerves. The primary cells of nervous tissue are **neurons**. The function of neurons is to receive electrical signals from sensory receptors or other neurons and to transmit **action potentials** to other neurons or effectors (muscles or glands). Action potentials are special electrochemical signals that only occur in neurons and muscle cells. In this way, the nervous system coordinates body functions almost instantly. **Neuroglia** are associated

Figure 5.27 Nervous Tissue (50×). (Right) Ed Reschke/Photolibrary/Getty Images

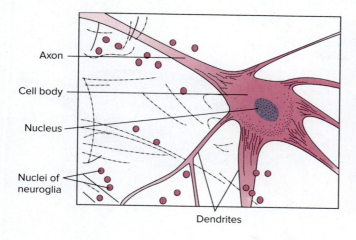
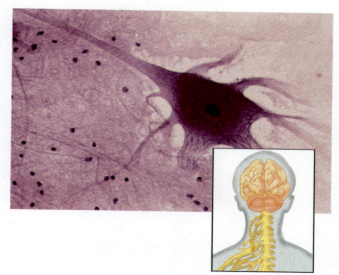

with the neurons and aid neuron functions. They provide support for the neurons and prevent contact between neurons except at particular sites. The structure of neurons varies in accordance with their particular roles, but all neurons possess common features. See Figure 5.27 illustrating a large neuron and surrounding neuroglia.

Neuron Structure

The basic structure of a neuron is shown in Figure 5.28. Obtain an anatomical model of a neuron and compare it to Figure 5.28 as you read this section. The **cell body** is the enlarged portion that contains the nucleus. The **nucleus** is centrally located with a well-defined **nucleolus**. The cytoplasm contains a cytoskeleton of microtubules and **neurofibrils** (label 5) composed of bundles of actin filaments. The meshwork of neurofibrils encompasses the dark-staining **chromatophilic substance** (label 2), which are composed of aggregations of rough endoplasmic reticulum (RER), and extends into the neuron processes. The cytoplasm also contains mitochondria, lysosomes, Golgi apparatus, and various inclusions, but these organelles are not shown in Figure 5.28.

Several rather short processes extend from the cell body and branch extensively to form vast numbers of **dendrites,** which are the primary sites for receiving action potentials from other neurons. The number of dendrites in different types of neurons will vary from one to thousands.

The **axon hillock,** a small mound on one side of the cell body, is where the axon originates. The **axon** is a single elongated process that carries action potentials from the dendrites to another neuron or an effector (muscle or gland). A neuron has only one axon. It may have side branches called **collaterals,** but they are usually few in number. The **axon terminals** (label 9) may be highly branched in order to contact multiple neurons or effectors. The tip of each terminal branch is a **terminal bouton,** which is in direct contact with a dendrite of another neuron or with an effector.

The axons of peripheral (i.e., outside the central nervous system) neurons are enclosed by a covering of **neurolemmocytes (Schwann cells)** (label 14), one type of neuroglia. In larger peripheral fibers, as shown in Figure 5.28, the inner multiple wrappings of neurolemmocyte membranes form the **myelin sheath** (label 16). The outer layer of the neurolemmocyte, the part containing the nucleus and most of the cytoplasm, constitutes the **neurilemma.** Axons with a myelin sheath are said to be **myelinated;** those enclosed in neurolemmocytes but lacking a myelin sheath are **unmyelinated.** Neurolemmocytes are essential for the regeneration of a damaged nerve fiber because they provide a pathway for the regrowth of the axon. Regeneration of nerve fibers is not possible in the brain and spinal cord because of the absence of neurolemmocytes in the central nervous system. In the brain and spinal cord, the myelin sheath of neurons is formed by **oligodendrocytes,** one of four types of neuroglia found only in the central nervous system. However, a neurilemma is absent; only neurolemmocytes form a neurilemma.

Figure 5.28 Neuron structure. (A) A motor neuron illustrates the basic structure of a neuron. (B) A neurolemmocyte forms the myelin sheath and neurilemma around axons of peripheral neurons.

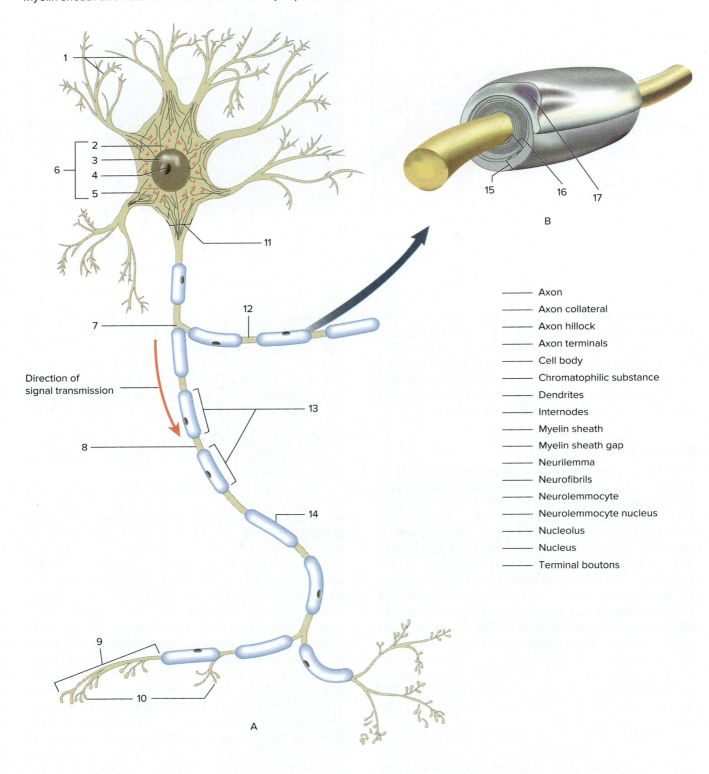

_____ Axon
_____ Axon collateral
_____ Axon hillock
_____ Axon terminals
_____ Cell body
_____ Chromatophilic substance
_____ Dendrites
_____ Internodes
_____ Myelin sheath
_____ Myelin sheath gap
_____ Neurilemma
_____ Neurofibrils
_____ Neurolemmocyte
_____ Neurolemmocyte nucleus
_____ Nucleolus
_____ Nucleus
_____ Terminal boutons

The myelin sheath and neurilemma are interrupted at regular intervals along the axon by the **myelin sheath gaps (nodes of Ranvier)**. These nodes are tiny spaces between the neurolemmocytes where the axon is exposed. They play an important role in increasing the speed of action potential transmission. The axon segments between the nodes, where neurolemmocytes are wrapped around the axon, are called **internodes**.

Structural Classes of Neurons

Neurons vary considerably in size, shape, and function. Most neurons belong to one of three basic structural types: multipolar neurons, bipolar neurons, and pseudounipolar neurons. See Figure 5.29.

Multipolar Neurons

Multipolar neurons have many dendrites and a single axon like the one in Figure 5.28. They are the most abundant neuron type and include most neurons in the brain and spinal cord. Spinal motor neurons are an example.

Bipolar Neurons

Bipolar neurons have a single dendrite and a single axon extending from opposite sides of the cell body. Olfactory neurons in the nasal cavity, some neurons in the retina of the eye, and sensory neurons in the inner ear are of this type.

Pseudounipolar (Unipolar) Neurons

Pseudounipolar neurons have a single process extending from the cell body. The process quickly branches in the shape of a T with the two branches extending in opposite directions. Spinal sensory neurons are an example.

Functional Classification of Neurons

There are three functional types of neurons: sensory neurons, interneurons, and motor neurons. An example of the interaction of the three functional types of neurons is shown in Figure 5.30.

Sensory Neurons

The bipolar neurons of the eyes, ears, and nose mentioned above are sensory neurons. All other sensory neurons are pseudounipolar neurons. Sensory neurons carry action potentials from **sensory receptors** (label 6) toward the central nervous system (brain and spinal cord). Thus, they are sometimes called **afferent neurons.** Note in Figure 5.30 that a **spinal nerve** is joined to the spinal cord by two roots: an **anterior** (ventral) **root** (label 10) and a **posterior** (dorsal) **root** (label 9). A nerve consists of bundles of axons. Spinal sensory neurons enter the spinal cord via the posterior root of a spinal nerve, and their cell bodies are located within a **posterior root ganglion** (dorsal root ganglion), a swollen region on the posterior root. A ganglion consists of a cluster of neuron cell bodies. Only the axon of the sensory neuron enters the spinal cord where it forms a synapse with interneurons. A **synapse** is the junction between an axon terminal bouton and either (1) a dendrite or cell body of another neuron or (2) an **effector,** such as a muscle or gland.

Interneurons

All interneurons are multipolar neurons that occur only in the central nervous system and compose about 99% of the neurons in the body. They carry action potentials from one part of the central nervous system to another. In Figure 5.30, the interneuron is shown linking the sensory neuron with the motor neuron within the gray matter of the spinal cord. The gray matter in the brain and spinal cord is composed mostly of interneuron cell bodies, and their myelinated axons compose the white matter. Myelin is responsible for the lighter coloration. A synapse is formed between the sensory neuron and interneuron and between the interneuron and motor neuron.

Figure 5.29 Structural types of neurons.

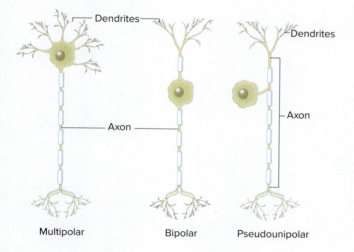

Histology

Figure 5.30 Interaction of the three functional types of neurons in the spinal cord and spinal nerves.

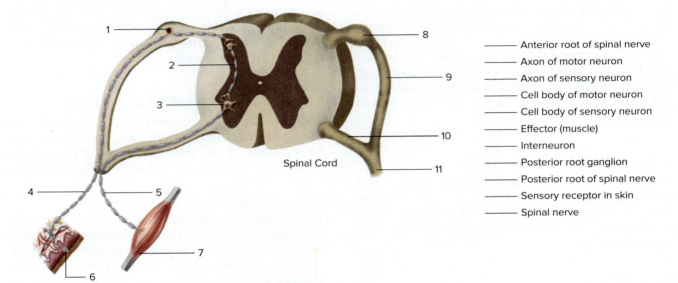

_____ Anterior root of spinal nerve
_____ Axon of motor neuron
_____ Axon of sensory neuron
_____ Cell body of motor neuron
_____ Cell body of sensory neuron
_____ Effector (muscle)
_____ Interneuron
_____ Posterior root ganglion
_____ Posterior root of spinal nerve
_____ Sensory receptor in skin
_____ Spinal nerve

Motor Neurons

Motor neurons are multipolar neurons that carry action potentials away from the central nervous system to an effector (a muscle or gland). Thus, they are also called **efferent neurons.** The cell body and dendrites of a motor neuron are located within the central nervous system, and only the myelinated axon is located outside the central nervous system. In Figure 5.30, note that the axon of a motor neuron exits the spinal cord via the anterior root of the spinal nerve and terminates at an effector.

Assignment

1. Label Figures 5.28 and 5.30.
2. Complete Sections E, F, G, and H of the laboratory report.
3. Examine slides of each type of tissue discussed in this lab. Make drawings for yourself to help you study.

Exercise 6

THE INTEGUMENTARY SYSTEM

Objectives

After completing this exercise, you should be able to

1. Identify the structures of the integumentary system.
2. List the functions of the integumentary system.
3. Identify the components of the skin when viewed microscopically.

Materials

Calipers (or pointed scissors)
Metric ruler
Model of human skin
Prepared slides of
   human, thick skin
   human, thin skin

The **integument** (skin) consists of epithelial and connective tissues with associated nervous and muscle tissues. The functions of the skin include (1) protection of underlying tissues from abrasion, dehydration, microorganisms, and ultraviolet radiation; (2) excretion; (3) maintenance of body temperature; (4) detection of certain stimuli; (5) participation in the synthesis of vitamin D; and (6) absorption of lipid soluble substances, such as topical creams. The skin consists of two layers: a superficial epidermis and an underlying dermis.

The integument is attached to underlying tissues by the **subcutaneous layer** (hypodermis). The subcutaneous layer is sometimes called the **superficial fascia.** The subcutaneous layer consists primarily of areolar connective tissue and adipose tissue, and it contains abundant blood vessels and nerves that service the integument. The adipose tissue provides cushioning and insulation for underlying tissues and serves as a storehouse for energy reserves.

The Epidermis

The **epidermis**, label 1 in Figure 6.1, consists of stratified squamous epithelium. In the *thin skin* covering the majority of the body's surface, the epidermis is organized into four layers. *Thick skin*, which is organized into five layers, is hairless and found only in areas subjected to high levels of abrasion, such as the palms and soles. Prepared slides of thick skin reveal five distinct layers: an outer **stratum corneum** (label 14), a thin translucent **stratum lucidum** (label 15), a darkly stained **stratum granulosum** (label 16), a multi-layered **stratum spinosum** (label 17), and an innermost **stratum basale** (label 18), a single layer of columnar cells. The stratum lucidum is apparent only in thick skin. Cells of all layers originate from cells produced by the **stem cells** (label 19) of the stratum basale. As the cells are pushed to the surface by new cells formed below them, they change in shape, structure, and composition.

The stratum basale consists of two other major cell types. **Tactile epithelial cells** (label 20) in the stratum basale work with nerve endings (label 21) in the dermis in touch sensation detection. The brown-black skin pigment **melanin** is produced by **melanocytes** (label 22) in the stratum basale and is incorporated into the cells of the stratum spinosum. Inherited differences in skin color are due to the amount of melanin produced rather than the number of melanocytes. Melanin protects underlying tissues from harmful ultraviolet radiation. Exposure to ultraviolet radiation increases the production of melanin, resulting in a "tan." Clustering of melanin results in freckles.

The granules in cells of the stratum granulosum are precursors of **keratin,** a water-repellent protein. As the cells are pushed toward the surface, they become the keratinized, flattened, dead cells of the stratum corneum. Due to this feature, the epidermis of the skin is classified as *keratinized stratified squamous epithelium*. Surface cells of this outer layer are constantly sloughed off and replaced by underlying cells.

Unit 3: Protection, Support, and Movement

Figure 6.1 Skin structure. (A) Basic components; (B) epidermal layers.

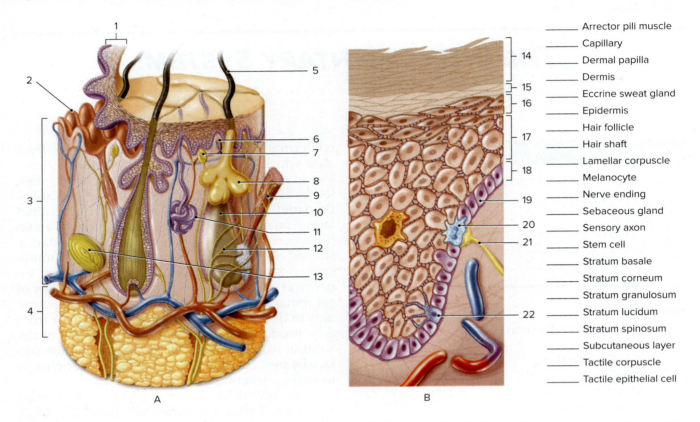

The Dermis

The **dermis** (label 3) varies in thickness from less than 1 mm to over 6 mm. It is composed mostly of dense irregular connective tissue. It contains numerous capillaries that provide nourishment not only for the dermis but also for the epidermis. Axons and sensory receptors are also present in the dermis.

The epidermis is attached to the dermis. The dermis has numerous projections called **dermal papillae** (label 2) that extend into the epidermis. These papillae are most abundant and pronounced in skin of the fingertips, palms of the hands, and soles of the feet, where they produce dermal ridges (fingerprints) that improve the frictional characteristics of these areas.

Abundant collagenous and elastic fibers are found in the dermis. They are responsible for the strength and elasticity of the skin.

Hair

A hair is formed of dead, heavily keratinized epidermal cells. The **hair shaft** extends above the skin, and the **hair root** is located below the skin surface. The root is enveloped by **internal** and **external root sheaths,** formed of epidermal cells, that compose the **hair follicle** (Figure 6.2).

The hair follicle extends inward into the dermis and sometimes into the subcutaneous layer. The tip of the hair root forms an enlargement called the **bulb.** The bulb is composed of stratum basale cells that produce the cells forming the hair. The bulb encompasses a dermal papilla known as a **hair papilla.** It contains capillaries that nourish the bulb and follicle.

An **arrector pili muscle,** formed of smooth muscle, extends diagonally from the hair follicle upward to the epidermis. Contraction of this muscle raises the hair into a more upright position and produces "goose bumps." In most mammals, contraction of the arrector pili muscle is believed to provide insulation, although this function may be insignificant in humans.

Glands

Two kinds of glands are present in the skin: sebaceous glands and sweat glands.

Figure 6.2 Hair follicle and associated structures.

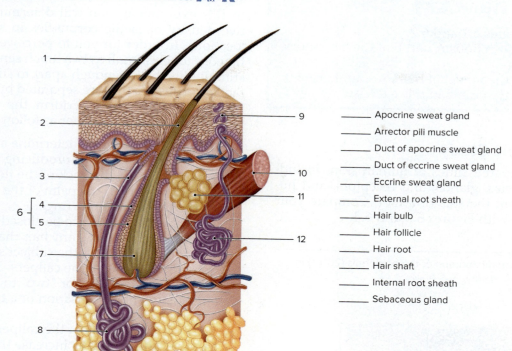

_____ Apocrine sweat gland
_____ Arrector pili muscle
_____ Duct of apocrine sweat gland
_____ Duct of eccrine sweat gland
_____ Eccrine sweat gland
_____ External root sheath
_____ Hair bulb
_____ Hair follicle
_____ Hair root
_____ Hair shaft
_____ Internal root sheath
_____ Sebaceous gland

Sebaceous Glands

Sebaceous glands (label 11 in Figure 6.2) are located within the epithelial tissue of the hair follicle. They secrete an oily secretion called **sebum** into the follicle, which serves to keep the hair soft and pliable and to kill bacteria on the skin. Sebum also helps to waterproof the skin.

Sweat Glands

There are two types of sweat glands: eccrine and apocrine. The smaller **eccrine sweat glands** (Figures 6.1 and 6.2) carry their secretions directly to the surface of the skin. These are simple tubular glands that have a coiled basilar portion located in the dermis. They are widely distributed over the body and are abundant on the head, neck, back, palms of the hands, and soles of the feet. They are absent on the lips, glans penis, and clitoris. Eccrine glands located in the axillae and palms may be stimulated by psychological factors such as fear and stress, but most eccrine glands are activated primarily by thermal stimuli. The watery perspiration secreted by these glands removes waste materials from the blood and cools the body surface by evaporation.

The larger **apocrine sweat glands** (label 8 in Figure 6.2) empty their secretions into a hair follicle. The coiled basilar portion is located in the subcutaneous layer. They occur primarily in the axillary, areolar, beard, and genital regions and become active at puberty under the influence of the sex hormones. These glands become stimulated during times of emotional stress and sexual excitement. Apocrine sweat is viscous and milky in color, due to the addition of lipid and proteins. Although the secretions are essentially odorless, decomposition by bacteria produces waste products that cause body odor.

Sensory Receptors

The skin plays an important role in sensory perception because it contains numerous sensory receptors for touch, pressure, heat, cold, and pain stimuli. As previously mentioned, the epidermis contains tactile epithelial cells important in touch sensations. Additionally, **tactile corpuscles** are located in the dermis near the epidermis (label 7 in Figure 6.1), often within dermal papillae. They are especially important receptors in hairless skin. Free nerve endings wrapped around hair follicles (label 12 in Figure 6.1) also serve as touch receptors. **Lamellar corpuscles** (label 13 in Figure 6.1) are spherical or ovoid, multilayered structures that lie deep in the dermis that are used to detect course touch and deep pressure.

Assignment

1. Label Figures 6.1 and 6.2.
2. Complete Sections A and B of the laboratory report.
3. Examine the model of human skin. Locate the parts described in this exercise.

Microscopic Study

Examine prepared slides of human skin, including hair follicles, glands, and receptors, and human skin from the sole of a foot. Compare your observations with Figures 6.3 and 6.4.

Figure 6.3 Photomicrograph of the epidermis (200×) in thick skin. APR Micrographs/Lutz Slomianka

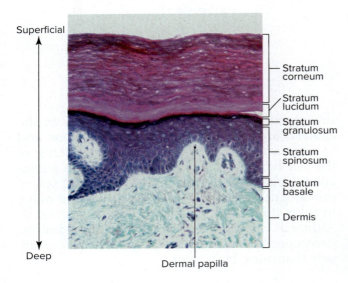

Distribution of Touch Receptors

In this experiment, you will determine the relative density of tactile corpuscles in skin at four locations. In order for you to perceive two simultaneous touch stimuli as two touch sensations, the stimuli must be far enough apart to stimulate two tactile corpuscles that are separated by an unstimulated touch receptor. Perform the experiment with your laboratory partner as follows.

1. Using this procedure, determine and record the minimum distance producing a two-point sensation on (1) the back of the neck, (2) the inner forearm, (3) the palm of the hand, and (4) the tip of the index finger.
2. The subject's eyes are to be closed during the experiment. Avoiding any hair that may be present, lightly touch the subject's skin with one or two points of the calipers. The subject responds either "one" or "two" to indicate either a one-point sensation or a two-point sensation.
3. Start with the points of the calipers close together and gradually increase the distance between the points until the subject reports a two-point sensation about 75% of the time. Measure and record this distance between the tips of the calipers as the minimum distance producing a two-point sensation.

Assignment

Complete the laboratory report.

Figure 6.4 Photomicrograph showing thin skin (40×). APR Micrographs/Victor P. Eroschenko

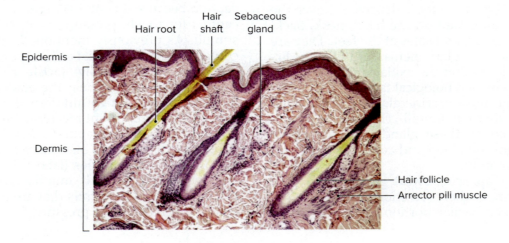

Protection, Support, and Movement

Exercise 7

SKELETAL OVERVIEW

Objectives

After completing this exercise, you should be able to

1. Identify the parts of a longitudinally sectioned long bone.
2. Identify the major types of bone fractures.
3. Identify the major bones of the body and their relationships.
4. Identify the surface features of bones.

Materials
Fresh beef bones sawed longitudinally
Human femur sawed longitudinally
Articulated human skeleton
Dissection instruments

In this exercise, you will study the structure of a typical long bone, certain types of bone fractures, and the skeleton as a whole.

Long Bone Structure

Figure 7.1 depicts a tibia that has been sectioned to reveal its internal structure. The bone consists of an elongated shaft, the **diaphysis,** and expanded terminal portions, the **epiphyses.** During the growing years, a plate of hyaline cartilage, the **epiphysial plate** (metaphysis), is present at the junction of the epiphyses and the diaphysis. The bone grows in length as new cartilage tissue is formed on the epiphysial side of the plate and cartilage tissue is replaced by bone tissue on the diaphyseal side. Longitudinal growth ceases when the cartilage tissue ceases to reproduce and is completely replaced by bone tissue. The lines of fusion that form between the epiphyses and the diaphysis are called **epiphysial lines.**

The inside of the epiphyses and ends of the diaphysis consist of **spongy bone. Compact bone** forms the rigid tube of the diaphysis and is reduced in thickness where it covers the underlying spongy bone.

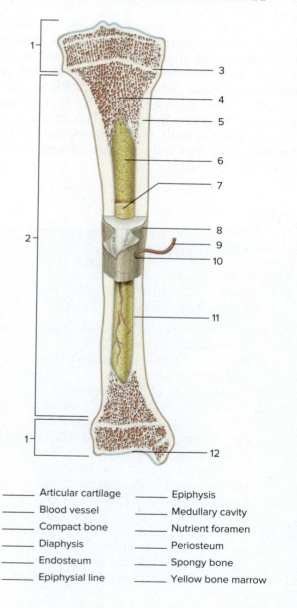

Figure 7.1 Anatomy of a long bone, the tibia.

_____ Articular cartilage   _____ Epiphysis
_____ Blood vessel   _____ Medullary cavity
_____ Compact bone   _____ Nutrient foramen
_____ Diaphysis   _____ Periosteum
_____ Endosteum   _____ Spongy bone
_____ Epiphysial line   _____ Yellow bone marrow

Unit 3: Protection, Support, and Movement

The chamber in the diaphysis is the **medullary cavity.** It is lined with a thin epithelial membrane, the **endosteum,** that extends into the spaces in spongy bone and into the central canals of osteons. The medullary cavity and the adjoining spaces in spongy bone are filled with fatty **yellow bone marrow. Red bone marrow,** which forms blood cells, occurs in spongy bone in the epiphyses of the femur and humerus, but the epiphyses of other long bones contain yellow bone marrow. Most red bone marrow in adults is found in spongy bone of the ribs, sternum, pelvic girdle, and vertebrae.

Except for the articular surfaces, the bone is covered with the **periosteum,** a tough, connective tissue membrane that tightly adheres to the bone surface. The periosteum is richly supplied with blood vessels and nerves. Nutrient blood vessels enter the bone via small channels called **nutrient foramina** (singular, **foramen**). The articular surfaces are covered with a smooth **articular cartilage** (hyaline type) that reduces friction and protects the ends of the bone.

Parts of the Skeleton

The adult skeleton consists of 206 named bones and several small unnamed ones. Bones vary considerably in size and shape and are categorized into two main groups: (1) bones of the **axial skeleton** and (2) bones of the **appendicular skeleton.** Only the major bones of the skeleton will be considered in this section. Refer to Figure 7.2.

The Axial Skeleton

The axial skeleton consists of the **skull, vertebral column** (spine), and **thoracic cage.** The skull is formed of 21 fused bones plus the movable **mandible** (lower jaw). The vertebral column includes 24 somewhat movable vertebrae plus the **sacrum** and **coccyx** (tail bone). The thoracic cage is formed by 12 pairs of **ribs,** the **sternum** (breastbone), and the **costal cartilages** (label 9), which join the ribs to the sternum. Not shown in Figure 7.2 are the **hyoid bone,** a U-shaped bone located under the lower jaw and associated with the larynx, and the **auditory ossicles,** which play a major role in hearing.

The Appendicular Skeleton

The appendicular skeleton is formed by the bones of the pectoral (shoulder) and pelvic girdles and the upper and lower limbs.

The **pectoral** (shoulder) **girdles** consist of the **clavicles** (collarbones) and **scapulae** (shoulder blades), which attach the upper limbs to the trunk. An upper limb consists of the **humerus** (arm bone), **radius** and **ulna** (forearm bones), **carpals** (wrist bones), **metacarpals** (bones of the palm), and **phalanges** (finger bones).

The **pelvic girdle** is formed by two **coxal bones** (hipbones) that join anteriorly at the **pubic symphysis.** The pelvic girdle unites with the **sacrum** of the vertebral column at the **sacroiliac joint.** Together the pelvic girdle and sacrum make up the **pelvis** (label 11). A lower limb consists of the **femur** (thighbone), **tibia** (shinbone), **fibula** (smaller bone of the leg), **patella** (kneecap), **tarsals** (ankle bones), **metatarsals** (bones of the instep), and **phalanges** (toe bones). Note that both finger and toe bones are called phalanges.

Surface Features of Bones

As you examine the bones of the skeleton, you will notice a variety of surface features, such as grooves and ridges. Study the major surface features of bones in Table 7.1. Surface features of bones will be emphasized in subsequent exercises on the skeletal system, so learn to recognize them as soon as possible.

Assignment

1. Label Figure 7.1.
2. Examine the cut surface of a split beef bone. Locate the parts shown in Figure 7.1.
   a. Remove a little yellow bone marrow from the medullary cavity. Note its texture.
   b. Determine the distribution of compact and spongy bone.
   c. Locate the red bone marrow.
   d. Use a scalpel to cut away a small portion of the periosteum. Determine if the periosteum is elastic or nonelastic and strong or weak.
   e. Muscles are attached to bones by tendons, and bones are bound to other bones by ligaments. Examine the attachment of a tendon or ligament to the bone. Determine how the periosteum is involved in the attachment.
   f. Examine the articular cartilages to determine how they aid movement at a joint.
3. Examine the split human femur. Compare its epiphysial lines with the epiphysial cartilages in the beef bone. Note the distribution of compact and spongy bone.
4. Complete Section B of the laboratory report.

Figure 7.2 Major bones of the body. Axial skeleton in brown, appendicular skeleton in gray.

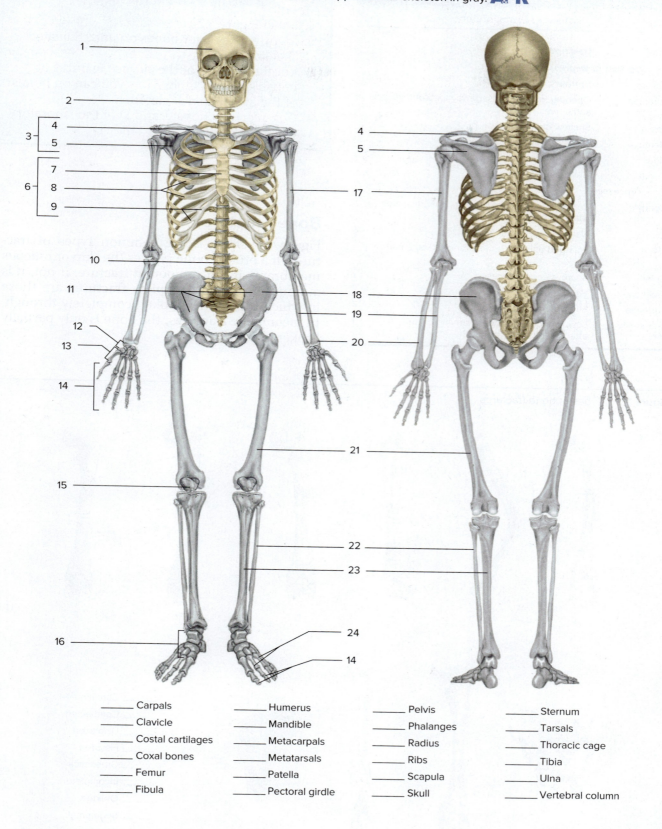

_____ Carpals
_____ Clavicle
_____ Costal cartilages
_____ Coxal bones
_____ Femur
_____ Fibula

_____ Humerus
_____ Mandible
_____ Metacarpals
_____ Metatarsals
_____ Patella
_____ Pectoral girdle

_____ Pelvis
_____ Phalanges
_____ Radius
_____ Ribs
_____ Scapula
_____ Skull

_____ Sternum
_____ Tarsals
_____ Thoracic cage
_____ Tibia
_____ Ulna
_____ Vertebral column

Skeletal Overview 59

TABLE 7.1
Surface Features of Bones

Term	Description
Cavities or Depressions	
Fissure	Narrow slit
Foramen	Opening that is a passageway for blood vessels and nerves
Fossa	Shallow depression
Meatus	Large canal or passageway
Sinus	Air-filled cavity
Sulcus	Groove or furrow
Processes	
Condyle	Rounded knucklelike process that articulates with another bone
Crest	Narrow ridge
Head	Enlarged end of a bone supported by a narrow neck
Spine	Sharp, slender process
Trochanter	Very large process
Tubercle	Small rounded process
Tuberosity	Low, roughened process serving as an attachment site for a muscle

Assignment

1. Label Figure 7.2.
2. Identify the major bones on an articulated skeleton.
3. Locate as many of the surface features of bones listed in Table 7.1 as you can on bones of the articulated skeleton.
4. Complete Sections C and D of the laboratory report.

Bone Fractures

Figure 7.3 shows some common types of fractures. If a broken bone pierces the skin or mucous membrane, it is a **compound fracture;** if not, it is a **simple fracture. Complete fractures** are those in which the bone is broken completely through. In **incomplete fractures,** the bone is only partially broken or splintered.

Figure 7.3 Types of bone fractures.

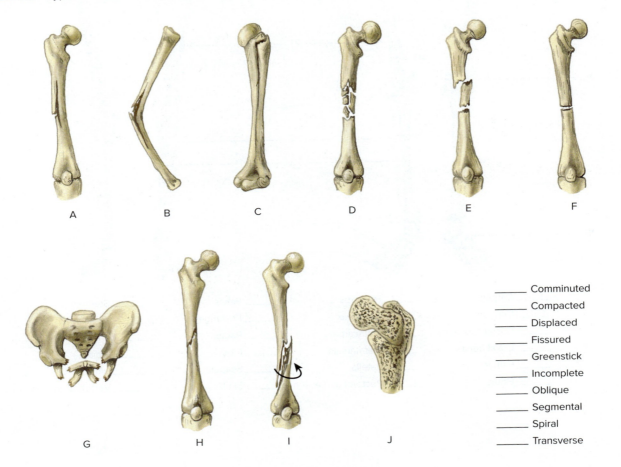

_____ Comminuted
_____ Compacted
_____ Displaced
_____ Fissured
_____ Greenstick
_____ Incomplete
_____ Oblique
_____ Segmental
_____ Spiral
_____ Transverse

60 Protection, Support, and Movement

There are several types of complete fractures. If the break is at right angles to the long axis of the bone, it is a **transverse fracture.** Fractures at some other angle are **oblique fractures.** A **spiral fracture** results from excessive twisting of the bone. Several pieces are broken out of the bone in a **comminuted fracture.** If only one piece is broken out of the bone, it is a **segmental fracture.** When bone fragments have been moved out of alignment as in illustration G, the fracture may also be said to be **displaced.**

There are two types of incomplete fractures. A **greenstick fracture** occurs on only one side of a bone when the bone is bent. A **fissured fracture** is a lengthwise split in a long bone.

When one piece of the bone is forced into another part of the same bone, a **compacted fracture** (illustration J) results. **Compression fractures** (not shown) of the vertebrae (crushed vertebrae) sometimes occur because of excessive vertical forces.

Assignment

1. Label Figure 7.3.
2. Complete the laboratory report.

Skeletal Overview **61**

Exercise 8

THE AXIAL SKELETON

Objectives

After completing this exercise, you should be able to

1. Identify the skull bones and their distinctive features, sutures, paranasal sinuses, and major foramina.
2. Describe the relationships of the bones forming the skull.
3. Identify the fontanelles and unossified sutures of the fetal skull.
4. Identify the bones, and their significant parts, that form the vertebral column and thorax.

Materials
- Human skulls with removable top of cranium
- Fetal skulls
- Skeletons, articulated and disarticulated
- Vertebral column, mounted
- Pipe cleaners

The major components of the **axial skeleton** are the skull, vertebral column, and thoracic cage. Bones of the axial skeleton are covered in this exercise.

The adult skull consists of 22 bones. Of those, 21 are firmly joined together by immovable joints called **sutures,** plus the movable lower jaw. Eight of the interlocked bones form the **cranium** encasing the brain, and 14 compose the face. Bones of the skull contain numerous **foramina,** passageways for cranial nerves and blood vessels. Only a few of the larger foramina will be considered here.

While studying the laboratory bones, be very careful not to damage them. Some parts are delicate and easily broken. *Never use a pencil or pen as a pointer or probe.* Use a pipe cleaner instead.

The Cranium

The cranium is formed of the following bones: one **frontal,** two **parietals,** one **occipital,** two **temporals,** one **sphenoid,** and one **ethmoid.** Locate the bones and their distinctive features in Figures 8.1 through 8.4 as you study this section.

Frontal Bone

The frontal bone forms the upper front portion of the cranium, including the forehead, upper parts of the orbital cavities, and the roof of the nasal cavity. Above each orbital cavity is a **supraorbital foramen** that is reduced in some skulls to a **supraorbital notch.**

Parietal Bones

The parietal bones form the roof and sides of the cranium behind the frontal bone, to which they are joined by **coronal sutures.** The parietal bones are joined to each other at the midline by the **sagittal suture.**

Occipital Bone

Most of the bottom of the skull consists of the occipital bone. It joins to the parietal bones by the **lambdoid suture.** As shown in Figure 8.2, the floor of the occipital bone contains the large **foramen magnum,** which surrounds the brainstem. When the brainstem passes through the foramen magnum it ends and the spinal cord begins. On either side of the foramen magnum are the **occipital condyles.** These knucklelike processes articulate with the atlas, the first vertebra of the vertebral column.

Temporal Bones

The temporal bones are located just below the parietal bones on each side of the skull. Each joins to the parietal bone above it by a **squamous suture** and behind it with the occipital bone by the lambdoid suture. The ear components are located within the temporal bones. The **external acoustic** (auditory) **meatus,** which leads inward toward the middle ear, is a large hole on the lateral surface. Just in front of the external acoustic meatus is the **mandibular fossa.** The mandibular condyle fits into this fossa to form the **temporomandibular joint.**

Unit 3: Protection, Support, and Movement

Figure 8.1 Lateral view of the skull.

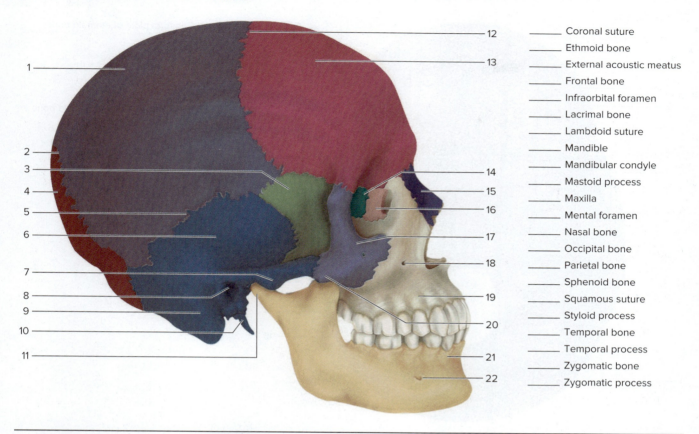

_____ Coronal suture
_____ Ethmoid bone
_____ External acoustic meatus
_____ Frontal bone
_____ Infraorbital foramen
_____ Lacrimal bone
_____ Lambdoid suture
_____ Mandible
_____ Mandibular condyle
_____ Mastoid process
_____ Maxilla
_____ Mental foramen
_____ Nasal bone
_____ Occipital bone
_____ Parietal bone
_____ Sphenoid bone
_____ Squamous suture
_____ Styloid process
_____ Temporal bone
_____ Temporal process
_____ Zygomatic bone
_____ Zygomatic process

Figure 8.2 Inferior view of the skull.

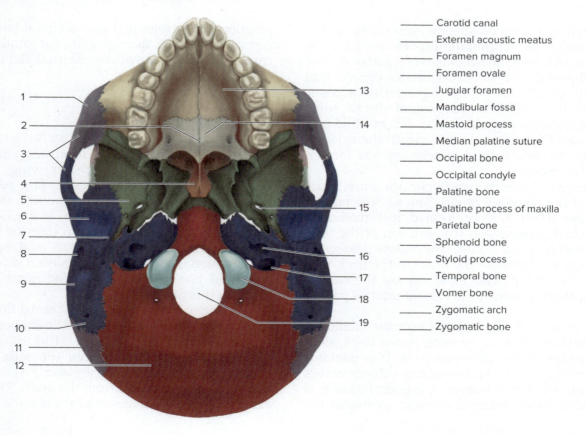

_____ Carotid canal
_____ External acoustic meatus
_____ Foramen magnum
_____ Foramen ovale
_____ Jugular foramen
_____ Mandibular fossa
_____ Mastoid process
_____ Median palatine suture
_____ Occipital bone
_____ Occipital condyle
_____ Palatine bone
_____ Palatine process of maxilla
_____ Parietal bone
_____ Sphenoid bone
_____ Styloid process
_____ Temporal bone
_____ Vomer bone
_____ Zygomatic arch
_____ Zygomatic bone

The Axial Skeleton

Figure 8.3 Internal view of the cranial floor.

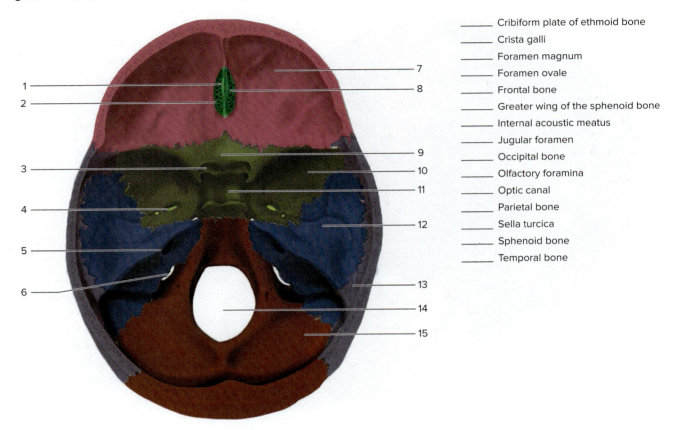

____ Cribiform plate of ethmoid bone
____ Crista galli
____ Foramen magnum
____ Foramen ovale
____ Frontal bone
____ Greater wing of the sphenoid bone
____ Internal acoustic meatus
____ Jugular foramen
____ Occipital bone
____ Olfactory foramina
____ Optic canal
____ Parietal bone
____ Sella turcica
____ Sphenoid bone
____ Temporal bone

There are three major processes on each temporal bone. The **zygomatic process** projects forward to join with the cheekbone (zygomatic). The **styloid process** is a slender spinelike process that extends downward below the external acoustic meatus. It serves as an attachment site for some tongue and pharyngeal muscles. The **mastoid process** is a rounded eminence beneath the external acoustic meatus. It is an attachment site for certain neck muscles.

Figure 8.2 depicts two important paired foramina of the temporal bones. The narrow **jugular foramina** (label 17) are located just medial to the styloid processes at the junction of the temporal and occipital bones. They allow passage of the jugular veins from the brain to the neck. Just in front of the jugular foramina are the **carotid canals,** through which the carotid arteries pass to the brain.

In Figure 8.3, the **internal acoustic** (auditory) **meatus** (label 5) is the opening on the thick, sloping portion of the temporal bone that contains the inner ear. Facial and vestibulocochlear nerves pass through this foramen. A jugular foramen is just behind each internal acoustic meatus at the junction of the temporal and occipital bones. Insert a pipe cleaner through a jugular foramen and carotid canal to clarify their external and internal openings.

Sphenoid Bone

The sphenoid is colored hunter green in the figures. It consists of a central *body* and two winglike structures, the greater wings of the sphenoid, that extend laterally. The sphenoid forms part of the floor of the cranium, the sides of the cranium in the "temple" areas, and the back of the orbital cavities. An inferior view of the skull (Figure 8.2) shows an oval opening, the **foramen ovale** (label 15), on each side of the sphenoid bone. The mandibular nerve passes through this opening.

As shown in the view of the cranial floor (Figure 8.3), the body forms a saddle-shaped structure called the **sella turcica** (Turkish saddle). The **hypophyseal fossa** within the sella turcica is occupied by the pituitary gland (hypophysis), which hangs downward from the brain. The **optic canals** (label 3), passageways for the optic nerves,

Figure 8.4 Anterior view of the skull. A&PR

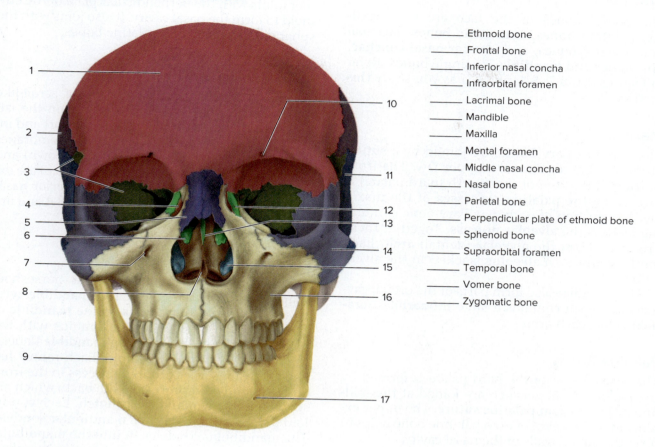

_____ Ethmoid bone
_____ Frontal bone
_____ Inferior nasal concha
_____ Infraorbital foramen
_____ Lacrimal bone
_____ Mandible
_____ Maxilla
_____ Mental foramen
_____ Middle nasal concha
_____ Nasal bone
_____ Parietal bone
_____ Perpendicular plate of ethmoid bone
_____ Sphenoid bone
_____ Supraorbital foramen
_____ Temporal bone
_____ Vomer bone
_____ Zygomatic bone

are located on each side of the sella turcica. The foramina ovale (label 4) are located near the rear lateral tips of the greater wings.

Figure 8.4 shows the greater wing of the sphenoid forming the back of the orbital cavities and the lateral walls of the cranium. The **superior orbital fissure** (not labeled) is the opening medial to the sphenoid. The **inferior orbital fissure** (not labeled) is below the sphenoid.

Ethmoid Bone

The ethmoid forms part of the roof of the nasal cavity and part of the medial surface of the orbit (label 14 in Figure 8.1) and closes the front of the cranium. The **perpendicular plate** of the ethmoid (label 13 in Figure 8.4 and label 7 in Figure 8.7) extends downward to form most of the nasal septum. It joins with the frontal bones in the front and the sphenoid and vomer in the back. Delicate, scroll-shaped processes, the **superior** (not shown), **middle,** and **inferior nasal conchae,** project from the lateral portions of the ethmoid toward the perpendicular plate. The superior and middle nasal conchae are parts of the ethmoid bone, but the inferior nasal conchae are separate bones attached to the ethmoid.

The upper portion of the ethmoid forms the roof of the nasal cavity and a small part of the anterior floor of the cranium joining with the frontal and sphenoid bones. This portion consists of two **cribriform plates** (label 2 in Figure 8.3), which contain numerous **olfactory foramina** (label 8) for the passage of olfactory nerves from the brain to the nasal cavity; and the **crista galli** (label 1 in Figure 8.3 and label 4 in Figure 8.7), which projects upward into the cranial cavity between the cribriform plates. Membranes enclosing the brain are attached to the crista galli.

Assignment

Label the cranial bones and their pertinent parts in Figures 8.1 through 8.4.

The Axial Skeleton 65

The Face

The paired bones of the face are the **maxillae, palatine bones, zygomatic bones, lacrimal bones, nasal bones,** and **inferior nasal conchae.** The **vomer** and **mandible** are single bones. Refer to Figures 8.1, 8.2, 8.4, and 8.5 as you study this section.

Maxillae

The two maxillary bones are joined by a suture at the midline to form the upper jaw. The front portion of the roof of the mouth (hard palate) is formed by the **palatine processes** of the maxillae. The portion of each maxillary bone containing teeth is the **alveolar process.** Together these processes form the **alveolar (dental) arch.** Each tooth occupies an **alveolus** (socket) in the alveolar arch.

The maxillae also form parts of the orbital cavities, and each contains a large **infraorbital foramen** below each orbit.

Palatine Bones

The rear portion of the hard palate is formed by the palatine bones, which are joined at the midline by the **median palatine suture.** The upper extension (not shown) of each palatine bone helps to form the lateral walls of the nasal cavity.

Zygomatic Bones

The zygomatic bones (cheekbones) form the prominences of the cheeks and the lateral walls of the orbits. The **temporal process** reaches backward to unite with the zygomatic process of the temporal bone to form the **zygomatic arch.**

Lacrimal Bones

A small lacrimal bone is located between the ethmoid and maxilla on the medial wall of each orbital cavity. Each lacrimal bone has a small groove for a tear duct that carries tears from the eye into the nasal cavity.

Nasal Bones

The nasal bones are thin bones fused at the midline to form the bridge of the nose.

Vomer

The vomer is a thin, flat bone (label 4 in Figure 8.2 and label 8 in Figures 8.4 and 8.7) located on the midline of the nasal cavity. Most of the upper vomer joints with the perpendicular plate of the ethmoid to form the nasal cavity. It also joins with the sphenoid, maxillae, and palatine bones.

Inferior Nasal Conchae

The nasal conchae are three pairs of scroll-like bony features that extend medially from the lateral walls of the nasal cavity. They support and increase the surface area of the mucous membranes of the nasal cavity. The superior (not shown) and middle nasal conchae (label 6 in Figure 8.4) are extensions of the ethmoid bone. The **inferior nasal conchae** (label 15) are shown in Figure 8.4 and are attached to the ethmoid bone.

Mandible

The lower jawbone consists of a horseshoe-shaped, horizontal **body of the mandible** with an upward-projecting **ramus of the mandible** at each end. The junction of each ramus with the body is called the **angle of the mandible** (label 8 in Figure 8.5) of the mandible. Each ramus has two extensions, the **coronoid process** in the front and the **condyloid process** in the back, which are separated by the **mandibular notch.** Each condyloid process terminates in a **mandibular condyle.** The mandibular condyles fit into the mandibular fossa of the temporal bones. Like the maxillae, the mandible has an *alveolar process* that contains the teeth.

The prominent **mandibular foramina** are located on the medial surfaces of the rami. Blood vessels and nerves supplying the teeth enter these openings and later emerge through the smaller **mental foramina,** which are located on the anterior body, to supply the lips and chin.

Assignment

Complete the labeling of Figures 8.1 through 8.5.

The Paranasal Sinuses

Some of the skull bones contain cavities, the **paranasal sinuses,** that reduce the weight of the skull without appreciably weakening it. The sinuses are

Figure 8.5 The mandible. **APR**

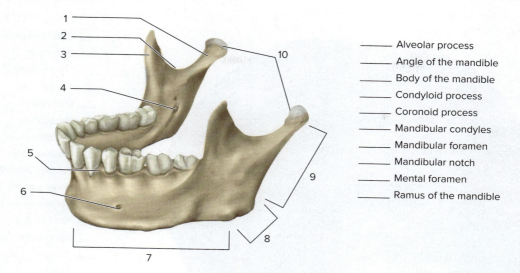

____ Alveolar process
____ Angle of the mandible
____ Body of the mandible
____ Condyloid process
____ Coronoid process
____ Mandibular condyles
____ Mandibular foramen
____ Mandibular notch
____ Mental foramen
____ Ramus of the mandible

Figure 8.6 The paranasal sinuses.

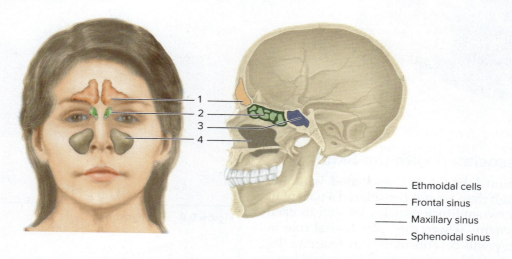

____ Ethmoidal cells
____ Frontal sinus
____ Maxillary sinus
____ Sphenoidal sinus

lined with a mucous membrane and have passageways leading into the nasal cavity. The sinuses are named after the bones in which they are located and are shown in Figure 8.6. The **frontal sinuses** (label 3 in Figure 8.7) occur in the forehead above the eyes. The large **maxillary sinuses** are below the eyes. The **sphenoidal sinus** (label 2 in Figure 8.7) is centrally located under the sella turcica. The **ethmoidal cells** consist of a number of small, air-filled spaces.

Assignment

1. Label Figures 8.6 and 8.7.
2. Complete Sections A (except for Figure 8.9), B, and C of the laboratory report.
3. Examine a human skull and locate the parts illustrated and described. *Remember to use only pipe cleaners as pointers.*

The Axial Skeleton

Figure 8.7 Sagittal view of skull. A&PR

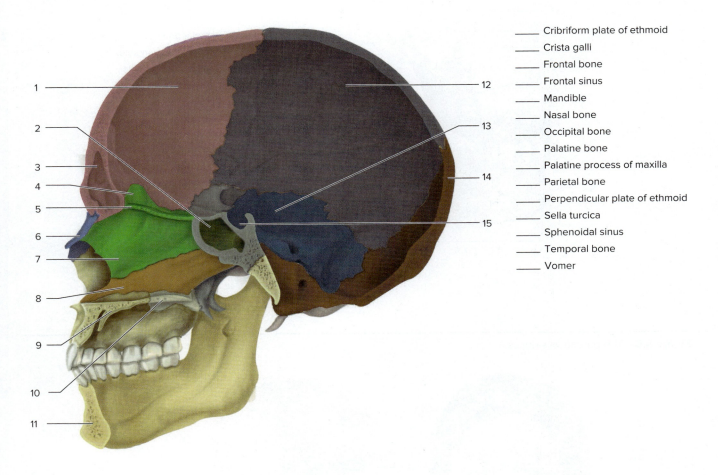

____ Cribriform plate of ethmoid
____ Crista galli
____ Frontal bone
____ Frontal sinus
____ Mandible
____ Nasal bone
____ Occipital bone
____ Palatine bone
____ Palatine process of maxilla
____ Parietal bone
____ Perpendicular plate of ethmoid
____ Sella turcica
____ Sphenoidal sinus
____ Temporal bone
____ Vomer

Bones Associated with the Skull

Seven additional bones are associated with the skull, although they are not considered a part of it. There are three **auditory ossicles** located in each middle ear cavity, where they play a vital role in hearing. They will be considered in Exercise 16, Senses. The **hyoid bone** is a small, U-shaped bone located below the mandible and just above the larynx. See Figure 8.8. It does not articulate with another bone. Instead, it is suspended from the styloid processes by slender stylohyoid muscles and ligaments. The hyoid is an attachment site for certain muscles controlling the tongue, larynx, and mandible.

The Fetal Skull

The incompletely ossified fetal skull is shown in Figure 8.9. The unossified sutures and **fontanelles,** membranous areas at the junction of several cranial bones, allow the compression of the skull during childbirth and growth of the brain after birth.

Figure 8.8 The hyoid bone.

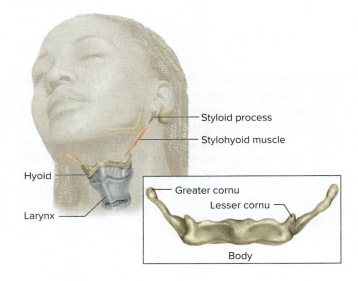

68   Protection, Support, and Movement

Figure 8.9 The fetal skull. (A) Lateral view; (B) superior view.

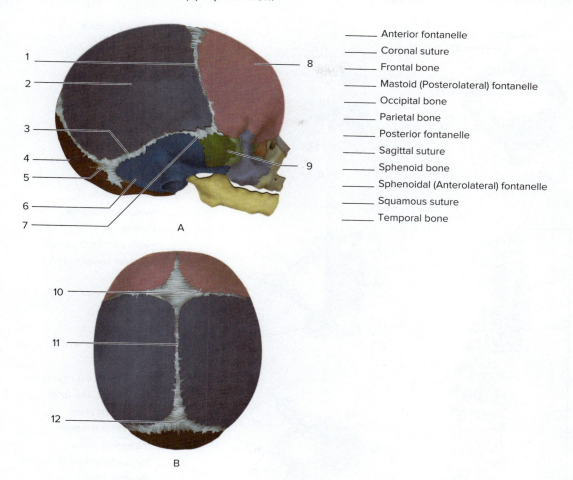

_____ Anterior fontanelle
_____ Coronal suture
_____ Frontal bone
_____ Mastoid (Posterolateral) fontanelle
_____ Occipital bone
_____ Parietal bone
_____ Posterior fontanelle
_____ Sagittal suture
_____ Sphenoid bone
_____ Sphenoidal (Anterolateral) fontanelle
_____ Squamous suture
_____ Temporal bone

There are six fontanelles. The large **anterior fontanelle** lies on the midline at the junction of the frontal and parietal bones. The smaller **posterior fontanelle** is on the midline at the junction of the parietal and occipital bones. A **sphenoidal (anterolateral) fontanelle** occurs on each side of the skull at the junction of the frontal, parietal, zygomatic, sphenoid, and temporal bones. A **mastoid (posterolateral) fontanelle** is on each side of the skull at the junction of the temporal, parietal, and occipital bones.

Assignment

1. Label Figure 8.9 and complete Section A of the laboratory report.
2. Complete Section D of the laboratory report.
3. Examine a fetal skull. Locate the skull bones and fontanelles shown in Figure 8.9.

The Vertebral Column

The vertebral column forms a flexible but sturdy vertical axis extending from the skull to the pelvis. It is composed of 24 movable vertebrae, the sacrum, and the coccyx. They are all bound together by ligaments and muscles to form a unified structure. Refer back to Figure 7.2 and note the relationship between the vertebral column and the rest of the skeletal system. The vertebral column exhibits four defined **spinal curvatures** that are shown in Figure 8.10. From top to bottom, they are the **cervical, thoracic, lumbar,** and **sacral curvatures.**

Vertebrae share many common features, although they vary in size and shape in different regions of the vertebral column. Each has a large structural mass, the **vertebral body,** that is the major load-bearing contact between adjacent vertebrae. Between the bodies of adjacent vertebrae are fibrocartilaginous **intervertebral discs** that serve as protective cushions.

The Axial Skeleton

Figure 8.10 The vertebral column. A&PR

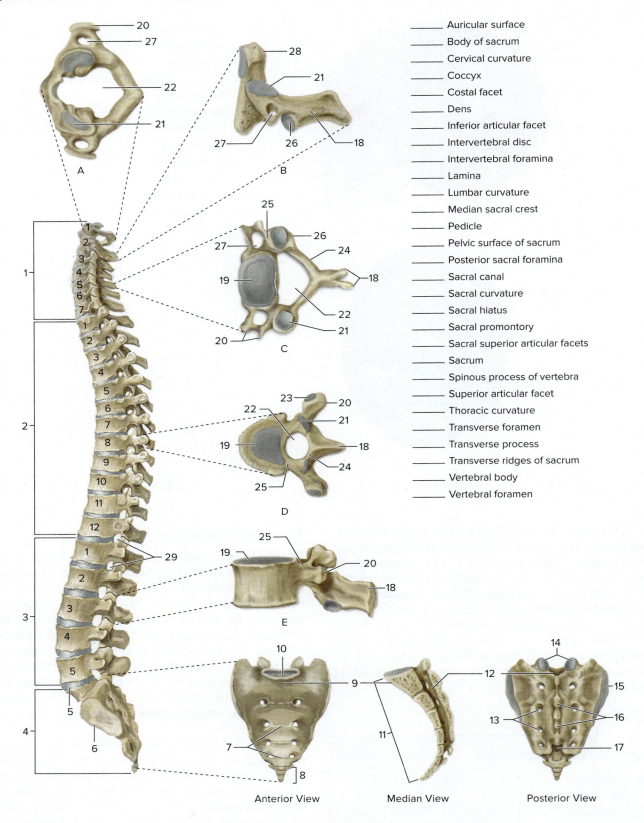

_____ Auricular surface
_____ Body of sacrum
_____ Cervical curvature
_____ Coccyx
_____ Costal facet
_____ Dens
_____ Inferior articular facet
_____ Intervertebral disc
_____ Intervertebral foramina
_____ Lamina
_____ Lumbar curvature
_____ Median sacral crest
_____ Pedicle
_____ Pelvic surface of sacrum
_____ Posterior sacral foramina
_____ Sacral canal
_____ Sacral curvature
_____ Sacral hiatus
_____ Sacral promontory
_____ Sacral superior articular facets
_____ Sacrum
_____ Spinous process of vertebra
_____ Superior articular facet
_____ Thoracic curvature
_____ Transverse foramen
_____ Transverse process
_____ Transverse ridges of sacrum
_____ Vertebral body
_____ Vertebral foramen

Projecting posteriorly from each side of the vertebral body are two processes, the **pedicles** (label 25), that form the anterolateral portion of the **vertebral arch.** The posterolateral portion of the neural arch is formed by the **laminae** (label 24), which join to form the backward projecting **spinous process.**

The **vertebral foramen** is located in the center of the vertebral arch. The spinal cord descends from the brain through the vertebral foramina, where it is protected by the surrounding vertebral arches of the vertebrae. Spinal nerves emerging from the spinal cord exit through the **intervertebral foramina** (label 29), small openings between pedicles of adjacent vertebrae.

In addition to the bodies, vertebrae contact each other by two pairs of articular surfaces located on the **transverse processes** (label 20). The **superior articular facets** (label 21) of one vertebra join with the **inferior articular facets** (label 26) of the adjacent vertebra above it.

Cervical Vertebrae

The first seven vertebrae are the **cervical vertebrae** of the neck. The first vertebra is the **atlas.** Illustration A shows its superior surface. Its vertebral foramen is larger than that of other vertebrae to accommodate the large spinal cord in the cervical region. The superior articular surfaces join with the occipital condyles of the skull. A small **transverse foramen** occurs in each transverse process. These foramina are found only in cervical vertebrae and serve as passageways for the vertebral arteries and veins.

The second cervical vertebra, the **axis,** is shown in illustration B. It is unique in having an upward-projecting **dens** arising from the vertebral body, which serves as a pivot for the rotation of the atlas. When the head is turned, the atlas rotates on the axis. This pivotal movement is enabled because the atlas lacks a vertebral body.

Illustration C shows the superior surface of a nonspecialized cervical vertebra. Note the Y-shaped spinous process.

Thoracic Vertebrae

Below the cervical vertebrae are the 12 **thoracic vertebrae.** A superior view of one is shown in illustration D. These vertebrae are distinguished in having ribs attached to them. They have **costal facets** (label 23) on their transverse processes and on the sides of their vertebral body. Also, their spinous processes project downward.

Lumbar Vertebrae

The next five vertebrae are the **lumbar vertebrae** (illustration E). They have a larger vertebral body than other vertebrae because of the greater stress that occurs in this region of the vertebral column.

The Sacrum

Below the fifth lumbar vertebra is the **sacrum,** which composes the rear wall of the pelvic cavity. It consists of five fused vertebrae. The **transverse ridges** where the vertebrae have fused are visible. The **pelvic surface of the sacrum** is concave, which causes an anterior protrusion of the body of the first sacral vertebra. This protrusion is the **sacral promontory.**

The back of the sacrum exhibits fused spinous processes forming the **median sacral crest.** On each side are rows of **posterior sacral foramina** through which nerves and blood vessels pass. The vertebral arches of the fused vertebrae form the **sacral canal,** which continues to a lower opening, the **sacral hiatus.**

The **sacral superior articular facets** (label 14) join with the inferior articular surfaces of the fifth lumbar vertebra. On each lateral surface is an auricular surface (label 15) that joins to the ilium of the pelvic girdle to form the *sacroiliac joint.*

The Coccyx

Four or five rudimentary vertebrae fuse together to form the **coccyx.** The triangular coccyx is joined to the sacrum by ligaments.

The Thoracic Cage

The thoracic vertebrae, ribs, costal cartilages, and sternum form the **thoracic cage.** It protects the heart and lungs and provides a support for the pectoral (shoulder) girdles. See Figure 8.11.

Figure 8.11 The thoracic cage. A&PR

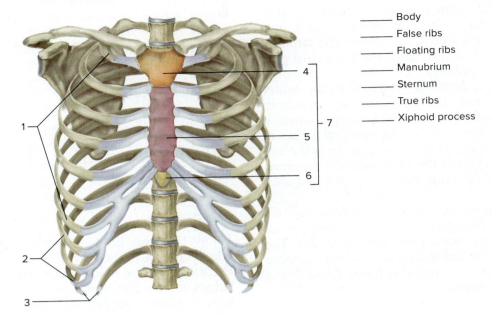

_____ Body
_____ False ribs
_____ Floating ribs
_____ Manubrium
_____ Sternum
_____ True ribs
_____ Xiphoid process

Sternum

The **sternum** is notched along the sides where the costal cartilages attach. It consists of three parts: an upper **manubrium**, a middle **body**, and a lower **xiphoid process.**

Ribs

There are 12 pairs of **ribs** attached to thoracic vertebrae. The first seven pairs of ribs attach directly to the sternum by **costal cartilages.** They are the **true ribs.** The remaining five pairs are **false ribs.** The cartilages of the first three pairs of false ribs are fused to the costal cartilages of ribs above them. The last two pairs of false ribs are not attached to bone nor cartilage in the front and are called **floating ribs.**

Figure 8.12 shows the structure of a thoracic vertebra, articulation of a rib with a thoracic vertebra, and the structure of a rib. The lateral view of the thoracic vertebra in illustration A shows, in addition to structures previously studied, two **costal facets** for attachment of a rib: one facet is near the vertebral body and the other facet is on the transverse process.

As shown in illustration C, a rib consists of the head, neck, tubercle, and body. The **head** (label 13) joins with the vertebral column via two **articular facets.** The **tubercle** (label 14) also has an **articular facet of the tubercle** (label 8). The **neck** is the short segment between the head and tubercle. The **body** is the flattened, curved remainder of the rib.

Illustration B shows a rib–vertebra articulation. The **articular facets** on the head of the rib connect to the *costal facet* near the vertebral body of its own vertebra and also with the vertebral body of the adjacent vertebra (not shown) above it. The articular facet of the tubercle articulates with the costal facet on the vertebra's transverse process.

Assignment

1. Label Figures 8.10, 8.11, and 8.12.
2. Complete the laboratory report.
3. Examine the skeletal material available in the laboratory and locate the parts discussed. *Use only pipe cleaners as pointers.*

Figure 8.12 Rib anatomy and its articulation.

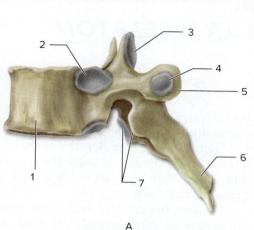

A
THORACIC VERTEBRA
(Lateral View)

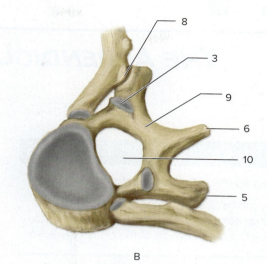

B
RIB-TO-VERTEBRA ARTICULATION
(Superior Anteroventral View)

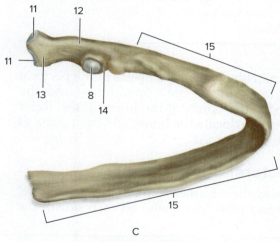

C
RIGHT CENTRAL RIB
(Posterior Ventral View)

_____ Articular facet of rib
_____ Articular facet of tubercle
_____ Body of rib
_____ Body of vertebra
_____ Costal facet (near body)
_____ Costal facet (on transverse process)
_____ Head of rib
_____ Inferior articular facets
_____ Lamina of vertebra
_____ Neck of rib
_____ Spinous process
_____ Superior articular facet
_____ Transverse process
_____ Tubercle of rib
_____ Vertebral foramen

The Axial Skeleton 73

Exercise 9

The Appendicular Skeleton

Objectives

After completing this exercise, you should be able to

1. Identify the bones, including their significant parts, that compose the appendicular skeleton.
2. Describe the relationships of the bones involved.

Materials
Skeletons, articulated and disarticulated
Pelves, male and female
Pipe cleaners

The **appendicular skeleton** consists of the pectoral girdles and bones of the upper limbs and the pelvic girdle and bones of the lower limbs. Refer to the figures to locate the bones and their parts as you study this exercise.

The Pectoral Girdles

A **pectoral girdle** is formed by a clavicle (collarbone) and a scapula (shoulder blade) on each side of the body. The right and left pectoral girdles support the upper limbs and allow a high range of motion.

Each **clavicle** is a slender, S-shaped bone. Its lateral end articulates with the shoulder blade, and its medial end articulates with the manubrium of the sternum.

The **scapula** is a flat, triangular bone that does not articulate directly with the axial skeleton. Instead, it is held in place by muscles, thus giving greater mobility to the shoulder. On its **lateral border** is the shallow **glenoid cavity,** which articulates with the head of the humerus. Above the glenoid cavity are two large processes. The **coracoid process** (label 8 in Figure 9.1) projects forward under the clavicle. The **acromion** projects laterally and articulates with the clavicle. On the back of the scapula, the **spine of the scapula** is a slender process that runs diagonally from the **medial border** to the acromion. A **suprascapular notch** occurs on the **superior border** at the base of the coracoid process.

The Upper Limb

The **upper limb** consists of the arm, forearm, and hand.

Figure 9.1 The scapula.

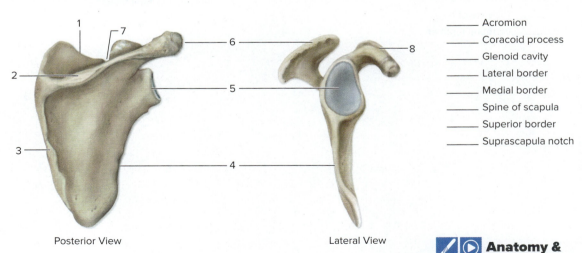

_____ Acromion
_____ Coracoid process
_____ Glenoid cavity
_____ Lateral border
_____ Medial border
_____ Spine of scapula
_____ Superior border
_____ Suprascapula notch

Posterior View   Lateral View

Anatomy & Physiology Revealed® 4.0
Unit 3: Protection, Support, and Movement

Figure 9.2 The upper limb and pectoral girdle.

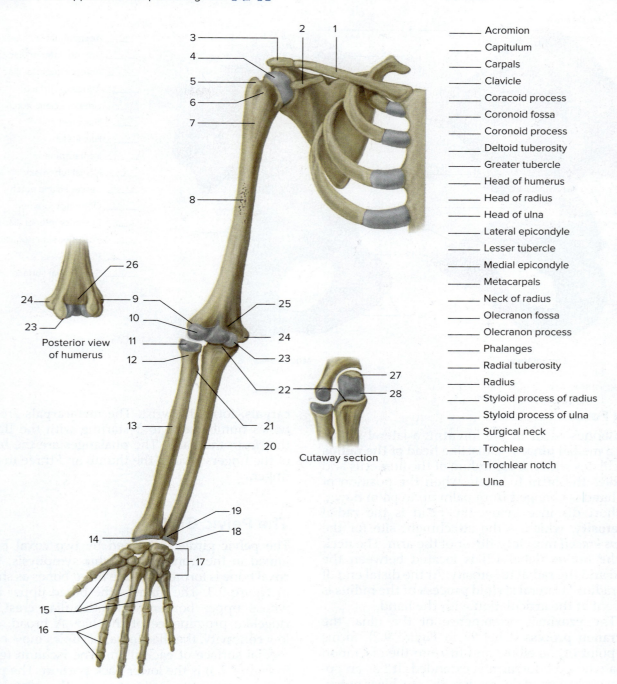

_____ Acromion
_____ Capitulum
_____ Carpals
_____ Clavicle
_____ Coracoid process
_____ Coronoid fossa
_____ Coronoid process
_____ Deltoid tuberosity
_____ Greater tubercle
_____ Head of humerus
_____ Head of radius
_____ Head of ulna
_____ Lateral epicondyle
_____ Lesser tubercle
_____ Medial epicondyle
_____ Metacarpals
_____ Neck of radius
_____ Olecranon fossa
_____ Olecranon process
_____ Phalanges
_____ Radial tuberosity
_____ Radius
_____ Styloid process of radius
_____ Styloid process of ulna
_____ Surgical neck
_____ Trochlea
_____ Trochlear notch
_____ Ulna

The Arm

The **humerus** is the bone of the arm. The rounded **head of the humerus** fits into the glenoid cavity of the scapula. Just distal to the head are two processes: a **greater tubercle** on the lateral surface and a **lesser tubercle** in front of it. The **surgical neck** is distal to the tubercles and is so named because of the frequency of fractures in this area. The **deltoid tuberosity** is a rough, raised area near the midpoint of the lateral surface of the shaft to which the _deltoid_ muscle attaches.

The distal end of the humerus has two condyles. The **capitulum** is the lateral condyle articulating with the radius. The **trochlea** is the medial condyle, which articulates with the ulna. Proximal to these condyles are the **lateral** and **medial epicondyles.** The depression on the front of the humerus just proximal to the trochlea is the **coronoid fossa.** The **olecranon fossa** (label 26 in Figure 9.2) is in a similar location on the back of the humerus.

The Appendicular Skeleton **75**

Figure 9.3 The coxal bone. A&PR

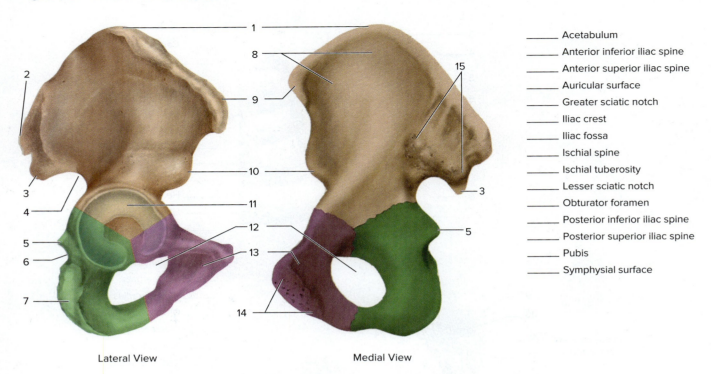

_____ Acetabulum
_____ Anterior inferior iliac spine
_____ Anterior superior iliac spine
_____ Auricular surface
_____ Greater sciatic notch
_____ Iliac crest
_____ Iliac fossa
_____ Ischial spine
_____ Ischial tuberosity
_____ Lesser sciatic notch
_____ Obturator foramen
_____ Posterior inferior iliac spine
_____ Posterior superior iliac spine
_____ Pubis
_____ Symphysial surface

Lateral View   Medial View

The Forearm

Two bones occur in the forearm: a lateral **radius** and a medial **ulna**. The disclike **head of the radius** articulates with the capitulum of the humerus and enables the head to rotate when the position of the hand is changed from palm up to palm down. A short distance below the head is the **radial tuberosity**, which is the attachment site for the *biceps brachii* muscle, a flexor of the arm. The **neck of the radius** (label 12) is located between the head and the radial tuberosity. At the distal end of the radius, a lateral **styloid process of the radius** is present at the articulation with the hand.

The proximal prominence of the ulna, the **olecranon process** (label 27 in Figure 9.2), forms the point of the elbow and fits into the olecranon fossa when the forearm is extended. It has an appropriately named depression, the **trochlear notch**, that articulates with the trochlea of the humerus. A small eminence in front of the trochlear notch is the **coronoid process** (label 22). At the distal end of the ulna, the knoblike **head of the ulna** articulates with the radius and a fibrocartilaginous disc that separates it from the hand. The **styloid process of the ulna** is the distal medial prominence.

The Hand

The skeleton of each hand consists of carpals, metacarpals, and phalanges. Eight small bones, the **carpals**, form the wrist. The **metacarpals** are five bones numbered 1 to 5 starting with the thumb that form the palm. The **phalanges** are the bones of the fingers: two in the thumb and three in each finger.

The Pelvic Girdle

The **pelvic girdle** is formed by two **coxal bones** joined in the front at the **pubic symphysis.** Each coxal bone is formed of three fused bones as shown in Figure 9.3. The **ilium** is the broad upper bone whose upper border forms the **iliac crest,** the ridgelike prominence of the hip. A broad, shallow concavity, the **iliac fossa,** occupies most of the medial surface of each ilium. The **ischium** (green in Figure 9.3) is the lower back portion. The **pubis** is the lower anterior part. Note the **symphysial surface** on the pubis and the **auricular surface** on the ilium. A large cup-shaped fossa, the **acetabulum,** is located at the junction of the ilium, ischium, and pubis on the lateral surface of each coxal bone. The acetabulum receives the head of the femur. The pelvic girdle joins with the sacrum at the **sacroiliac joints** forming the **pelvis.** See Figure 9.4.

The iliac crest extends between the **anterior superior iliac spine** (label 9 in Figure 9.3) and the **posterior superior iliac spine** (label 2). Just below each of these prominences are smaller iliac spines:

Figure 9.4 The lower limb and pelvis.

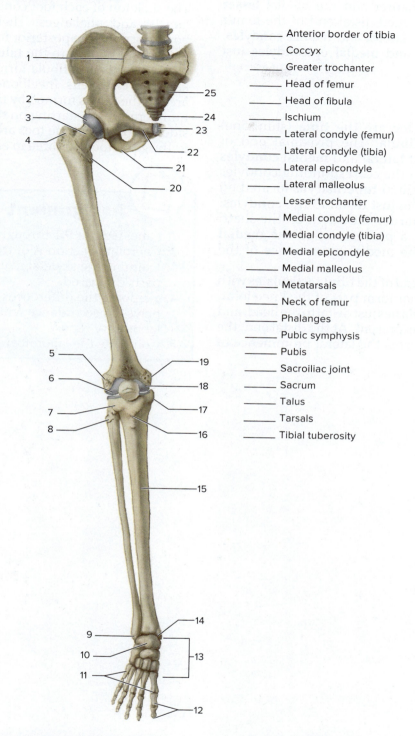

___ Anterior border of tibia
___ Coccyx
___ Greater trochanter
___ Head of femur
___ Head of fibula
___ Ischium
___ Lateral condyle (femur)
___ Lateral condyle (tibia)
___ Lateral epicondyle
___ Lateral malleolus
___ Lesser trochanter
___ Medial condyle (femur)
___ Medial condyle (tibia)
___ Medial epicondyle
___ Medial malleolus
___ Metatarsals
___ Neck of femur
___ Phalanges
___ Pubic symphysis
___ Pubis
___ Sacroiliac joint
___ Sacrum
___ Talus
___ Tarsals
___ Tibial tuberosity

the **anterior inferior iliac spine** and the **posterior inferior iliac spine.** Between the posterior inferior iliac spine and the **ischial spine** (label 5) is the **greater sciatic notch.** The **lesser sciatic notch** is just below the ischial spine. The **ischial tuberosity** (label 7) is located on the back of the ischium. The large opening surrounded by the ischium and pubis is the **obturator foramen.**

The Lower Limb

The **lower limb** consists of the thigh, leg, and foot.

The Thigh

The **femur** is the bone of the thigh. The rounded **head of the femur** fits into the acetabulum of a coxal bone. Two large processes occur at the base

The Appendicular Skeleton 77

of the **neck of the femur** (label 3 in Figure 9.4): the lateral **greater trochanter** and the medial **lesser trochanter.** The enlarged distal end of the femur terminates with the **lateral** and **medial condyles,** which have **lateral** and **medial epicondyles** just proximal to them.

The Leg

The bones of the leg are the **tibia** (shinbone) and the smaller **fibula.** The proximal end of the tibia consists of **lateral** and **medial condyles** that articulate with the corresponding condyles of the femur. The **tibial tuberosity** is located on the front of the tibia just distal to the condyles. A sharp **anterior border** is evident on the shaft. At the distal end, a process called the **medial malleolus** forms the medial prominence of the ankle.

The proximal **head of the fibula** articulates with the tibia but does not form part of the knee joint. The **neck of the fibula** lies just distal to the head, and an anterior border is present. At the distal end, the **lateral malleolus** forms the lateral prominence of the ankle.

The Foot

The skeleton of each foot consists of tarsals, metatarsals, and phalanges. There are seven **tarsal** bones forming the posterior foot. The most prominent tarsal bones are the **talus,** which articulates with the tibia and fibula forming the ankle joint, and the **calcaneus** (heelbone, not shown). The arch of the foot is formed by five **metatarsal** bones numbered 1 to 5 starting on the medial (great toe) side. The bones of the toes are the **phalanges:** two in the great toe and three in each of the other toes.

Assignment

1. Label Figures 9.1 through 9.4.
2. Complete Section A of the laboratory report.
3. Examine the skeletal material and locate the parts discussed.
4. Compare the differences in male and female pelves in accordance with Section B of the laboratory report.
5. Complete the laboratory report.

Exercise 10

ARTICULATIONS AND BODY MOVEMENTS

Objectives

After completing this exercise, you should be able to

1. State the basic functional types of joints and describe their characteristics.
2. Classify the joints of the body by functional type.
3. Identify the components of diarthrotic joints.
4. Identify the various types of body movements possible at diarthrotic joints.

Materials
Skeleton, articulated
Fresh knee joint of cow or lamb, sectioned longitudinally
Models of various joints

Before You Proceed

Consult with your instructor about using protective disposable gloves when performing portions of this exercise.

The Basic Functional Types

All joints of the body can be classified according to three functional categories as shown in Figure 10.1.

Immovable Joints

Synarthrotic joints are immovable because the bones forming the joint are tightly bonded to each other by dense connective tissue or cartilage tissue. There are two types of immovable joints: sutures and synchondroses.

Sutures are irregular joints between the immovable bones of the skull (illustration A in Figure 10.1). The bones are bonded together by dense connective tissue that is continuous with the **periosteum** (label 2) on the outer surface of the bones and with the **dura mater** on the inner surface of the bones.

Synchondroses have cartilage tissue as the bonding tissue. An example is the bonding of the epiphyses to the diaphysis by the epiphysial cartilages in the long bones of children.

Slightly Movable Joints

The bones of **amphiarthrotic** (slightly movable) **joints** are also bound by dense connective tissue or cartilage tissue but not as tightly as in synarthrotic joints. The two types of slightly movable joints are symphyses and syndesmoses.

Symphyses have a cushioning pad of **fibrocartilage** between the bones. The articulations of the bodies of vertebrae and the joining of coxal bones at the pubic symphysis are examples. Adjacent to the cartilaginous pads, the articular surfaces of the bones are covered with **articular cartilages** (label 4 in Figure 10.1) that reduce friction in the joint. The joint is wrapped in a capsule formed of **ligaments** (label 5).

Syndesmoses lack fibrocartilage, but the bones are held together by dense connective tissue forming interosseous ligaments. The attachment of the fibula to the tibia, as shown in Figure 10.3, is an example.

Freely Movable Joints

Bones forming **diarthrotic** (freely movable) **joints** are bound together by an **articular capsule** formed of ligaments. A **synovial membrane** lines the inside of the articular capsule and secretes **synovial fluid** that lubricates the joint. The articular surfaces of the bones are covered by protective, friction-reducing **articular cartilages.** These joints may also contain **bursae,** sacs of synovial fluid that reduce friction. There are six kinds of diarthrotic joints.

Unit 3: Protection, Support, and Movement

Figure 10.1 Types of articulations.

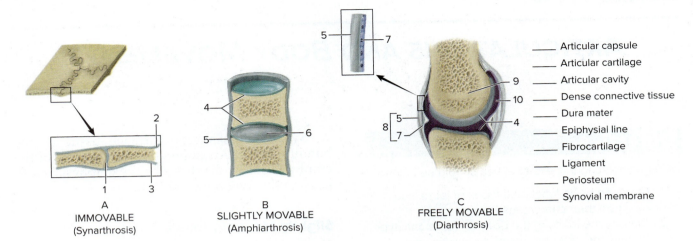

A IMMOVABLE (Synarthrosis)
B SLIGHTLY MOVABLE (Amphiarthrosis)
C FREELY MOVABLE (Diarthrosis)

___ Articular capsule
___ Articular cartilage
___ Articular cavity
___ Dense connective tissue
___ Dura mater
___ Epiphysial line
___ Fibrocartilage
___ Ligament
___ Periosteum
___ Synovial membrane

Gliding joints occur between small bones with flat or slightly convex surfaces, such as the carpal and tarsal bones.

Hinge joints allow movement in only one plane, such as in the elbow and knee.

Condyloid joints allow movement in two planes. They are formed by a rounded condyle articulating with an elliptical depression, such as between the carpals and the radius.

Saddle joints occur where the ends of both bones are saddle-shaped, convex in one direction and concave in the other. An example is the joint between the trapezium (a carpal bone) and the metacarpal bone of the thumb, which permits a variety of movements.

Pivot joints allow rotational movement around a pivot point, such as the atlas around the dens of the axis.

Ball-and-socket joints allow angular movement in all directions, such as in the shoulder and hip joints.

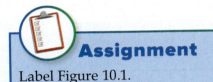

Assignment

Label Figure 10.1.

The Shoulder Joint

The shoulder joint (Figure 10.2) is a ball-and-socket joint. It is the most freely movable joint in the body, due to the shallowness of the glenoid cavity and the looseness of the articular capsule.

Figure 10.2 The shoulder joint.

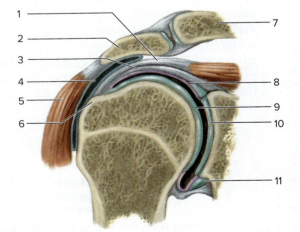

___ Acromion
___ Articular capsule
___ Articular cartilage of the glenoid cavity
___ Articular cartilage of the humerus
___ Clavicle
___ Deltoid muscle
___ Glenoid labrum
___ Greater tubercle of the humerus
___ Subdeltoid bursa
___ Supraspinatus tendon
___ Synovial membrane

Protection, Support, and Movement

Figure 10.3 The knee joint. (A) Anterior view; (B) posterior view. **APR**

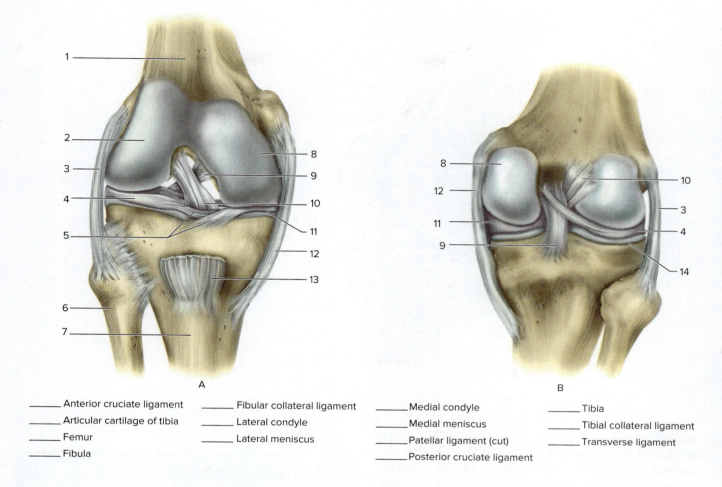

_____ Anterior cruciate ligament
_____ Articular cartilage of tibia
_____ Femur
_____ Fibula
_____ Fibular collateral ligament
_____ Lateral condyle
_____ Lateral meniscus

_____ Medial condyle
_____ Medial meniscus
_____ Patellar ligament (cut)
_____ Posterior cruciate ligament
_____ Tibia
_____ Tibial collateral ligament
_____ Transverse ligament

A fibrocartilage ring, the **glenoid labrum** (label 11), around the edge of the glenoid cavity slightly increases the depth of the cavity. Ligaments of the articular capsule and rotator cuff tendons of associated muscles support the joint. The **acromion** and the lateral end of the **clavicle** are shown above the joint. The tendon of **deltoid muscle** is attached to the acromion and the **tendon of the supraspinatus muscle** inserts on the **greater tubercle of the humerus**. Note the **subdeltoid bursa** under the deltoid muscle and tendon. A portion of the **articular capsule** (label 3), lined with **synovial membrane,** is shown just below the supraspinatus tendon. Both the head of the humerus and the glenoid cavity are coated with **articular cartilages.**

The Knee Joint

The structure of the knee joint, a complex hinge joint, is shown in Figure 10.3 in anterior and posterior views. The knee is probably the most highly stressed joint in the body. Some of this stress is absorbed by two fibrocartilage pads, the **lateral** and **medial menisci** (labels 4 and 11). They are thicker at the periphery and thinner at the center of the joint, providing a recess for the condyles. The menisci are wrapped together peripherally by a **transverse ligament** (label 5). **Articular cartilages** cover the articular surfaces of the femur (not labeled) and the tibia (label 14).

The joint is enveloped by an articular capsule composed of a complex of several layers of ligaments and muscle tendons. Within the capsule are the **anterior cruciate ligament** (label 10) and the **posterior cruciate ligament** (label 9). They are named for their anterior or posterior attachment to the tibia and because they cross to form an X. The **fibular collateral ligament** is the lateral ligament extending from the head of the fibula

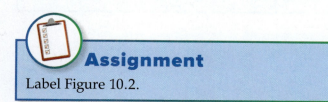

Assignment
Label Figure 10.2.

Articulations and Body Movements

Figure 10.4 A sagittal section of the knee joint.

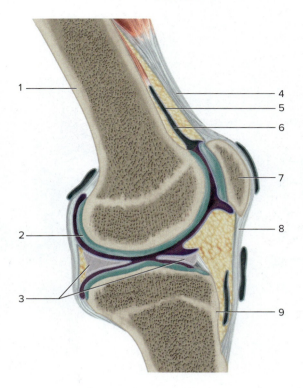

_____ Articular cartilage
_____ Femur
_____ Menisci
_____ Patella
_____ Patellar ligament
_____ Quadriceps tendon
_____ Suprapatellar bursa
_____ Synovial membrane
_____ Tibia

to the lateral epicondyle of the femur. The **tibial collateral ligament** extends from the tibia to the medial epicondyle of the femur. In the anterior view, a stub of the **patellar ligament** (cut) is shown attached to the tibial tuberosity. Other ligaments are not shown.

Figure 10.4 shows a sagittal section of the knee joint. Note the articular cartilages and menisci within the joint. The **quadriceps tendon** from the large group of muscles on the anterior of the thigh, and the **patellar ligament,** which attaches to the tibia, hold the patella in place. A **suprapatellar bursa** lies beneath the quadriceps tendon. There are a dozen or so bursae associated with muscle tendons of the knee joint but not shown here. **Synovial membranes** line the bursae and the articular capsule.

As you can see, there is no bony reinforcement to prevent dislocation of the knee in any direction. The knee is especially vulnerable to lateral and rotational forces, so it is often injured in vigorous sport activities. A torn ligament is either a collateral ligament or a cruciate ligament. Damage to a meniscus is called a torn cartilage.

Assignment

1. Label Figures 10.3 and 10.4.
2. Complete Sections A through C of the laboratory report.
3. Examine the joint models and the major structures of each.
4. Examine an articulated skeleton and classify the joints in accordance with Section D of the laboratory report.
5. Examine the diarthrotic joints on your body and correlate the degree and type of movement with the type of freely movable joint.
6. Examine a fresh cow or sheep hinge joint that has been sectioned longitudinally. Compare it with Figures 10.3 and 10.4 and identify as many structures as possible.

82 Protection, Support, and Movement

Muscle Arrangements

Muscles usually function in groups to bring about body movements. The **prime mover** is the primary muscle in the group that produces the desired action. The assisting muscles, the **synergists,** impart steadiness and prevent unnecessary action. Other muscles that hold structures in position for a movement to occur are called **fixators.**

Muscles are usually arranged in **antagonistic groups** so that the contraction of one muscle, the **agonist** (prime mover), causes a specific action and the contraction of the other muscle, the **antagonist,** causes the opposite action. Smooth movement requires that each muscle relaxes while the opposing muscle contracts.

Body Movements

The arrangement of muscles in the body allows them to produce specific body movements upon contraction. The common types of movements are noted in the following paragraphs. It is important to note that body movements can be described correctly in two different formats. One format is to describe the **movement** *of the* **body part** being moved, such as "flexion of the forearm." The other format describes the **movement** *at the* **joint** involved in the movement, such as "flexion at the elbow." For simplicity, all movements in this laboratory textbook use the *movement of the body part* format.

Flexion and Extension

Flexion is the decrease of the angle formed by the bones at a joint. **Extension** is the increase of this angle. Flexion of the forearm occurs at the elbow when the hand is brought toward the shoulder. Straightening of the elbow produces extension.

Specific terms are applied to the movement of the foot. Flexing the foot upward is called **dorsiflexion.** Moving the sole of the foot downward is called **plantar flexion.** Extension of the foot at the ankle is the same as plantar flexion. When extension is excessive, as when leaning backward, it is called **hyperextension.**

Abduction and Adduction

The movement of a limb away from the midline of the body is **abduction.** Movement toward the midline is **adduction.** These terms may be applied to fingers and toes by using the axis of the limb as the point of reference.

Pronation and Supination

The palm of the hand faces forward in the anatomical position. Rotating the hand so the palm faces backward is **pronation. Supination** rotates the palm to again face forward. These terms also apply when the upper limb is not in the anatomical position. Turning the palm downward is pronation, and turning it upward is supination.

Elevation and Depression

Movement of structures upward is **elevation,** whereas movement downward is **depression.** For example, a shoulder shrug elevates the scapula.

Inversion and Eversion

These terms apply to foot movements. **Inversion** is when the sole of the foot is turned inward toward the midline of the body. **Eversion** is turning the sole outward away from the midline.

Rotation and Circumduction

The movement of a body part around its longitudinal axis is **rotation.** If the body part can flex, extend, abduct, and adduct so that it traces a circle, the movement is called **circumduction.** For example, the arm can circumduct at the shoulder joint.

Constriction and Dilation

Sphincter muscles are circular muscles that surround openings of the body. Contraction causes **constriction** of the opening; relaxation causes **dilation** of the opening.

Assignment

1. Identify the types of movements in Figure 10.5. Some movements are illustrated more than once.
2. Complete the laboratory report.

Figure 10.5 Body movements. A&PR

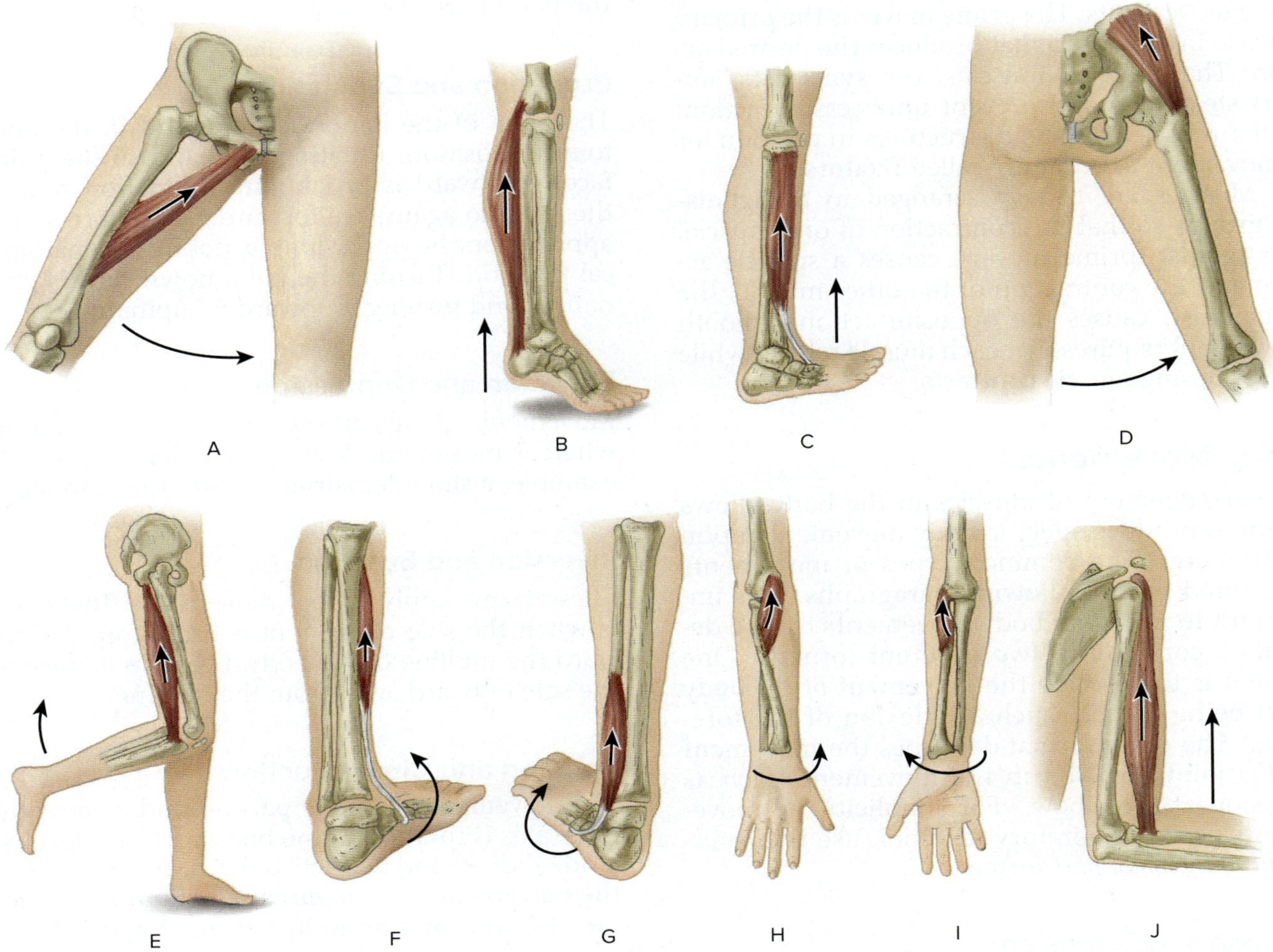

A B C D

E F G H I J

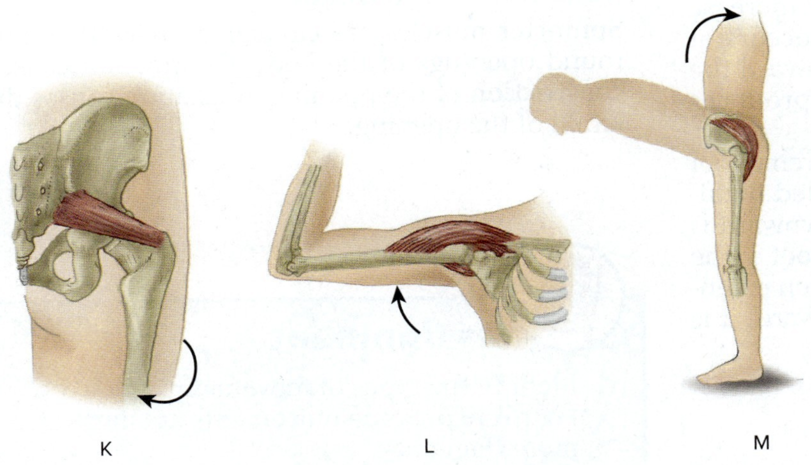

_____ Abduction (2 places)
_____ Adduction (2 places)
_____ Dorsiflexion
_____ Eversion
_____ Extension
_____ Flexion (2 places)
_____ Inversion
_____ Plantar flexion
_____ Pronation
_____ Rotation
_____ Supination

K L M

84 Protection, Support, and Movement

Exercise 11

MUSCLE ORGANIZATION AND CONTRACTIONS

Objectives

After completing this exercise, you should be able to

1. Identify and describe the structure of a skeletal muscle.
2. Describe the basic structure and function of the neuromuscular junction.
3. Describe the orientation of thick and thin myofilaments in resting and contracted muscle fibers.
4. Describe the basic mechanism of muscle contraction.

Materials

Test tube of glycerinated muscle
Dropping bottles of
   glycerol
   0.25% ATP in triple distilled water
   0.05 M KCl
   0.001 M $MgCl_2$
Microscope slides and cover glasses, three each
Dissecting instruments
Plastic ruler, clear and flat
Microscopes, compound and dissecting

Before You Proceed

Consult with your instructor about using protective disposable gloves when performing portions of this exercise.

Movement of limbs and other body parts by the contraction (shortening) of skeletal muscles results from the contraction of muscle fibers (cells) composing the muscles involved. The type of movement is determined by the sites of muscle attachment and the type of articulation involved. Typically, a muscle is attached to the skeleton at each end and spans across a joint where the movement occurs. In this exercise, you will study the structure of skeletal muscles and the interaction of a motor neuron and a muscle fiber in the process of contraction, as well as the mechanism of contraction.

Figure 11.1 shows the orientation of two muscles of the arm as examples. The **triceps brachii** extends the forearm, and the **biceps brachii** and **brachialis** flex the forearm.

Muscle Attachments

A muscle is attached to a bone by the fusion of its dense connective tissue with the periosteum. This attachment may be direct or indirect. In a **direct attachment,** the muscle's connective tissue does not extend beyond the muscle but fuses directly to the periosteum. In an **indirect attachment,** the muscle's connective tissue extends beyond the muscle as a narrow bandlike or ropelike **tendon** or a broad sheetlike **aponeurosis** (not shown). Typically when a muscle contracts one end remains stationary and the other end moves. Older terminology referred to the stationary end of a muscle as the *origin* and the movable end as the *insertion.* However, the origin and insertion of a muscle are not consistent nor always easy to determine and new terminology is beginning to replace origin and insertion. Generally, muscles will be described as having either a **superior attachment** or **inferior attachment** if they are muscles of the head or trunk. Muscles of the upper and lower limbs are referred to as having a **proximal attachment** and **distal attachment.** When the new terminology is not clear, origin and insertion are still used.

Figure 11.1A shows the orientation of three muscles of the arm that span the elbow joint: the **triceps brachii, brachialis,** and **biceps brachii.** The triceps brachii has three proximal attachments (label 1). Two proximal attachments fuse by direct attachment to the back of the humerus, and one proximal attachment is attached by a tendon to

Unit 3: Protection, Support, and Movement

Figure 11.1   Muscle anatomy and attachments.

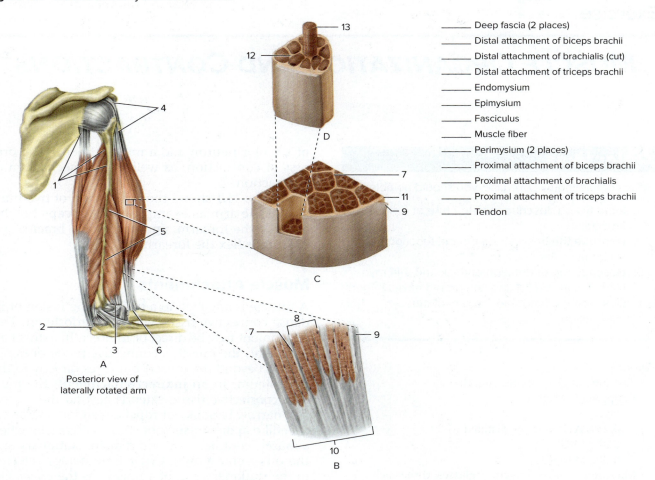

_____ Deep fascia (2 places)
_____ Distal attachment of biceps brachii
_____ Distal attachment of brachialis (cut)
_____ Distal attachment of triceps brachii
_____ Endomysium
_____ Epimysium
_____ Fasciculus
_____ Muscle fiber
_____ Perimysium (2 places)
_____ Proximal attachment of biceps brachii
_____ Proximal attachment of brachialis
_____ Proximal attachment of triceps brachii
_____ Tendon

the scapula. Its distal attachment is on the olecranon process of the ulna. The proximal attachment of the brachialis is fused by direct attachment (label 5) to the lower half of the front of the humerus, and its distal attachment is on the coronoid process of the ulna. The biceps brachii has two proximal attachments (label 4) attached by tendons to the scapula, and its distal attachment is on the radial tuberosity. The distal attachments of all three muscles are attached to bones by tendons.

When a muscle contracts, it exerts a pull between the two attachments. Contraction of the triceps brachii exerts a pull on the olecranon, which leads to extension of the forearm. Contraction of the brachialis and biceps brachii exerts a pull on the ulna and radius, respectively, which leads to flexion of the forearm.

Microscopic Structure

Illustrations B, C, and D in Figure 11.1 show the detailed structure of a skeletal muscle. A muscle is composed of many **muscle fibers** that are bound together in small bundles called **fasciculi** (singular, *fasciculus*). Only one of three fasciculi is labeled (label 8) in illustration B, which shows the fusion of connective tissue to form a tendon.

Illustration C shows how adjacent fasciculi are separated from and bound to each other by connective tissue called **perimysium** (label 7). A coarser connective tissue, the **epimysium** (label 11), surrounds and binds together all of the fasciculi in a muscle. Surrounding the epimysium, the **deep fascia** (label 9) also envelops the entire muscle.

Illustration D shows a single fasciculus. Note that each muscle fiber is separated from and bound to adjacent muscle fibers by a thin layer of connective tissue called **endomysium**. A single muscle fiber is extended for easy recognition.

Assignment

Label Figure 11.1.

Introduction to Muscle Contraction

Contraction of a muscle fiber is a complex process that involves both chemical changes and microscopic structural changes within the muscle fiber. Furthermore, contraction of skeletal muscle is dependent upon action potentials from a motor neuron to initiate the process.

Skeletal muscle fibers are innervated by motor neurons. A motor neuron and the muscle fibers it innervates constitute a **motor unit.** A large motor unit involves the innervation of many (e.g., 100–2,000) muscle fibers by a single motor neuron. Large motor units activate muscles for which precise control is unimportant. A small motor unit involves innervation of only a few (e.g., 5–30) muscle fibers activated by a single motor neuron. Small motor units activate muscles for which precise control is important, such as in muscles controlling finger movements.

The junction between a motor neuron and its target cell, such as another neuron, is called a **synapse.** When the target cell is a muscle fiber, the synapse is called a **neuromuscular junction.** Figure 11.2 shows (1) the terminal boutons of a motor neuron's axon attached to a muscle fiber forming a neuromuscular junction, and (2) the structure of a neuromuscular junction.

Contraction of a skeletal muscle fiber involves the interaction of (1) a motor neuron, (2) the neuromuscular junction, and (3) a muscle fiber. As an action potential passes along an axon to the neuromuscular junction, it causes chemical changes at the neuromuscular junction that form an action potential in the muscle fiber. This action potential, in turn, initiates changes in the muscle fiber that produce the contraction.

The Neuromuscular Junction

The specialized structures of a neuromuscular junction perform specific functions in the transmission of an action potential from a motor neuron to a muscle fiber. Each **terminal bouton** (label 2) fits into a depression in the **sarcolemma** (label 3) called the **motor end plate.** The space between the terminal bouton and the sarcolemma is the **synaptic cleft** (label 7). Numerous folds of the sarcolemma adjacent to the synaptic cleft greatly increase its surface area.

When an action potential reaches the terminal bouton, **synaptic vesicles** (label 8) release the neurotransmitter **acetylcholine** into the synaptic cleft by exocytosis (label 9). Acetylcholine quickly binds to acetylcholine receptors in the sarcolemma, producing an action potential in the muscle fiber. The action potential moves inward along the **transverse tubule** (T tubule) (label 5) to the **sarcoplasmic reticulum** (label 6), causing it to release calcium ions that enable contraction of the myofibrils. After initiating muscle contraction, acetylcholine is quickly (within 1/500 of a second) inactivated by an enzyme, **acetylcholinesterase,** on the sarcolemma, and the sarcolemma is returned to its resting state. It is now capable of receiving another action potential.

Energy for both neuron function and muscle contraction is supplied by adenosine triphosphate (ATP). ATP is produced by aerobic respiration in numerous **mitochondria** present in neurons and muscle fibers.

Myofibril Structure

Figure 11.3 shows the structure of a muscle fiber. Illustration A shows a single muscle fiber composed of several myofibrils. Note that the sarcoplasmic reticulum and T tubules envelop each myofibril, that mitochondria are located between myofibrils, and that the entire muscle fiber is enveloped by the sarcolemma. Nuclei are located on the periphery of the muscle fiber.

Illustration B shows a single myofibril and its striations. The dark and light striations of a myofibril are due to the arrangement of two kinds of **myofilaments** that are composed of proteins. The myofilaments not only create the striations of skeletal muscle, but the repeating pattern also forms **sarcomeres,** the contractile units of skeletal muscle. Illustrations C and D show the structure and arrangement of the myofilaments when the myofibril is relaxed and when it is contracted.

Thick myofilaments are composed of the protein *myosin.* They are the larger of the myofilaments and are located at the center of the sarcomeres. The length of the thick myofilaments corresponds to the **A band,** which is the darker striation formed by overlap of the thick myofilaments with the thin myofilaments. The **thin myofilaments** are made up mostly of the protein *actin.* Thin myofilaments extend toward the center of the sarcomere from the **Z disc,** a system of proteins that makes up the border between adjacent sarcomeres. The lighter striation, the **I band,** is composed only of thin myofilaments.

Compare the position of the Z discs and the thick and thin myofilaments in the relaxed state (illustration C) and in the contracted state (illustration D).

Mechanics of Contraction

The shortening of a muscle fiber by contraction results from the interaction of thick and thin myofilaments in the presence of calcium

Figure 11.2 The neuromuscular junction.

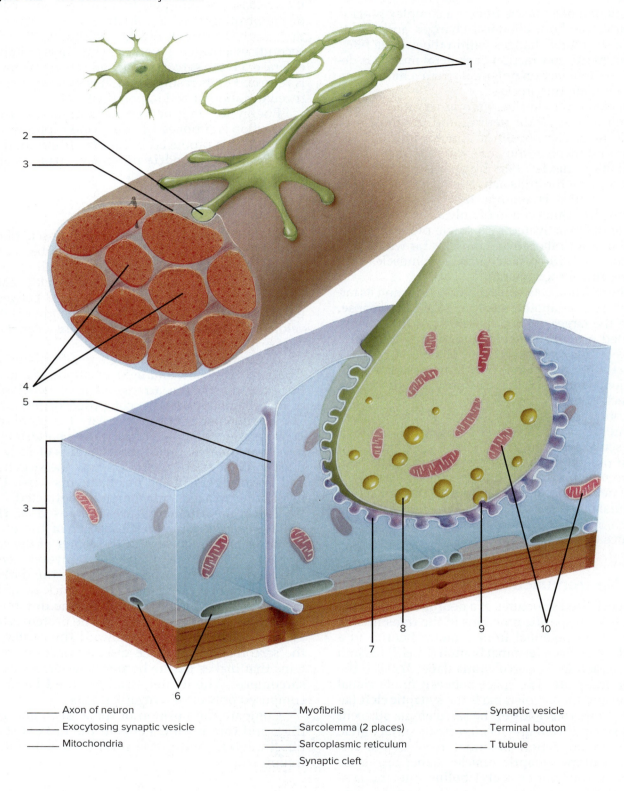

_____ Axon of neuron
_____ Exocytosing synaptic vesicle
_____ Mitochondria
_____ Myofibrils
_____ Sarcolemma (2 places)
_____ Sarcoplasmic reticulum
_____ Synaptic cleft
_____ Synaptic vesicle
_____ Terminal bouton
_____ T tubule

Figure 11.3 The structure of a muscle fiber.

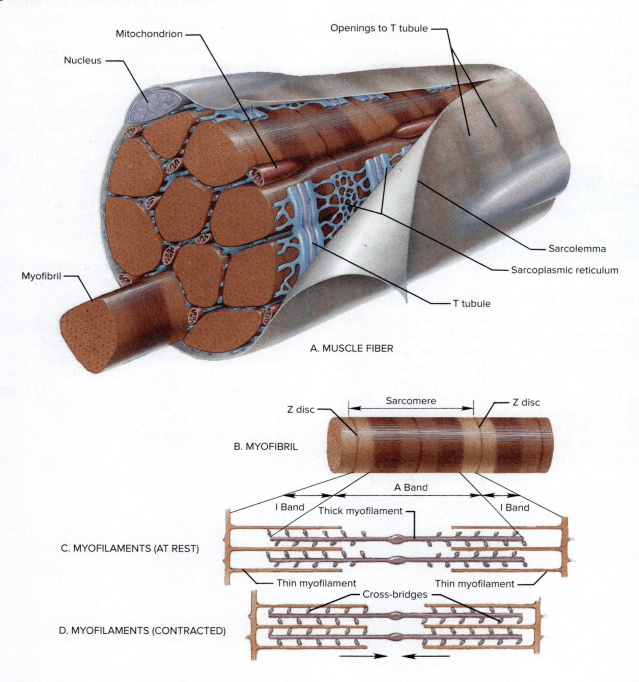

and magnesium ions. ATP provides the necessary energy. The most accepted explanation for myofibril contraction is known as the sliding-filament model. The sequence of events is described below.

1. As the action potential moves along the sarcolemma, T tubules, and sarcoplasmic reticulum, large quantities of calcium ions are released.

2. Calcium ions cause changes that expose the chemically active sites on the thin myofilaments.

3. The *myosin heads* of the thick myofilaments attach to the actin active sites on the thin myofilament forming cross-bridges. The heads bend, exerting a power stroke that pulls the thin myofilaments toward the center of the A band, thus

Muscle Organization and Contractions

shortening the sarcomere by pulling Z discs closer together.
4. After the first power stroke, the heads detach from the first actin active sites, bond with next active sites, and produce another power stroke.
5. This ratchetlike process repeats itself until nervous stimulation stops, maximal contraction is attained, or fatigue occurs.

Assignment

1. Label Figure 11.2.
2. Complete Sections B through F of the laboratory report.

Experimental Muscle Contraction

Perform the following experiment in which a glycerinated muscle will be induced to contract using ATP, magnesium ions, and potassium ions.

The muscle has already been removed from its test tube and cut it into 2 cm segments that have been placed in a Petri dish of glycerol. These muscle segments have been teased apart to obtain thin muscle strands composed of very few fibers. The strands should be no more than 0.2 mm thick for use in the experiment.

1. Place a muscle strand in a small drop of glycerol on a clean microscope slide and add a cover glass. Examine it microscopically with the high-dry or oil immersion objective. Draw the pattern of striations of the relaxed muscle fibers in Section G of the laboratory report.
2. Place three or four of the thinnest strands on another slide in a minimal amount of glycerol. Arrange the strands straight, parallel, and close to each other.
3. Using a dissecting microscope, measure the length of the strands by placing a clear plastic ruler under the slide. Record the lengths on the laboratory report. Estimate the width of the strands.
4. While observing through the dissecting microscope, add one drop from each of the solutions: ATP, magnesium chloride, and potassium chloride. Observe the contraction.
5. After 30 seconds, remeasure and record the length of the strands. Calculate and record the percentage of contraction. Has the width of the strands changed?
6. Remove one of the contracted strands to another slide, add a cover glass, and observe the pattern of the striations using a compound microscope. Draw the pattern of striations of the contracted muscle fibers in Section G on the laboratory report. Complete the laboratory report.
7. Clean your workstation.

Exercise 12

AXIAL MUSCLES

Objectives

After completing this exercise, you should be able to
1. Identify the major axial muscles on charts or models.
2. Describe the action of the muscles studied.
3. Describe the location and attachments of the muscles studied.

Materials
Charts or models of muscles

Organization of Muscles

The muscles of the body are organized like the skeleton into **axial muscles** and **appendicular muscles.** In this exercise you will study the major axial muscles of the body including their locations, actions, and attachments. Axial muscles are defined as having both of their attachments on the axial skeleton or within the axial region of the body. Axial muscles are described as having either a *superior attachment* or an *inferior attachment* when attaching two bones. However, sometimes axial muscles have one attachment point on the skin or another structure. In these cases, the less moveable end of the muscles is called the *origin* and the more movable end of the muscle is called the *insertion*.

Muscle movements were previously discussed as having the format or "movement of the body part." With some muscles the movement format creates difficult to understand movements and therefore simpler terminology will be used. For example, instead of describing a muscle action as "depressing the corners of the mouth," we can simply say "frowning."

Head and Neck Muscles

Refer to Figure 12.1 as you study this section.

Orbicularis Oris This sphincter muscle encircles the mouth. Its *origin* is on several facial muscles, maxilla, mandible, and nasal septum. The *insertion* is on the lips.

  Action: Closes and puckers the lips.

Orbicularis Oculi This is the sphincter muscle encircling the eye. Its *origin* is on the frontal and maxilla bones. Its *insertion* is on the eyelids.

  Action: Closes the eyelids.

Occipitofrontalis This muscle is composed of two muscular parts that lie over the frontal and occipital bones. They are joined by the **epicranial aponeurosis,** which covers the top of the skull. The **frontal belly** has its *origin* on the aponeurosis and its *insertion* on the soft tissue under the eyebrows. The **occipital belly** has its *origin* on the mastoid process and occipital bone. Its *insertion* is on the aponeurosis.

  Action: The frontal belly elevates the eyebrows and wrinkles the forehead. The occipital belly pulls the scalp backward.

Zygomaticus These two muscles (major and minor parts) have their *origin* on the zygomatic arch and extend diagonally to their *insertion* on the orbicularis oris at the corner of the mouth.

  Action: Smiling.

Depressor Anguli Oris This muscle has its *origin* on the mandible and its *insertion* on the orbicularis oris at the corner of the mouth. It is an antagonist of the zygomaticus.

  Action: Frowning.

Unit 3: Protection, Support, and Movement

Figure 12.1 Head and neck muscles.

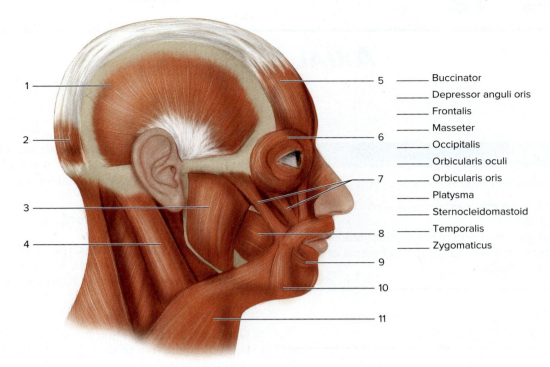

Platysma This is a broad sheetlike muscle that covers the front and side of the neck. Its *origin* is on the fascia of the upper chest and shoulder. The *insertion* is on the mandible and the muscles around the mouth. It is involved in many facial expressions.

  Action: Depresses the lower lip and assists in opening the mouth.

Buccinator This horizontal muscle is located in the walls of the cheeks. Its *origin* is on the mandible and maxilla, and its *insertion* is on the orbicularis oris at the corner of the mouth.

  Action: Compresses the cheek, holds food between teeth during chewing, and interacts with other muscles in facial expressions.

Masseter This is the primary chewing muscle. Its *superior attachment* is on the zygomatic arch, and its *inferior attachment* is on the lateral surface of the ramus and angle of the mandible.

  Action: Elevates the mandible as in chewing.

Temporalis This is a large, fan-shaped muscle on the side of the head. Its *superior attachment* is on the temporal, parietal, and frontal bones. The *inferior attachment* is on the coronoid process of the mandible.

  Action: Acts synergistically with the masseter to elevate the mandible as in chewing.

Sternocleidomastoid This muscle is located on the side of the neck and is partially covered by the platysma. Its *inferior attachment* is on the sternum and medial end of the clavicle, and its *superior attachment* is on the mastoid process.

  Action: Contraction of both muscles flexes the head down toward the chest. Contraction of one muscle laterally rotates the head away from the side of the contracting muscle.

Assignment
Label Figure 12.1.

Respiratory Muscles

Refer to Figure 12.2 as you study this section.

External Intercostals Intercostal muscles are found between the ribs. The external intercostals (label 3) have a *superior attachment* on the lower border of the upper rib and an *inferior attachment*

Figure 12.2 Muscles of the anterior trunk.

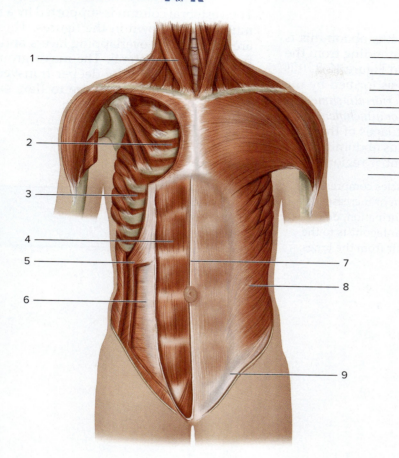

_____ External intercostal
_____ External oblique
_____ Inguinal ligament
_____ Internal intercostal
_____ Internal oblique
_____ Linea alba
_____ Rectus abdominis
_____ Sternocleidomastoid
_____ Transverse abdominis

on the upper edge of the lower rib. The fibers are directed obliquely anterior.

Action: Elevate thoracic cage and increase volume of the thorax, resulting in inspiration of air in breathing.

Internal Intercostals These muscles (label 2) have an *inferior attachment* on the upper border of the lower rib and a *superior attachment* on the lower edge of the upper rib. The muscle fibers are arranged in a direction opposite to the external intercostals.

Action: Depress thoracic cage and draw ribs closer together. This decreases volume of the thorax, causing expiration of air in breathing.

Abdominal Muscles

The abdominal wall consists of four pairs of thin muscles that have a common collective action. The right side of the abdominal wall has been removed in Figure 12.2 to show the three layers composed of the first three muscles described next, starting with the outermost muscle.

External Oblique The *superior attachment* of this superficial muscle is on the lower eight ribs. Its *inferior attachment* is on the iliac crest and the **linea alba** (label 7), the white line at the midline of the abdomen where the aponeuroses of the right and left external obliques meet. The lower margin of each aponeurosis forms the **inguinal ligament,** which extends from the anterior superior iliac spine of the ilium to near the pubic symphysis. The muscle fibers run diagonally from the ribs toward the linea alba.

Internal Oblique This muscle (label 5) lies just under the external oblique. Its *inferior attachment* is on the iliac crest and inguinal ligament. Its *superior attachment* is on the costal cartilages of the lower three ribs, the linea alba, and the pubic bone. The muscle fibers also run diagonally.

Transversus Abdominis This innermost muscle *originates* from the inguinal ligament, iliac crest, and the costal cartilages of the lower six ribs. It *inserts* on the linea alba and the pubic bone.

Axial Muscles   93

The muscle fibers run horizontally across the abdomen.

Rectus Abdominis The right rectus abdominis is the narrow segmented muscle extending from the thoracic cage to the pubic bone in Figure 12.2. It is embedded within the aponeurosis formed by the preceding three muscles. Its *inferior attachment* is on the pubic bone, and its *superior attachment* is on the xiphoid process and the cartilages of the fifth, sixth, and seventh ribs. The rectus abdominis aids in flexion of the trunk in the lumbar region.

> *Collective Action:* These four muscles compress the abdominal organs and maintain or increase intra-abdominal pressure, aiding in urination, defecation, and childbirth. They are antagonists to the diaphragm and aid in forcing air from the lungs.

Muscles of the Vertebral Column

The vertebral column is supported by a network of muscles not shown in the figures. These muscles are arranged in overlapping layers and in groups from medial to lateral. The major groups are the **erector spinae** and the deeper **transversospinalis** group. These muscles serve to flex, extend, and stabilize the vertebral column.

Assignment
Label Figure 12.2.

Exercise 13

APPENDICULAR MUSCLES

Objectives

After completing this exercise, you should be able to

1. Identify the major appendicular muscles on charts or models.
2. Describe the action of the muscles studied.
3. Describe the location and attachments of the muscles studied.

Materials
Charts or models of muscles

Organization of Muscles

The muscles of the body are organized like the skeleton into **axial muscles** and **appendicular muscles.** In this exercise you will study the major appendicular muscles of the body including their locations, actions, and attachments. Appendicular muscles are defined as having one or more of their attachments on the appendicular skeleton. Appendicular muscles are usually described as having either a *proximal attachment* or a *distal attachment* when attaching two bones. However, when appendicular muscles have one attachment point in the axial skeleton the axial end of the muscles is called the *origin* and the appendicular end of the muscle is called the *insertion*.

Muscles Moving the Pectoral Girdle

Refer to Figure 13.1 as you study this section.

Serratus Anterior This muscle covers the upper lateral surface of the ribs. Its *origin* is on the upper eight or nine ribs, and its *insertion* is on the medial border of the scapula.

Action: Pulls scapula downward and forward toward the chest.

Trapezius This large triangular muscle of the upper back has its *origin* on the occipital bone and the spinous processes of the cervical and thoracic vertebrae. It *inserts* on the spine and acromion of the scapula and on the lateral third of the clavicle.

Action: Adducts and elevates scapula; pulls head backward (hyperextension) if scapulae are fixed.

Muscles Moving the Arm

Refer to Figure 13.1 as you study this section.

Pectoralis Major This fan-shaped muscle occupies the upper quadrant of the chest. Its *proximal attachment* is on the clavicle, sternum, costal cartilages, and aponeurosis of the external oblique. Its *distal attachment* is along a groove between the greater and lesser tubercles of the humerus.

Action: Adduction and medial rotation of the arm.

Deltoid This muscle of the shoulder has a *proximal attachment* on the lateral third of the clavicle and on the acromion and spine of the scapula. The *distal attachment* is on the deltoid tuberosity of the humerus.

Action: Flexion, abduction, and extension of the arm.

Coracobrachialis This muscle covers the upper medial surface of the humerus. Its *proximal attachment* is on the coracoid process of the scapula, and its *distal attachment* is on the middle medial surface of the humerus. Coracobrachialis is shown in Figure 13.2.

Action: Flexes and adducts arm.

Latissimus Dorsi This large muscle covers the lower back. Its *proximal attachment* is a large aponeurosis attached to thoracic and lumbar vertebrae, spine of the sacrum, iliac crest, and lower

Unit 3: Protection, Support, and Movement

Figure 13.1 Muscles moving the pectoral girdle.

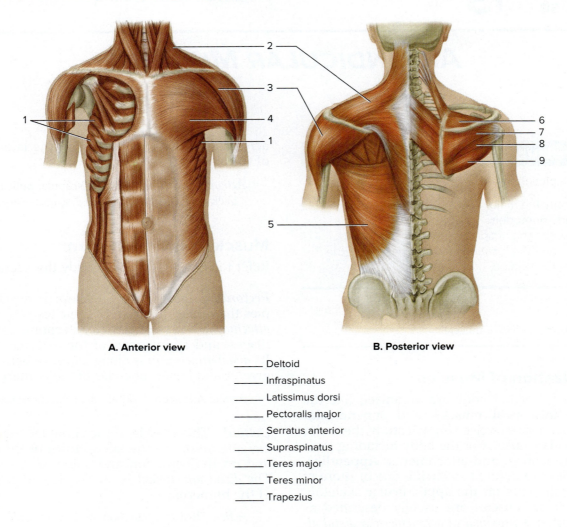

A. Anterior view B. Posterior view

_____ Deltoid
_____ Infraspinatus
_____ Latissimus dorsi
_____ Pectoralis major
_____ Serratus anterior
_____ Supraspinatus
_____ Teres major
_____ Teres minor
_____ Trapezius

ribs. Its *distal attachment* is on the groove between the greater and lesser tubercles of the humerus.

Action: Extends, adducts, and rotates the arm medially; pulls shoulder downward and backward.

Infraspinatus This muscle (label 7) has a *proximal attachment* on the inferior margin of the spine and back of the scapula. Its *distal attachment* is on the greater tubercle of the humerus. It is partially covered by the trapezius and deltoid in Figure 13.1. The small muscle just below it, the **teres minor**, synergistically assists the infraspinatus.

Action: Rotates arm laterally.

Supraspinatus This muscle (label 6) is covered by the trapezius and deltoid. It has a *proximal attachment* above the spine of the scapula and a *distal attachment* on the greater tubercle of the humerus.

Action: Assists deltoid in abducting arm.

Teres Major This muscle is the bottom muscle of the three muscles with a *proximal attachment* on the back of the scapula. It has a *distal attachment* on the lesser tubercle of the humerus.

Action: Rotates arm medially.

Assignment

1. Label Figure 13.1.
2. Complete Sections A and B of the laboratory report.
3. Locate these muscles on a wall chart or model.

96 Protection, Support, and Movement

Figure 13.2 Muscles attached to the humerus.

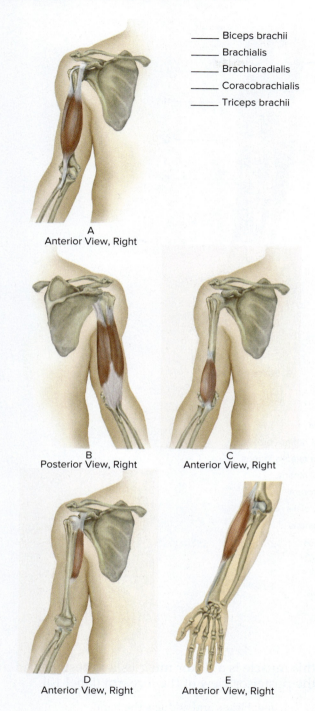

_____ Biceps brachii
_____ Brachialis
_____ Brachioradialis
_____ Coracobrachialis
_____ Triceps brachii

A Anterior View, Right
B Posterior View, Right
C Anterior View, Right
D Anterior View, Right
E Anterior View, Right

Muscle Moving the Forearm

The first four muscles described are attached to the humerus and are shown in Figure 13.2. The last two are forearm muscles and are depicted in Figure 13.3.

Biceps Brachii This large muscle on the front of the arm bulges when the forearm is flexed. It has *proximal attachments* at two sites on the scapula: the coracoid process and superior to the glenoid cavity. Its *distal attachment* is on the radial tuberosity.

Action: Flexes forearm; supination.

Brachialis This muscle lies just under the biceps brachii. Its *proximal attachment* is a direct attachment on the lower half of the humerus. Its *distal attachment* is on the coronoid process of the ulna.

Action: Flexes the forearm.

Brachioradialis This is the most superficial muscle on the lateral side of the forearm. Its *proximal attachment* is on the lateral epicondyle of the humerus. Its *distal attachment* is on the lateral surface of the radius just above the styloid process.

Action: Flexes the forearm.

Triceps Brachii This muscle covers the posterior surface of the upper arm and is the antagonist of the brachialis. Its *proximal attachment* occurs at three sites: the tubercle below the glenoid cavity, and the posterior and medial surfaces of the humerus. It has a *distal attachment* on the olecranon process of the ulna.

Action: Extends the forearm.

Supinator This small muscle (label 3, Figure 13.3) near the elbow has a *proximal attachment* on the lateral epicondyle of the humerus and proximal ulna. It curves around the radius and has a *distal attachment* on the lateral margin of the radial tuberosity.

Action: Supination.

Pronator Teres This muscle has a *proximal attachment* on the medial epicondyle of the humerus and a *distal attachment* on the proximal lateral surface of the radius.

Action: Pronation.

Muscles Moving the Hand

Muscles moving the hand are shown in Figure 13.3. All illustrations are of the right forearm.

Appendicular Muscles

Figure 13.3 Muscles moving the forearm and hand.

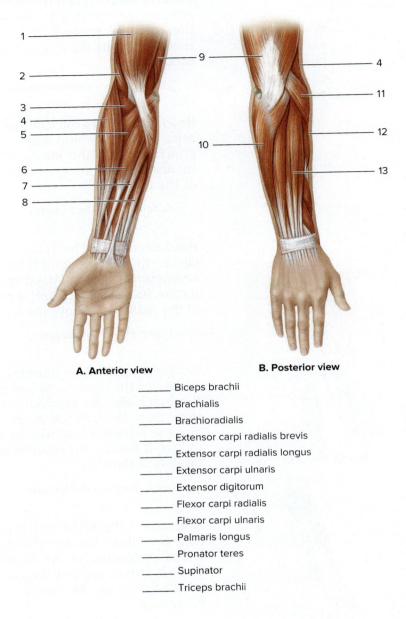

A. Anterior view   B. Posterior view

_____ Biceps brachii
_____ Brachialis
_____ Brachioradialis
_____ Extensor carpi radialis brevis
_____ Extensor carpi radialis longus
_____ Extensor carpi ulnaris
_____ Extensor digitorum
_____ Flexor carpi radialis
_____ Flexor carpi ulnaris
_____ Palmaris longus
_____ Pronator teres
_____ Supinator
_____ Triceps brachii

Flexor Carpi Radialis This muscle has a *proximal attachment* on the medial epicondyle of the humerus and extends diagonally across the arm where its tendon divides to form *distal attachments* on the proximal ends of the second and third metacarpals.

Action: Flexes and abducts the hand.

Flexor Carpi Ulnaris This muscle also has a *proximal attachment* on the medial epicondyle of the humerus, but it has a *distal attachment* on the proximal end of the fifth metacarpal. Even though this muscle is a flexor muscles it is best viewed in the posterior view of the forearm (label 10).

Action: Flexes and adducts the hand.

Flexor Digitorum This muscle has a *proximal attachment* on the humerus, ulna, and radius. Its tendon divides to form *distal attachments* on the middle phalanges of fingers 2–5. It is beneath the flexor carpi radialis and flexor carpi ulnaris and therefore not visible in any of the figures of this exercise.

Action: Flexes fingers 2–5.

98   Protection, Support, and Movement

Extensor Carpi Radialis Longus and Brevis These two muscles are shown in Figure 13.3 (labels 12—longus, and 13—brevis). Both have a *proximal attachment* on the lateral epicondyle of the humerus; the longus is attached in the more proximal position. The longus has a *distal attachment* on the proximal end of the second metacarpal; the brevis has a *distal attachment* on the proximal end of the third metacarpal.

Action: Both muscles extend and abduct the hand.

Extensor Carpi Ulnaris This muscle lies on the medial side of the posterior forearm (label 11). It has a *proximal attachment* on the lateral epicondyle of the humerus, crosses over, and has a *distal attachment* on the base of the fifth metacarpal.

Action: Extends and adducts the hand.

Extensor Digitorum This muscle (label 14) also has a *proximal attachment* from the lateral epicondyle of the humerus. Its tendon divides to form *distal attachments* on the posterior surfaces of the distal phalanges of fingers 2–5.

Action: Extends fingers 2–5.

Assignment
1. Label Figures 13.2 and 13.3.
2. Complete Sections D and E of the laboratory report.
3. Locate the muscles on a wall chart or model.

Muscles Moving the Thigh

The muscles that move the thigh have proximal attachments on the pelvis or vertebral column and distal attachments on the femur. See Figure 13.4.

Psoas Major This muscle has a *proximal attachment* on the lumbar vertebrae, passes through the pelvis, and a *distal attachment* on the lesser trochanter of the femur.

Action: Flexes the thigh.

Iliacus This muscle has a direct *proximal attachment* in the iliac fossa and merges with the psoas major to form a *distal attachment* on the lesser trochanter of the femur. The term **iliopsoas** is used to refer to the psoas major and iliac muscles together since they merge and share a synergistic function.

Action: Flexes the thigh.

Gluteus Maximus This superficial buttocks muscle is covered by the deep fascia that invests the thigh muscles. A broad tendon, the **iliotibial tract** (label 5), extends downward from the deep fascia and attaches to the tibia. The gluteus maximus has *proximal attachments* on the ilium, sacrum, and coccyx and *distal attachments* on the back of the femur and the iliotibial tract.

Action: Extends and laterally rotates the thigh.

Gluteus Medius and Gluteus Minimus These muscles lie under the gluteus maximus, with the minimus being the innermost and is thus not seen in Figure 13.4. They have *proximal attachments* from the ilium and *distal attachments* on the greater trochanter of the femur.

Action: Abduct and medially rotate the thigh.

Tensor Fasciae Latae This muscle has a *proximal attachment* on the anterior superior iliac crest and its *distal attachment* is into the iliotibial tract.

Action: Flexes and abducts thigh.

Adductor Longus and Pectineus Both muscles have *proximal attachments* on the pubis and *distal attachments* on the back of the femur. The pectineus (label 6) is shorter and inserts above the adductor longus.

Action: Adduct, flex, and laterally rotate the thigh.

Adductor Magnus This muscle, the strongest of the adductors, has *proximal attachments* on the ischium and pubis and *distal attachment* on the medial epicondyle of the femur. Adductor magnus (label 8) is mostly hidden behind other muscles in Figure 13.4.

Action: Adduct, flex, and laterally rotate the thigh.

Assignment
Label Figure 13.4.

Appendicular Muscles

Figure 13.4 Muscles that move the thigh. **APR**

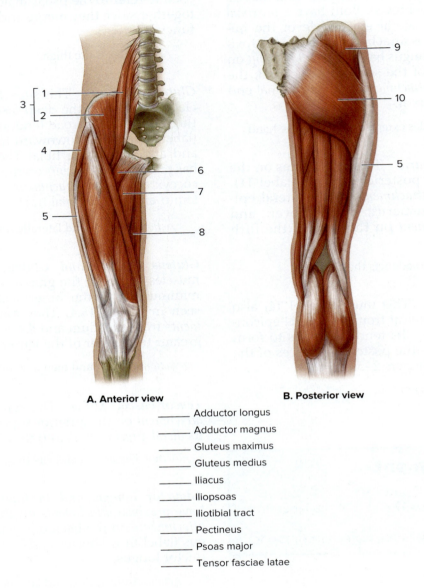

A. Anterior view B. Posterior view

_____ Adductor longus
_____ Adductor magnus
_____ Gluteus maximus
_____ Gluteus medius
_____ Iliacus
_____ Iliopsoas
_____ Iliotibial tract
_____ Pectineus
_____ Psoas major
_____ Tensor fasciae latae

Thigh and Leg Movements

The following muscles, shown in Figure 13.5, are primarily involved in the flexion and extension of the leg while some also act on the thigh.

Hamstrings The following three muscles are collectively known as the hamstrings.

Biceps Femoris This is the most lateral of the three hamstring muscles (label 7). It has *proximal attachments* on the ischial tuberosity and the back of the femur and *distal attachments* on the head of the fibula and lateral epicondyle of the tibia.

Semimembranosus This is the most medial muscle of the three hamstrings. Its *proximal attachment* is on the ischial tuberosity. The *distal attachment* is on the back of the medial epicondyle of the tibia.

Semitendinosus Located between the other two hamstrings, this muscle has a *proximal attachment* on the ischial tuberosity and *distal attachment* on the upper medial surface of the tibia.

Figure 13.5 Muscles that move the leg. A&PR

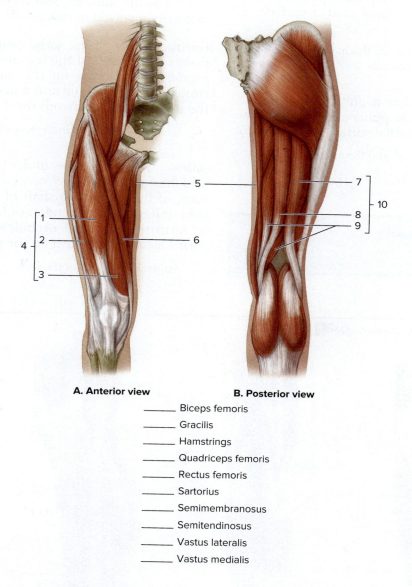

A. Anterior view   B. Posterior view

_____ Biceps femoris
_____ Gracilis
_____ Hamstrings
_____ Quadriceps femoris
_____ Rectus femoris
_____ Sartorius
_____ Semimembranosus
_____ Semitendinosus
_____ Vastus lateralis
_____ Vastus medialis

Collective Action: Flex lower leg; extend thigh; biceps femoris laterally rotates thigh; the other two rotate the thigh medially.

Quadriceps Femoris The quadriceps femoris is a large muscle that forms the front of the thigh. It is composed of the following four smaller muscles that unite to form a common tendon. The tendon envelops the patella and continues to form a *distal attachment* on the tibial tuberosity.

Rectus Femoris This is the superficial middle portion of the quadriceps. It has a *proximal attachment* on the anterior inferior iliac spine.

Vastus Lateralis This lateral part of the quadriceps has *proximal attachments* on the greater trochanter and the back of the femur.

Vastus Medialis This medial portion has a *proximal attachment* on the posterior surface of the femur.

Vastus Intermedius This portion is beneath the other portions of the quadriceps and therefore is not visible in Figure 13.5. It has *proximal attachments* on the anterior and lateral surfaces of the femur.

Collective Action: Extend the lower leg; the rectus femoris also flexes the thigh.

Appendicular Muscles   **101**

Sartorius This long muscle has a *proximal attachment* on the anterior superior iliac spine and *distal attachment* on the proximal medial surface of the tibia.

> *Action:* Flexes the leg and the thigh; rotates the thigh laterally and the leg medially.

Gracilis This muscle has a *proximal attachment* on the lower edge of the pubis and *distal attachment* on the proximal medial surface of the tibia.

> *Action:* Adducts the thigh and flexes the leg.

Muscle Moving the Foot

The following muscles, shown in Figure 13.6, are primarily involved in movements of the foot.

Gastrocnemius This large superficial muscle makes up the calf of the leg (sural region). It has *proximal attachments* on the lateral and medial condyles of the femur and a *distal attachment* on the back of the calcaneous (heel bone).

> *Action:* Flexes leg; plantar flexes the foot.

Soleus This muscle lies under the gastrocnemius and has *proximal attachments* on the head and back of the fibula and shaft of the tibia. Its tendon unites with the tendon of the gastrocnemius to form the **calcaneal (Achilles) tendon,** which forms a *distal attachment* on the calcaneus.

> *Action:* Plantar flexes the foot.

Assignment
Label Figure 13.5.

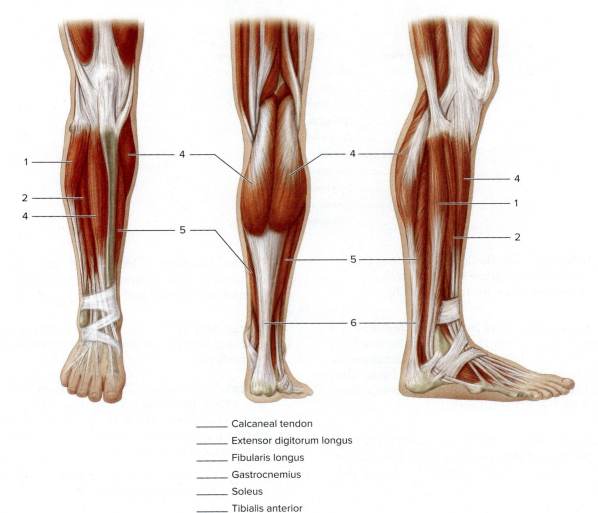

Figure 13.6 Muscle moving the foot. **APR**

_____ Calcaneal tendon
_____ Extensor digitorum longus
_____ Fibularis longus
_____ Gastrocnemius
_____ Soleus
_____ Tibialis anterior

Figure 13.7 Major surface muscles of the body. (A) Anterior view; (B) posterior view.

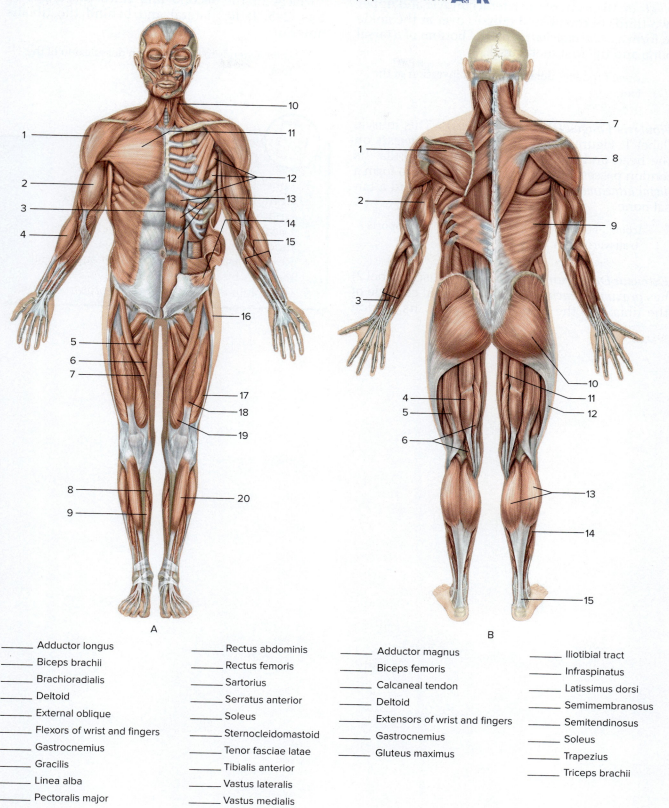

A

_____ Adductor longus
_____ Biceps brachii
_____ Brachioradialis
_____ Deltoid
_____ External oblique
_____ Flexors of wrist and fingers
_____ Gastrocnemius
_____ Gracilis
_____ Linea alba
_____ Pectoralis major

_____ Rectus abdominis
_____ Rectus femoris
_____ Sartorius
_____ Serratus anterior
_____ Soleus
_____ Sternocleidomastoid
_____ Tenor fasciae latae
_____ Tibialis anterior
_____ Vastus lateralis
_____ Vastus medialis

B

_____ Adductor magnus
_____ Biceps femoris
_____ Calcaneal tendon
_____ Deltoid
_____ Extensors of wrist and fingers
_____ Gastrocnemius
_____ Gluteus maximus

_____ Iliotibial tract
_____ Infraspinatus
_____ Latissimus dorsi
_____ Semimembranosus
_____ Semitendinosus
_____ Soleus
_____ Trapezius
_____ Triceps brachii

Appendicular Muscles   103

Tibialis Anterior This muscle has *proximal attachments* on the front of the lateral condyle and upper two-thirds of the tibia. It crosses over at the ankle to form a *distal attachment* on the bottom of a tarsal bone and the first metatarsal.

Action: Produces dorsiflexion and inversion of the foot.

Fibularis Longus (Peroneus Longus) This muscle (label 1, Figure 13.6) has *proximal attachments* on the head and proximal portion of the fibula. Its tendon passes under the arch of the foot to form a *distal attachment* on the first metatarsal and a tarsal bone.

Action: Plantar flexes and everts the foot; supports transverse arch.

Extensor Digitorum Longus This muscle (label 2) has *proximal attachments* on the lateral condyle of the tibia and the front of the fibula. Its tendon divides to form *distal attachments* on the upper surfaces of the second and third phalanges of toes 2–5. It lies lateral and behind the tibialis anterior.

Action: Extends toes 2–5; aids in dorsiflexion of the foot.

Assignment

1. Label Figure 13.6.
2. Complete Sections F through I of the laboratory report.
3. Locate the muscles on a wall chart or model.
4. Review your understanding of superficial muscles by labeling Figure 13.7.
5. Complete Section J of the laboratory report.

Exercise 14

THE SPINAL CORD AND REFLEX ARCS

Objectives

After completing this exercise, you should be able to

1. Identify the spinal cord, its protective coverings, and spinal nerves on charts or models.
2. Identify parts of the spinal cord when viewed microscopically in cross section.
3. Describe the pathways of somatic and visceral reflex arcs.
4. Describe the reflexes performed in this exercise.

Materials
Model of spinal cord in cross section
Prepared slides of spinal cord, x.s.
Reflex hammers

Automatic stereotyped responses to stimuli enable rapid adjustment to environmental conditions. These responses are called **reflexes**. Reflexes play an important role in maintaining **homeostasis**. There are two types of reflexes. **Somatic reflexes** involve skeletal muscle responses. **Visceral reflexes** involve responses by smooth muscle, cardiac muscle, or glands. For example, visceral reflexes control heart rate, blood pressure, body temperature, water balance, and other vital physiological conditions. The neural pathway involved in a reflex is known as a **reflex arc,** and it always involves the central nervous system—either the brain or the spinal cord. For example, a spinal reflex involves receptors, spinal nerves, spinal cord, and effectors.

The Spinal Cord

The spinal cord extends downward from the brain, within the vertebral canal. In infants, the spinal cord extends the full length of the vertebral canal, but in adults it ends near the second lumbar vertebra. This difference results from the differences in growth rates of the spinal cord and the axial skeleton. The back of the axial skeleton has been removed in Figure 14.1 to reveal the spinal cord and spinal nerves.

Extending downward from the end of the spinal cord are the proximal ends of spinal nerves serving body regions below the cord. This aggregation of nerves is called the **cauda equina,** which translates to "horse tail." The **dural sac** (dura mater) ends within the sacrum; only a portion of it is shown (label 11) for better observation of the nerves.

The Spinal Nerves

The spinal nerves emerge from the spinal cord through the intervertebral foramina on each side of the vertebral column. There are eight cervical, twelve thoracic, five lumbar, and five sacral pairs. These 30 pairs of nerves, plus one pair of coccygeal nerves, make a total of 31 pairs of spinal nerves.

As shown in the sectional view in Figure 14.1, each spinal nerve is joined to the spinal cord by **anterior** (ventral) and **posterior** (dorsal) **roots.** The enlargement on the posterior root is a **spinal ganglion** that contains cell bodies of sensory neurons. Note in the sectional view that the **gray matter** (label 19), which is composed of unmyelinated axons and neuron cell bodies, is encompassed by **white matter,** which is composed of myelinated nerve fibers.

Some spinal nerves combine to form complex networks called **plexuses** on each side of the spinal cord. In a plexus, the axons are rearranged and combined so that axons innervating a particular body part occur in the same nerve, although the axons may originate in different spinal nerves.

Cervical Nerves

There are eight pairs of cervical nerves. The first pair (C1) exits the spinal cord between the skull and the atlas. The eighth pair (C8) exits between the seventh cervical vertebra and the first thoracic vertebra. The upper cervical nerves unite to form the **cervical plexus** (label 16), which serves the muscles and skin of the neck. The lower cervical nerves unite with the first thoracic nerve (T1) to form the **brachial plexus,** which is located in the

Unit 4: Control and Integration

Figure 14.1 The spinal cord and spinal nerves.

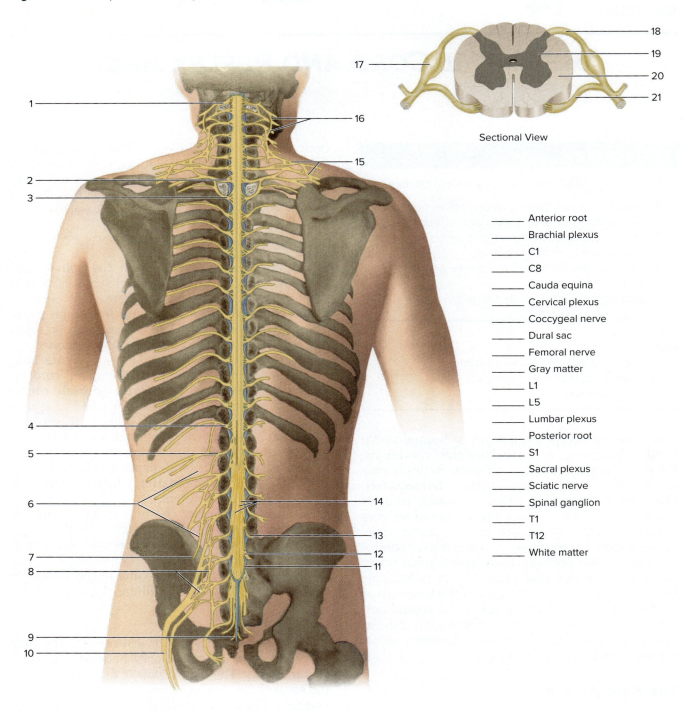

Sectional View

_____ Anterior root
_____ Brachial plexus
_____ C1
_____ C8
_____ Cauda equina
_____ Cervical plexus
_____ Coccygeal nerve
_____ Dural sac
_____ Femoral nerve
_____ Gray matter
_____ L1
_____ L5
_____ Lumbar plexus
_____ Posterior root
_____ S1
_____ Sacral plexus
_____ Sciatic nerve
_____ Spinal ganglion
_____ T1
_____ T12
_____ White matter

shoulder region. Nerves serving the shoulders, back, and upper limbs arise from this plexus.

Thoracic Nerves

There are 12 pairs of thoracic nerves, T1 through T12. Each nerve emerges from the spinal cord just below its corresponding vertebra. Most of the thoracic nerves do not mix to participate in plexuses and are located between ribs. For this reason, these spinal nerves are called **intercostal nerves.**

Lumbar Nerves

There are five pairs of lumbar nerves. Most of the lumbar nerves unite to form the **lumbar plexus** (label 6), from which the large **femoral**

106 Control and Integration

Figure 14.2 The spinal cord and meninges.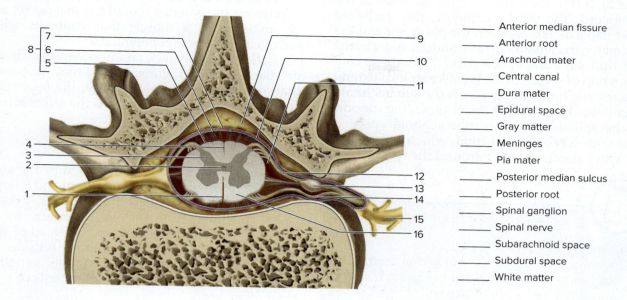

___ Anterior median fissure
___ Anterior root
___ Arachnoid mater
___ Central canal
___ Dura mater
___ Epidural space
___ Gray matter
___ Meninges
___ Pia mater
___ Posterior median sulcus
___ Posterior root
___ Spinal ganglion
___ Spinal nerve
___ Subarachnoid space
___ Subdural space
___ White matter

nerve (label 7) forms to supply the thigh and anterior leg.

Sacral and Coccygeal Nerves

There are five pairs of sacral nerves and one pair of coccygeal nerves (label 9). The **sacral plexus** (label 8) is formed by branches from the lumbar plexus and the sacral nerves. The large nerve arising from this plexus is the **sciatic nerve,** which serves the entire posterior lower limb.

The Spinal Cord in Cross Section

The cross-sectional structure of the spinal cord within the vertebral canal is shown in Figure 14.2. The H-shaped pattern of the gray matter is surrounded by white matter that is composed mostly of myelinated axons. Figure 14.3 illustrates the differences between gray matter and white matter and the regional variation of their proportions. Two grooves occur on the midline. The **anterior median fissure** is wider than the **posterior median sulcus.** The small **central canal** lies on the median line in the gray matter. It extends the length of the spinal cord and is continuous with the ventricles of the brain.

The Meninges

Surrounding the spinal cord (and also the brain) are three membranes, the **meninges** (singular, *meninx*). The outermost membrane, the **dura mater,** also covers the spinal nerve roots. It consists of dense connective tissue and is the toughest of the meninges. Outside the dura mater, between the dura mater and the vertebrae, is an **epidural space** that contains areolar connective tissue, adipose tissue, and blood vessels.

Figure 14.3 Cross-sectional histology of the thoracic and sacral spinal cord. Jose Luis Calvo/Shutterstock

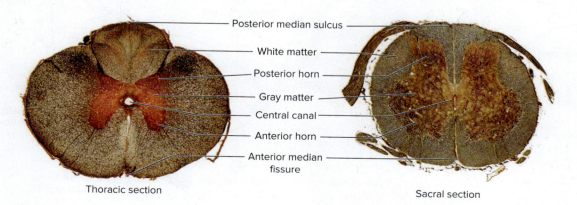

The Spinal Cord and Reflex Arcs   107

The innermost membrane is the **pia mater** (label 5), a thin membrane tightly adhered to the spinal cord. The third membrane, the **arachnoid mater** (label 6), lies between the pia mater and the dura mater. It is separated from the dura mater by the very thin **subdural space** (label 10), which contains a film of serous fluid. Weblike strands extend from the arachnoid mater across the **subarachnoid space** (label 11) to the pia mater providing support for the spinal cord. The subarachnoid space is filled with **cerebrospinal fluid,** which serves as a protective shock absorber around the spinal cord.

Assignment

1. Label Figures 14.1 and 14.2.
2. Examine a prepared slide of spinal cord, x.s., and locate the parts shown in Figure 14.2. Compare your observations with Figure 14.3. Use high power to locate neuron cell bodies in the gray matter.

The Somatic Reflex Arc

The components of a somatic reflex arc involving the spinal cord are shown in Figure 14.4. Arrows indicate the path of action potential transmission. These components are (1) a **receptor** that receives stimuli and forms action potentials, such as a touch receptor in the skin; (2) a **sensory neuron** that carries action potentials into the spinal cord via a spinal nerve and its posterior root; (3) an **interneuron** that transmits action potentials to a motor neuron within the gray matter of the spinal cord; (4) a **motor neuron** that transmits action potentials to an effector via the anterior root of a spinal nerve; and (5) an **effector,** a muscle that contracts when activated by the action potentials.

Reflex arcs may lack an interneuron, as in the patellar (knee-jerk) reflex, or they may have more than one interneuron. In general, the fewer synapses involved, the more rapid is the reflex action.

The Visceral Reflex Arc

Reflexes of the viscera are controlled by the **autonomic division** of the nervous system and follow a different pathway than somatic reflexes. Visceral reflex arcs differ from somatic reflex arcs by the presence of two efferent neurons instead of one. The two efferent neurons synapse outside the central nervous system in an autonomic ganglion. Figure 14.5 shows the major components of a visceral reflex arc.

There are two types of autonomic ganglia. **Paravertebral autonomic ganglia** or **sympathetic chain ganglia** (label 5) are united to form two chains; one chain lies along each side of the vertebral column. **Prevertebral ganglia** (label 2) are located in front of the vertebral bodies closer to the visceral organs.

Stimulation of a receptor in a visceral organ initiates action potentials that are carried by a **visceral afferent** (sensory) **neuron** (label 3). This afferent neuron passes through a paravertebral autonomic ganglion and enters the spinal cord via the posterior root of a spinal nerve. It synapses with a **preganglionic efferent neuron** (label 4) within the gray

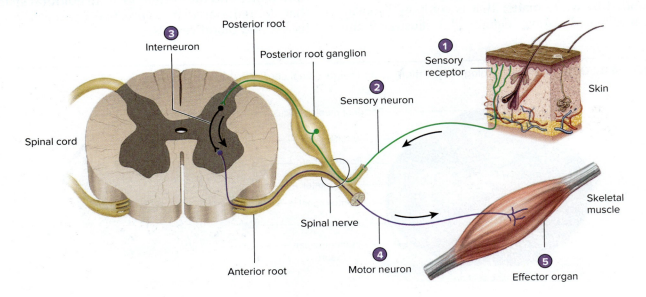

Figure 14.4 The spinal somatic reflex arc.

108   Control and Integration

Figure 14.5 The visceral reflex arc.

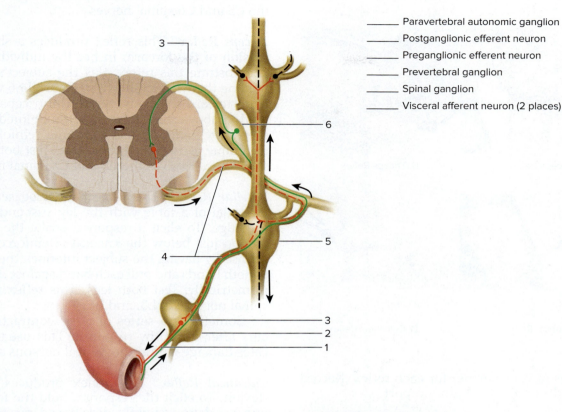

_____ Paravertebral autonomic ganglion
_____ Postganglionic efferent neuron
_____ Preganglionic efferent neuron
_____ Prevertebral ganglion
_____ Spinal ganglion
_____ Visceral afferent neuron (2 places)

matter of the spinal cord. This first efferent neuron exits the spinal cord via the anterior root of a spinal nerve and synapses with a **postganglionic efferent neuron** (label 1) at either a paravertebral autonomic ganglion or a prevertebral ganglion. This second efferent neuron carries action potentials to the visceral organ where the response occurs.

The autonomic division consists of two parts. The **sympathetic part** is involved with spinal nerves of the thoracic and lumbar regions. The **parasympathetic part** involves the cranial and sacral nerves. Most viscera are innervated by nerves from both parts, which enables a dynamic functional balance to be maintained through stimulation by one part and inhibition by the other.

Assignment
1. Label Figure 14.5.
2. Complete Sections A through D of the laboratory report.

Diagnostic Somatic Reflexes

Reflex tests are standard clinical procedures used by physicians in evaluating neural functions. Abnormal reflex responses may indicate pathology of the involved portion of the nervous system. Interpretation of the results requires considerable experience. Our purpose in performing three common tests is not to diagnose but to observe the reflexes and to understand why they are used.

In a clinical setting, the first thing clinicians look for is equal responses on the right and left sides of the body. Secondly, grossly decreased or increased responses are noted even if they are equal bilaterally.

Reflex Responses

Abnormal responses may be either **hypoflexia** (hypotonic or hypoactive), a diminished response, or **hyperflexia** (hypertonic or hyperactive), an exaggerated response. Hypoflexia may result from several causes such as malnutrition, neuronal lesions, aging, or deliberate control. Hyperflexia may be due to a loss of inhibition in motor control areas. Some reflexes occur only if pathological conditions exist.

Figure 14.6 shows the procedures to be used in performing the reflex tests. Each involves striking a tendon to stimulate a muscle's stretch receptors, which typically causes a quick contraction in response. Use the following scale in interpreting the

Figure 14.6 Methods used for testing four types of somatic reflexes.

A. Biceps Reflex   B. Triceps Reflex

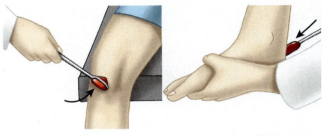

C. Patellar Reflex   D. Calcaneal Reflex

responses of your partner for each reflex. Record your results on the laboratory report.

++++ very strong response; often indicative of pathology
  +++ stronger than average response
    ++ average response
      + weaker than average response
      0 no response

Biceps Reflex This reflex causes flexion of the forearm. It is elicited by holding the subject's elbow, with the thumb placed on the tendon of the biceps brachii, and striking a sharp blow to the first joint of your thumb with the reflex hammer. Reinforce the response by asking the subject to squeeze his or her thigh with the other hand during the test. Test both arms. This reflex involves the C5 and C6 spinal nerves.

Triceps Reflex This reflex produces a slight extension of the forearm in healthy individuals. To demonstrate this reflex, flex the subject's elbow, holding the wrist as shown in Figure 14.6 with the palm toward the body. Strike the triceps brachii tendon above the elbow with the pointed end of the reflex hammer. Use the same reinforcement technique as in the biceps reflex. Test both arms. This reflex involves the C7 and C8 spinal nerves.

Patellar Reflex The subject must be seated on the edge of a table with the leg suspended over the edge. To elicit a response, strike the patellar tendon just below the kneecap. Reinforce the response by having the subject interlock the fingers of both hands and pull each hand against the other isometrically. Test both legs. This reflex involves spinal nerves L2, L3, and L4.

Sometimes a series of jerky contractions occurs after a healthy response. This usually indicates damage within the central nervous system.

Calcaneal Reflex This reflex produces plantar flexion. To elicit the response, hold the foot with one hand in a slightly dorsiflexed position and strike the calcaneal tendon with the reflex hammer. Reinforce the response as in the patellar reflex. Test both feet. This reflex involves spinal nerves S1 and S2.

Assignment

Complete the laboratory report.

Exercise 15

THE BRAIN

Objectives

After completing this exercise, you should be able to

1. Identify the structures of sheep and human brains on preserved specimens, charts, or models.
2. Describe the circulation of cerebrospinal fluid.
3. Identify the major functional areas of the human brain.
4. Describe the function of the cranial nerves.

Materials
- Colored pencils
- Human brain models
- Sheep brains, preserved
- Protective, disposable gloves
- Dissecting instruments and trays
- Long sharp knife

Before You Proceed

Consult with your instructor about using protective disposable gloves when performing portions of this exercise.

The Meninges

As introduced in Exercise 14, the brain and spinal cord are covered by three protective membranes, the **meninges.** Figure 15.1 shows the relationships of the meninges in a frontal section through the cranium and brain.

The **pia mater** (label 6) is the thin, innermost meninx adhered to the brain surface. The **arachnoid mater** is the intermediate meninx with delicate fibers extending from its inner surface through the **subarachnoid space** (label 8) to the pia mater. This space contains **cerebrospinal fluid,** which serves as a protective shock absorber for the brain. The outermost meninx is the tough, fibrous **dura mater** (label 4), which is attached to the inner surface of the cranial bones. Within the dura mater at the upper midline is the **superior sagittal sinus** that receives venous blood from the brain as it returns to the heart. The cerebrospinal fluid diffuses into the venous blood from the **arachnoid granulations** that project into the dural sinus. Note that the meninges extend into the longitudinal fissure that separates the cerebral hemispheres. Also notice that the section of cerebrum reveals that **gray matter** (label 9) is located superficial to the **white matter.**

The Ventricles and Cerebrospinal Fluid

There are four ventricles in the brain, as shown in Figure 15.2, which are filled with cerebrospinal fluid (CSF). The two **lateral ventricles** are located within the cerebral hemispheres and are separated by a thin membrane, the septum pellucidum (see Figure 15.5). The third and fourth ventricles are on the midline. The **third ventricle** (label 17 in Figure 15.6) lies between the lateral masses of the thalamus; the **fourth ventricle** (label 10) is between the brainstem and the cerebellum.

Each ventricle contains a **choroid plexus,** a mass of specialized capillaries, that secretes the cerebrospinal fluid. The choroid plexuses of the third (label 2) and fourth (label 11) ventricles are shown in Figure 15.6. Locate these plexuses in Figure 15.2, which depicts the circulation of the cerebrospinal fluid.

The CSF passes from each lateral ventricle into the third ventricle through the **interventricular foramen** (label 10 in Figure 15.2). From the third ventricle, the cerebrospinal fluid flows through the narrow **cerebral aqueduct** (label 9 in Figure 15.2) that passes through the midbrain into the fourth ventricle. The fourth ventricle is continuous with the central canal of the spinal cord.

Unit 4: Control and Integration

Figure 15.1 The meninges. (A) The meninges are protective membranes that envelop the brain and spinal cord; (B) detail of the meninges enveloping the brain.

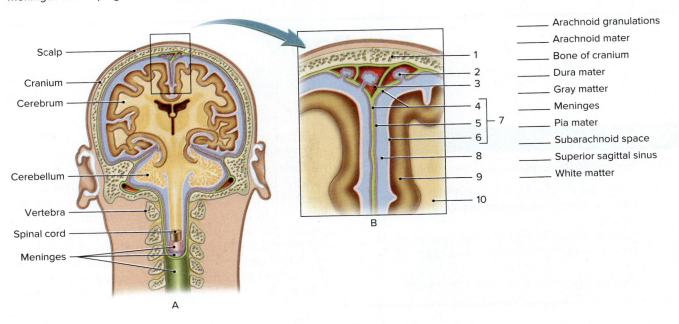

_____ Arachnoid granulations
_____ Arachnoid mater
_____ Bone of cranium
_____ Dura mater
_____ Gray matter
_____ Meninges
_____ Pia mater
_____ Subarachnoid space
_____ Superior sagittal sinus
_____ White matter

Figure 15.2 Origin and circulation of cerebrospinal fluid.

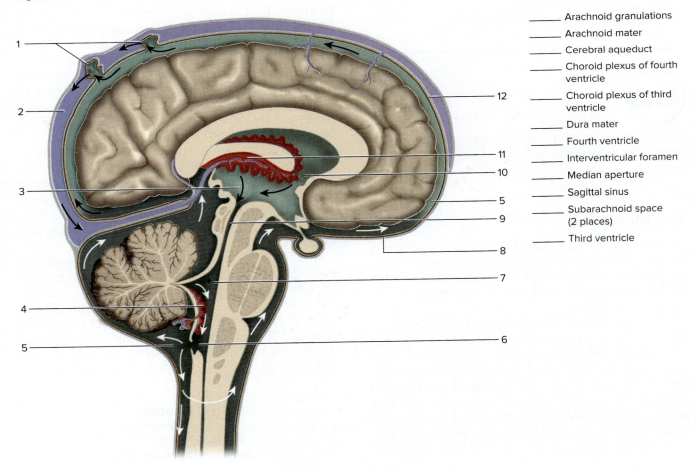

_____ Arachnoid granulations
_____ Arachnoid mater
_____ Cerebral aqueduct
_____ Choroid plexus of fourth ventricle
_____ Choroid plexus of third ventricle
_____ Dura mater
_____ Fourth ventricle
_____ Interventricular foramen
_____ Median aperture
_____ Sagittal sinus
_____ Subarachnoid space (2 places)
_____ Third ventricle

The cerebrospinal fluid then passes from the fourth ventricle through the **median aperture** and two other small openings into the **subarachnoid space** below the cerebellum. Some of the fluid moves upward within the subarachnoid space around the cerebellum and cerebrum and diffuses into the superior sagittal sinus through **arachnoid granulations** (label 1 in Figure 15.2). The remaining fluid flows downward in the subarachnoid space on the back of the spinal cord, upward in front of the spinal cord, then continues upward around the cerebrum, and finally diffuses into the superior sagittal sinus.

Assignment

1. Label Figures 15.1 and 15.2.
2. Use colored pencils to color code: the parietal lobe of the cerebrum, the cerebellum, and the diencephalon.
3. Complete Sections A through C of the laboratory report.

Major Parts of the Brain

Locate the parts of the sheep and human brains shown in Figure 15.3 as you study this section.

The Cerebrum

In both sheep and human, the cerebrum is the largest portion of the brain. The surface of the cerebrum has numerous ridges or convolutions called **gyri** (singular, *gyrus*) that increase the surface area of the gray matter. Gyri are bordered by shallow furrows called **sulci** (singular, *sulcus*). Deeper furrows are called **fissures.**

The cerebrum is divided into left and right **cerebral hemispheres** by the **longitudinal cerebral fissure** (label 20, Figure 15.4), which extends along the median line. Each hemisphere consists of four lobes (see Figure 15.3) that are named for the cranial bones covering them.

The **frontal lobe** makes up nearly the front half of the cerebrum. It is separated from the **parietal lobe** by the **central sulcus** (label 11, Figure 15.3), and its lower border is the **lateral sulcus** (label 9). The central sulcus divides the **precentral gyrus** of the frontal lobe from the **postcentral gyrus** of the parietal lobe. The **occipital lobe** is the most posterior lobe. Its anterior border lies at the **parieto-occipital sulcus,** which is distinct on the medial, but not on the lateral, cerebral surface. Therefore, its position is indicated by a dotted line in Figure 15.3. The **temporal lobe** lies below the lateral sulcus and in front of the parieto-occipital sulcus. If the temporal and frontal lobes are retracted at the lateral sulcus (lower right illustration in Figure 15.3), a small mass of cortex, the **insula,** is exposed. Its function involves memory and taste interpretation.

The cerebrum integrates and interprets incoming sensory action potentials as sensations and initiates conscious motor responses. It is also responsible for intelligence, will, and memory.

Each cerebral hemisphere controls the opposite side of the body. The right cerebral hemisphere controls the left side of the body, and the left cerebral hemisphere controls the right side of the body. This pattern of control results because nerve axons passing between the cerebral hemispheres and the spinal cord cross to the opposite side in the brainstem.

The internal structure of the brain can best be observed by examining median and frontal sections cut through it. First use the labeled illustration of the sheep brain in Figure 15.5 and the written descriptions to locate and label the structures in the human brain in Figure 15.6.

Note in the figures that the convolutions on the medial surface of the cerebral hemisphere have not been disturbed by the median section. The only part of the cerebrum that has been cut is the **corpus callosum** (label 20 in Figure 15.6). This large, curved band of white matter consists of axons connecting the two cerebral hemispheres. Locate the **superior sagittal sinus** in the dura mater along the midline and the **parieto-occipital sulcus** (label 1). The additional features labeled in Figures 15.5 and 15.6 are not part of the cerebrum and will be discussed later. First we will take a closer look at the functional regions of the cerebrum.

Assignment

1. Label Figures 15.3, 15.4, and 15.6.
2. Complete Sections A and D of the laboratory report.

Functional Areas of the Cerebrum
The functional areas of the cerebrum are known largely through studies of patients with brain lesions—local injuries of various types—plus some electrochemical experiments on living subjects. The major functional areas are shown in Figure 15.7 and discussed below. The gray matter on the surface of the cerebrum is called the **cerebral cortex.** Areas of the cerebral cortex that have a primary function are designated as parts

Figure 15.3 Lateral aspects of sheep and human brains. **APR**

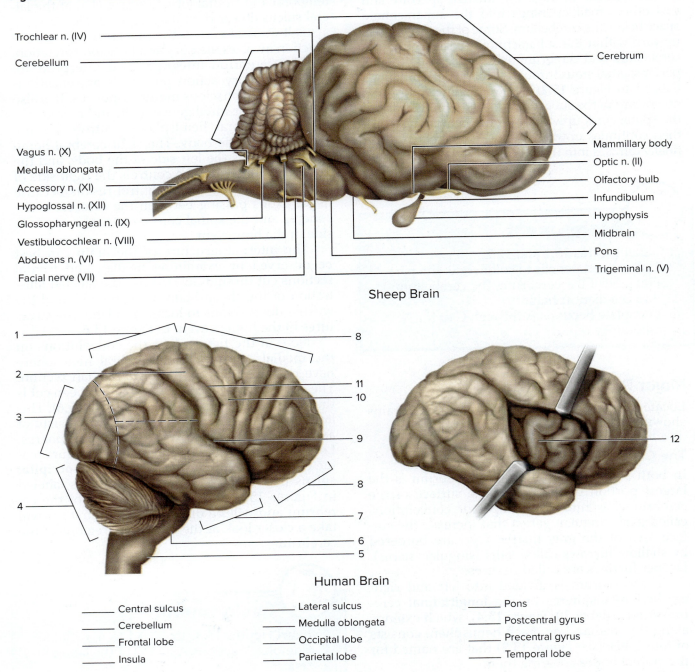

of the primary cortex, such as the primary motor cortex. Keep in mind that the brain functions as a whole unit with great interaction among its parts. The major functional areas noted below do not function in isolation but are always integrated with other brain areas to produce the appropriate sensation or action. **Association areas** are parts of the cerebral cortex that integrate inputs from other parts of the cerebrum to determine appropriate interpretations of sensory input and to determine appropriate action for motor output. Association areas occupy about 75% of the cerebrum. Refer to Figure 15.7 as you study this section.

Frontal Lobe The **primary motor cortex** (label 1) is on the precentral gyrus, which is located just in front of the central sulcus. It enables conscious control of skeletal muscles in the movement of

114 Control and Integration

Figure 15.4 Inferior aspects of the sheep and human brains. **APR**

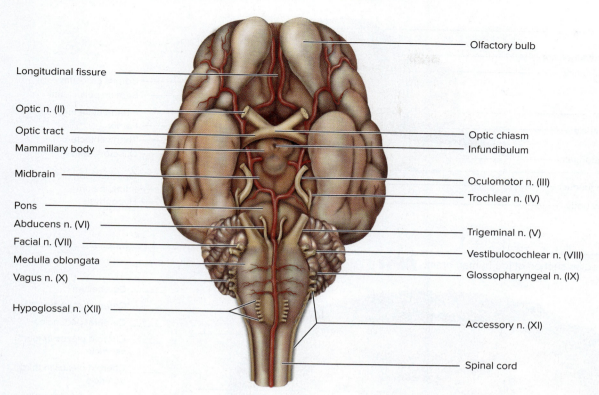

Sheep Brain

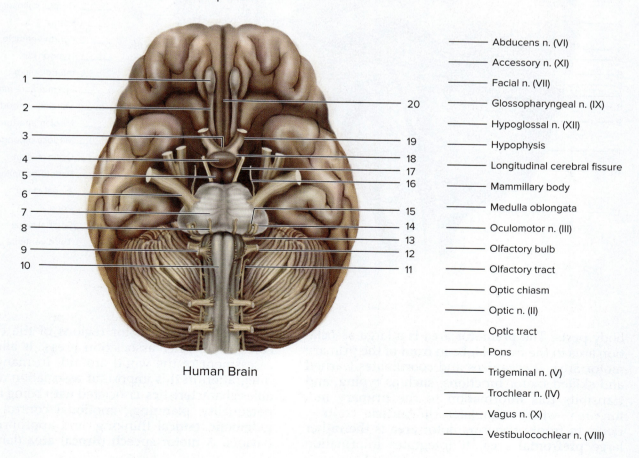

Human Brain

- Abducens n. (VI)
- Accessory n. (XI)
- Facial n. (VII)
- Glossopharyngeal n. (IX)
- Hypoglossal n. (XII)
- Hypophysis
- Longitudinal cerebral fissure
- Mammillary body
- Medulla oblongata
- Oculomotor n. (III)
- Olfactory bulb
- Olfactory tract
- Optic chiasm
- Optic n. (II)
- Optic tract
- Pons
- Trigeminal n. (V)
- Trochlear n. (IV)
- Vagus n. (X)
- Vestibulocochlear n. (VIII)

The Brain 115

Figure 15.5 Median section of the sheep brain.

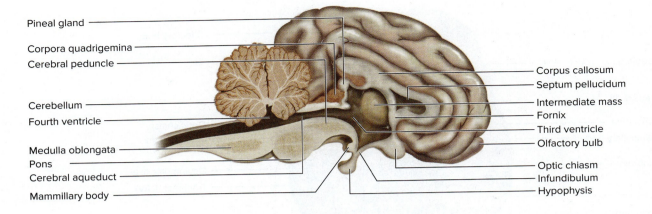

Figure 15.6 Median section of the human brain.

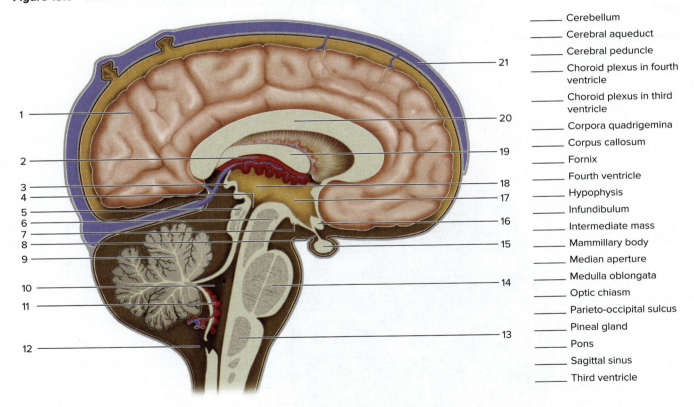

___ Cerebellum
___ Cerebral aqueduct
___ Cerebral peduncle
___ Choroid plexus in fourth ventricle
___ Choroid plexus in third ventricle
___ Corpora quadrigemina
___ Corpus callosum
___ Fornix
___ Fourth ventricle
___ Hypophysis
___ Infundibulum
___ Intermediate mass
___ Mammillary body
___ Median aperture
___ Medulla oblongata
___ Optic chiasm
___ Parieto-occipital sulcus
___ Pineal gland
___ Pons
___ Sagittal sinus
___ Third ventricle

body parts. The **premotor area** is a large association area in the frontal lobe in front of the primary motor area. It organizes and coordinates learned and skilled motor functions, such as typing, and transmits this information to the primary motor area, which stimulates the muscle contractions. In front of the premotor area is the rather large **prefrontal area.** It integrates information from sensory and motor regions of the cerebral cortex and other association areas. It allows an assessment of the world around. In many ways, integration in this important association area enables characteristics associated with being human: personality, planning, emotional control, moral judgment, critical thinking, and appropriate behaviors. A **motor speech (Broca) area** (label 4) is

116 Control and Integration

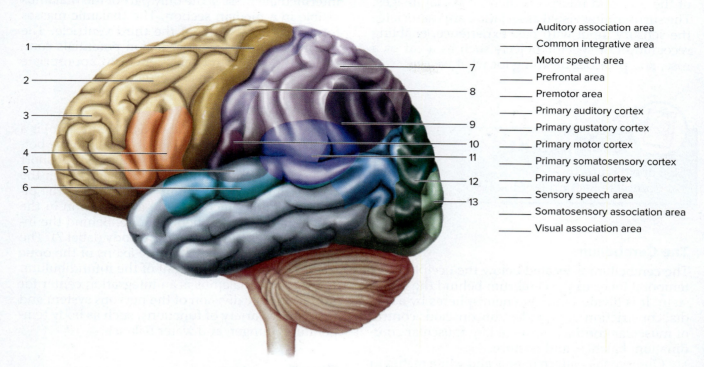

Figure 15.7 Major functional areas of the cerebrum.

_____ Auditory association area
_____ Common integrative area
_____ Motor speech area
_____ Prefrontal area
_____ Premotor area
_____ Primary auditory cortex
_____ Primary gustatory cortex
_____ Primary motor cortex
_____ Primary somatosensory cortex
_____ Primary visual cortex
_____ Sensory speech area
_____ Somatosensory association area
_____ Visual association area

located in the left frontal lobe just above the lateral sulcus and in front of the primary motor area. It receives input from the sensory speech area in the parietal lobe and generates programs for the muscles of the larynx, tongue, and lips to produce speech. These programs are transmitted to the primary motor area, which executes them as speech.

Parietal Lobe The **primary somatosensory cortex** lies on the postcentral gyrus, which is just behind the central sulcus. It receives sensory action potentials from receptors in skin and skeletal muscles and determines the location of the stimulated receptors. A **primary gustatory cortex** is located at the lower end of the primary somatosensory cortex, and it is responsible for taste sensation. Sensory input to the primary somatosensory area is processed and analyzed by the **somatosensory association area,** which is located just behind the primary somatosensory cortex. It enables comprehension of the sensory input. The **sensory speech (Wernicke) area** (label 11) is located at the junction of the parietal, temporal, and occipital lobes in one cerebral hemisphere, usually the left. It is responsible for understanding written and spoken language and transmits this information to the **common integrative area** (label 9), an area without definite borders above and behind the sensory speech area. The common integrative area involves parietal, temporal, and occipital lobes and usually occurs only in the left hemisphere. It integrates input from all sensory areas, interprets the information, and transmits the interpretation to the frontal lobe, where appropriate action is determined and taken. Lesions in the sensory speech or common integrative area may prevent understanding of written or spoken words and appropriate speech.

Temporal Lobe The **primary auditory cortex** (label 5) lies on the upper portion of the temporal lobe just inferior to the lateral sulcus. It receives action potentials from sound receptors of the cochlea in the internal ear and interprets them as to pitch, loudness, and rhythm. The surrounding **auditory association area** allows further perception of specific sounds, such as speech and music, and it enables us to remember specific music and the sound of a person's voice. It is an essential area for understanding speech. The **primary olfactory cortex** is located on the medial surface of the temporal lobe (not shown in Figure 15.7). It receives action potentials from olfactory receptors in the nasal epithelium and interprets them as specific odors.

Occipital Lobe The **primary visual cortex** is located in the back of the occipital lobe. It receives action potentials from light receptors in the retinas of the eyes and interprets them as visual images. The surrounding **visual association area** associates the sensory input with past experience, enabling recognition of specific objects such as a cat or a rose, and provides meaning for what we see.

Assignment
1. Label Figure 15.7.
2. Complete Section F of the laboratory report.

The Cerebellum

The cerebellum is located below the occipital and temporal lobes of the cerebrum behind the brainstem. It is divided into two hemispheres by a medial constriction. It provides subconscious control of muscular contractions enabling muscular coordination, balance, and posture.

Observe the pattern of gray and white matter in the cut surface of the cerebellum. The functions of the cerebellum are at the subconscious level. It receives action potentials from motor and visual centers in the brain, balance receptors in the internal ear, and muscles of the body. This information is integrated and action potentials are sent to muscles to maintain posture, balance, and muscle coordination. Unlike the control provided by the cerebrum, each hemisphere of the cerebellum coordinates muscle function on its own side of the body.

The Diencephalon

Structures between the cerebrum and the brainstem constitute the **diencephalon.** It contains the fornix, thalamus, hypothalamus, and third ventricle.

The Fornix

The band of white matter below the corpus callosum is the **fornix** (label 19 in Figure 15.6). It contains axons associated with the sense of smell and is much better developed in sheep than in humans. The small **pineal gland** (label 3) lies on the midline. It secretes a hormone, melatonin, which is involved in sleep–wake cycles. The pineal gland can also be seen in Figure 15.8.

The Thalamus

The **thalamus** consists of two ovoid masses of gray matter, one in each cerebral hemisphere, that are joined by an isthmus of nervous tissue called the **intermediate mass of the thalamus** (label 18). The intermediate mass is the only part of the thalamus visible in a median section. The thalamic masses form the lateral walls of the third ventricle. The thalamus receives sensory action potentials coming to the brain and relays them to appropriate sensory centers of the cerebrum.

The Hypothalamus

The **hypothalamus** (not labeled in Figure 15.6) is a small region that extends about 2 cm into the bottom of the brain in front of the brainstem. A short stalk, the **infundibulum** (label 8), hangs from the bottom of the hypothalamus and supports the **pituitary gland,** which fits into a depression in the sella turcica of the sphenoid bone. Behind the infundibulum is the **mammillary body** (label 7). The optic chiasm, where half of the axons of the optic tracts cross, lies just in front of the infundibulum.

The hypothalamus is an integration center for the autonomic division of the nervous system and regulates a variety of functions, such as body temperature, hunger, and water balance.

The Brainstem

The midbrain, pons, and medulla oblongata form the brainstem.

The Midbrain

The midbrain is a short section of the brainstem above the pons and is not visible in an intact human brain. It is necessary to depress the cerebellum as shown in Figure 15.8 to expose four rounded bodies known as the **corpora quadrigemina** of the midbrain. The larger pair is the **superior colliculi,** a visual reflex center, and the smaller pair is the **inferior colliculi,** an auditory reflex center. In addition, the midbrain is a pathway for nerve axons between lower and higher brain centers. The small median projection above the superior colliculi is the **pineal gland,** which is part of the diencephalon.

The short **cerebral aqueduct** (label 4, Figure 15.6) runs longitudinally through the midbrain and connects the third and fourth ventricles. In front of this duct is a **cerebral peduncle.** The cerebral peduncles contain the main motor tracts between the cerebrum and lower parts of the brainstem.

The Pons

The pons is a rounded bulge on the front of the brainstem between the midbrain and medulla oblongata. It contains axons that connect the cerebellum and medulla oblongata with higher parts of the brain.

Figure 15.8 Exposing structures of the diencephalon and midbrain. J and J Photography.

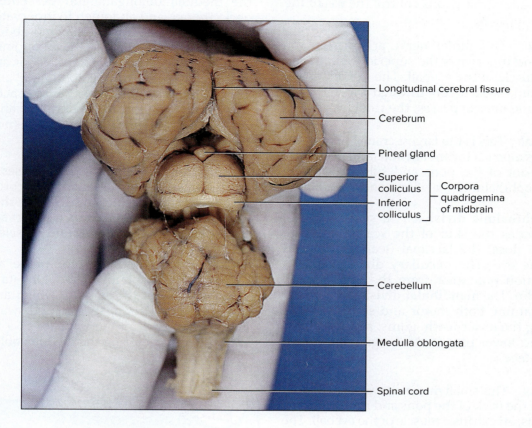

The Medulla Oblongata

The medulla oblongata is the bottom portion of the brainstem and is continuous with the spinal cord. The medulla oblongata contains centers that control the heart rate, breathing rate and depth, and diameter of blood vessels.

Assignment

1. Complete Sections E through G of the laboratory report.

The Cranial Nerves

There are 12 pairs of cranial nerves that arise from various parts of the brain and pass through foramina to innervate parts of the head and trunk. Each is designated by name and Roman numeral. Most cranial nerves are **mixed nerves** in that they carry both sensory and motor neuron processes. A few carry only sensory axons.

As you study this section, locate the cranial nerves first on the sheep brain and then on the human brain in Figure 15.4.

I Olfactory This cranial nerve contains only sensory axons for the sense of smell. It is not shown on the illustrations because it passes from the olfactory epithelium to the **olfactory bulb** (label 1) of the brain. It synapses with neurons in the olfactory bulb that carry the action potentials on to the primary olfactory cortex on the medial surface of the temporal lobe. In humans, an **olfactory tract** connects the bulb with the brain.

II Optic This sensory nerve functions in vision and contains axons of sensory neurons in the retina. Half of the axons cross over to the other side of the brain as they pass through the **optic chiasm** (label 3). They continue through the **optic tracts** (label 18) to the thalamus and on to the primary visual cortex in the occipital lobe.

III Oculomotor This motor nerve emerges from the midbrain and innervates the levator palpebrae muscle, which raises the eyelid, and four extrinsic muscles of the eyeball: the superior, medial, and

The Brain   119

inferior rectus muscles and the inferior oblique muscle. Its autonomic axons control the iris of the eye and the focusing of the lens.

IV Trochlear This motor nerve arises from the midbrain and innervates the superior oblique, an extrinsic muscle of the eyeball. The nerve derives its name from the structure through which the superior oblique muscle passes, the trochlea.

V Trigeminal This is the largest cranial nerve, and it innervates much of the mouth and face. It emerges from the front of the pons and subsequently divides into ophthalmic, maxillary, and mandibular branches.

The **ophthalmic division** is a sensory nerve that innervates the skin of the scalp, forehead, eyelid, and nose; the lacrimal (tear) gland; and parts of the eye. The **maxillary division** carries sensory action potentials from the upper teeth, gums, and lip. The **mandibular division** is a mixed nerve containing both motor and sensory axons innervating the lower teeth, gums, and lip; chewing muscles; lower part of the tongue; and lower part of the face.

VI Abducens This small motor nerve (label 8) originates from the back of the pons and innervates the lateral rectus, an extrinsic muscle of the eyeball. The lateral rectus muscle serves to abduct the eyeball.

VII Facial This mixed nerve (label 15) arises from the lower part of the pons. It innervates facial muscles, salivary glands, and the anterior two-thirds of the tongue, including taste buds.

VIII Vestibulocochlear This sensory nerve arises from the medulla oblongata and consists of two branches innervating the internal ear. The **vestibular branch** carries action potentials from the receptors contained within the internal ear that detect balance. The **cochlear branch** carries action potentials from the receptors contained within the internal ear that detect sound.

IX Glossopharyngeal This mixed nerve (label 13) emerges from the medulla oblongata just behind the vestibulocochlear nerve. It innervates much of the mouth lining, root (back) of the tongue, and pharynx. It carries motor action potentials for swallowing reflexes, salivary secretions, and respiration. It carries sensory action potentials to the brain from chemo- and baro-(pressure) receptors in carotid arteries, touch, pressure, and pain receptors of the tongue and pharynx, plus taste buds from the rear third of the tongue.

X Vagus This mixed nerve (label 12) arises from the medulla oblongata and descends into the neck, thorax, and abdomen to innervate the pharynx, larynx, and the thoracic and abdominal viscera. It is a major nerve of the autonomic division. It carries motor action potentials for swallowing, speech, and regulation of abdominal and thoracic visceral functions. It carries sensory action potentials from receptors in abdominal and thoracic viscera, including chemo- and baro-(pressure) receptors in the aortic arch.

XI Accessory This motor nerve arises from both the medulla oblongata and spinal cord. The **cranial branch** innervates muscles of the pharynx and larynx. The **spinal branch** innervates the trapezius and sternocleidomastoid muscles.

XII Hypoglossal This motor nerve arises from three origins on the medulla oblongata and innervates muscles of the tongue, which are involved in food manipulation, speech, and swallowing. In Figure 15.4, each origin appears as a tree stump with the roots entering the medulla oblongata.

Assignment

1. Obtain a preserved sheep brain and place it on a dissecting tray for study. Locate the external features as shown in Figures 15.3, 15.4, and 15.8.
2. Locate the external features studied in this exercise on the model of a human brain.
3. Complete Section H of the laboratory report.

Sheep Brain Dissection

Obtain a preserved sheep brain and review its external features shown in Figures 15.3 and 15.4. To view structures shown in Figure 15.5, you will make a median section of the brain. Place the brain dorsal surface up on your dissecting tray. Place a long, sharp knife (not a scalpel) in the longitudinal fissure and slice through the brain to form equal left and right halves. Examine the cut surface and locate all of the structures labeled in Figure 15.5.

To better understand the relationships of the thalamus and ventricles, you will make and examine a frontal section through the infundibulum of a whole sheep brain. Obtain a sheep brain and place it ventral surface up on your dissecting tray.

Use the knife to cut a frontal section by starting at the infundibulum and cutting perpendicularly through the longitudinal axis of the brain. Note the distribution of gray and white matter in the cerebrum. How do their functions differ?

When finished with your study, dispose of the brain as directed by your instructor. Clean your instruments, tray, and workstation.

Assignment

1. Locate the structures shown in Figures 15.2 and 15.6 on models of the human brain.

Exercise 16

SENSES

Objectives

After completing this exercise, you should be able to

1. Identify the parts of the eye and its accessory structures on a preserved eye, chart, or model.
2. Perform and interpret selected visual tests.
3. Identify the parts of the ear on charts and models, and describe their functions.
4. Perform and interpret simple tests of ear function.

Materials
- Beef eye, preferably fresh
- Dissecting instruments and tray
- Meterstick or tape measure
- Ishihara color blindness test plates
- Laboratory lamp
- Snellen eye chart
- Protective disposable gloves
- Tuning fork
- Cotton earplugs
- Prepared slides of cochlea, x.s.

Before You Proceed

Consult with your instructor about using protective disposable gloves when performing portions of this exercise.

In this exercise, you will study the anatomy of the human eye and ear. The eye is a complex special sense organ and includes several associated structures. The ear is involved in the special senses of hearing and equilibrium. The ear is anatomically divided into three parts: the external ear, the middle ear, and the internal ear. We will begin our discussion with the associated structures of the eye.

The Lacrimal Apparatus

The eye, lacrimal gland, and extrinsic eye muscles lie protected in the orbital cavity. The eye is protected by the **eyelids.** A thin membrane, the **conjunctiva,** lines the inner surface of the eyelids and covers the surface of the eye. It contains many blood vessels and pain receptors except where it covers the cornea.

The lacrimal apparatus is shown in Figure 16.1. The **lacrimal gland** is located in the upper lateral portion of the orbital cavity. It secretes a lubricating fluid that keeps the eye and the conjunctiva moist. The fluid reaches the eye surface via 6 to 12 tiny **lacrimal ducts** and flows downward and medially over the surface of the eye. At the medial corner of the eye, the fluid enters two tiny openings, the **lacrimal puncta,** and passes into the **superior** and **inferior lacrimal** canals. Then, the

Figure 16.1 The lacrimal apparatus of the eye.

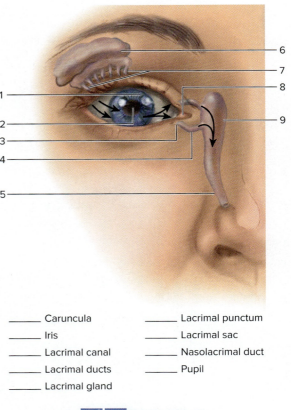

_____ Caruncula   _____ Lacrimal punctum
_____ Iris   _____ Lacrimal sac
_____ Lacrimal canal _____ Nasolacrimal duct
_____ Lacrimal ducts _____ Pupil
_____ Lacrimal gland

Unit 4: Control and Integration

122

Figure 16.2 The extrinsic muscles of the eye.

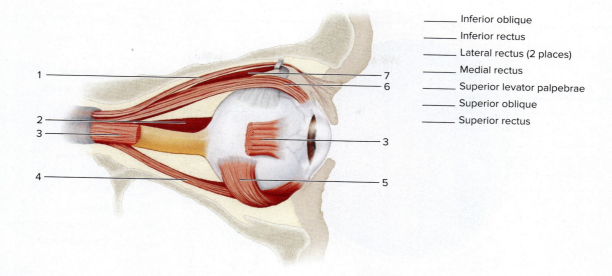

_____ Inferior oblique
_____ Inferior rectus
_____ Lateral rectus (2 places)
_____ Medial rectus
_____ Superior levator palpebrae
_____ Superior oblique
_____ Superior rectus

fluid flows into the **lacrimal sac** and passes through the **nasolacrimal duct** into the nasal cavity.

The small, red, conical body in the medial corner of the eye is the **caruncula**. It secretes a whitish substance that collects in this region.

Extrinsic Eye Muscles

The orbital cavity contains seven muscles: six extrinsic muscles that move the eye and the **superior levator palpebrae** muscle, which raises the upper eyelid. The attachments of these muscles are shown in Figure 16.2. The extrinsic eye muscles begin on bones at the back of the orbit.

The **lateral rectus** has been partially removed in the illustration. Its opposing muscle is the **medial rectus** (label 2). The **superior rectus** attaches on the upper surface of the eyeball and is opposed by the **inferior rectus** (label 4) attaching on the lower surface. The **superior oblique** passes through a cartilaginous loop, known as the trochlea, allowing it to exert an oblique pull on the eyeball. The **inferior oblique** is oriented in a similar manner, but only a portion (label 5) is shown. The interaction of these muscles enables movement of the eye in all directions.

Note in Figure 16.2 how the optic nerve (yellow) exits the back of the eye.

Internal Anatomy

A transverse section of the right eye as seen looking down on it is shown in Figure 16.3. The wall of the eyeball consists of three tunics or layers: an outer, fibrous **sclera**; a middle, pigmented **choroid** containing blood vessels; and an inner **retina** containing the light receptors.

The **optic nerve** (label 15) contains axons leading from the retina to the brain. The point where the optic nerve leaves the retina lacks light receptors and is known as the **optic disc** or **blind spot**. An artery and vein (not shown in Figure 16.3), which supply the retina and inside of the eye, enter and leave the eye at the optic disc and continue down the center of the optic nerve. Note that the optic nerve is wrapped in an extension of the **dura mater** (label 13).

Lateral to the optic disc is a small pit, the **fovea centralis,** the area of sharpest vision that is located in the center of a round yellow spot, the **macula lutea.** The macula is 0.5 mm in diameter and contains only cones.

The **cornea** covers the front of the eye as an extension of the sclera. It is transparent, and its greater curvature bends the light rays that pass through it. Light rays are precisely focused on the retina by the transparent, crystalline **lens** that is supported by **suspensory ligaments** attached to the **ciliary body.** Contraction or relaxation of smooth muscle in the ciliary body changes the shape of the lens to allow focusing of objects at different distances. Just in front of the lens is the colored portion of the eye, the circular **iris,** that regulates the amount of light entering the eye through the **pupil.**

The large cavity between the lens and retina is filled with a transparent, jellylike material, the **vitreous humor.** The cavity in front of the lens is filled with a watery liquid, the **aqueous humor** (label 7). Aqueous humor is constantly produced from capillaries behind the attachment of the iris

Figure 16.3 Internal anatomy of the eye.

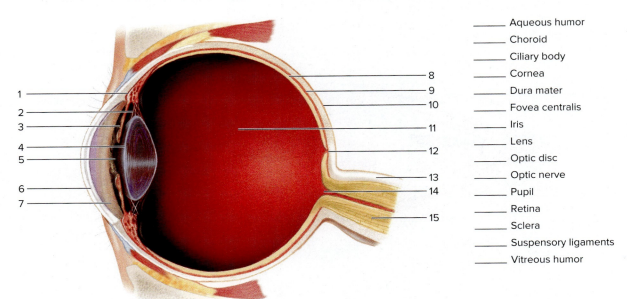

_____ Aqueous humor
_____ Choroid
_____ Ciliary body
_____ Cornea
_____ Dura mater
_____ Fovea centralis
_____ Iris
_____ Lens
_____ Optic disc
_____ Optic nerve
_____ Pupil
_____ Retina
_____ Sclera
_____ Suspensory ligaments
_____ Vitreous humor

to the ciliary body and is absorbed into small veins in front of the attachment of the iris to the sclera.

Assignment

1. Label Figures 16.1, 16.2, and 16.3.
2. Complete Sections A through D of the laboratory report.

Eye Dissection

An understanding of eye anatomy can best be obtained from an eye dissection. The procedures are described below.

1. Refer to Figure 16.4 as you examine the outer surface of the eyeball. Locate the optic nerve and any remnants of extrinsic muscles. Identify the cornea, iris, and pupil. Is the shape of the pupil the same as in humans?

Figure 16.4 The external structure of the eye. Cynthia Prentice-Craver/McGraw-Hill Education

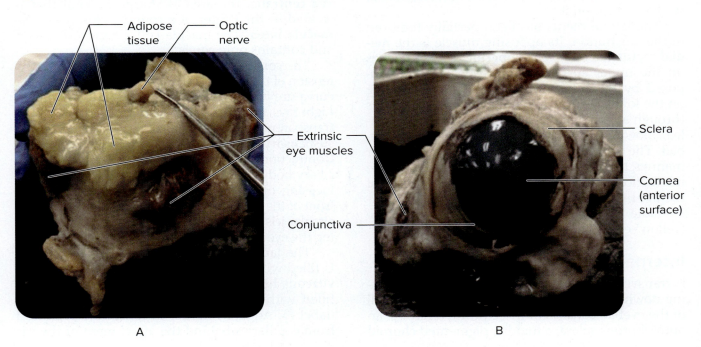

124 Control and Integration

2. Hold the eyeball as shown in Figure 16.5 and make a small incision in the sclera about 0.5 cm from the cornea. Insert scissors into the incision and cut around the eye while holding the cornea upward. Note the watery aqueous humor that exudes.
3. Refer to Figure 16.6 for the remaining steps. Gently lift off the front of the eye and place it on a tray with the inner surface upward. The lens usually remains with the vitreous humor in fresh eyes, but it may remain with the cornea in preserved eyes.
4. Look at the inner surface of the portion of the eye you removed in the previous step. Identify the thickened black ring and the ciliary body, and locate the iris. Are radial and circular muscle fibers evident?
5. Gently pour the vitreous humor and lens from the back section of the eye. Use a dissecting needle to carefully separate the lens from the vitreous humor.
6. Hold the lens by the edges and look through it at a distant object. What is unusual about the image? Place it on printed matter. Does it magnify the letters? Compare the consistency of the center and periphery of the lens using a probe or forceps.

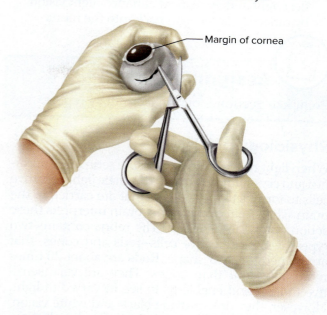

Figure 16.5 Making a frontal section of the eye.

7. Observe the thin, pale retina in the back of the eye. It separates easily from the choroid and forms wrinkles once the vitreous humor is removed. Note the optic disc.
8. Observe the black choroid and the iridescent nature of part of it. This reflective surface, called

Figure 16.6 The dissected eye. J and J Photography

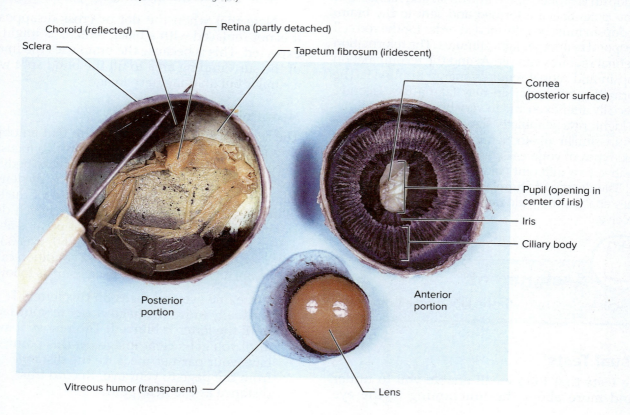

Senses 125

the *tapetum fibrosum* causes animals' eyes to reflect light and appears to enhance night vision by reflecting some light back into the retina.

Assignment
Complete Section E on the laboratory report.

Physiology of Vision

When light rays are focused on the retina by the cornea and crystalline lens, they stimulate light-sensitive cells to form action potentials that are carried to the brain. The visual center in the brain interprets these action potentials as vision. The retina contains two types of photosensitive cells—rods and cones—that are named for their shape. **Rods** are about 20 times more abundant than cones. They are very sensitive to light and enable us to see in very dim light. However, they only provide black and white vision. **Cones** require much more light to function but enable us to see in color. The fovea centralis, the area of the retina where light rays are focused for the sharpest, direct vision, contains only cones. The density of cones decreases and the density of rods increases with distance from the fovea centralis.

Rods contain **rhodopsin,** a photosensitive pigment composed of **opsin** and **retinal** in a loose chemical combination. When light strikes a rod cell, rhodopsin is broken down into opsin and retinal, and action potentials are formed and sent to the brain. Rhodopsin must be reformed in order for the rod cell to respond to another light stimulus. The reformation of retinal requires vitamin A, and the recombination of opsin and retinal requires energy from ATP. The reformation of rhodopsin may take a few seconds, especially after exposure to very bright light.

Light rays stimulate signal formation in cone cells in similar photochemical reactions. There are three types of cone cells, each with an opsin that is sensitive to a different color of light: blue, green, or red. The degree of stimulation of the three types of cones enables the wide range of color vision that we perceive.

Assignment
Complete Section F on the laboratory report.

Visual Tests

The tests that follow will enable you to understand more about the functioning of the eye.

Figure 16.7 The blind spot test.

Record the data for your eyes on your laboratory report.

The Blind Spot

Neither rods nor cones are present in the optic disc. Thus, light rays focused on this area produce no image. Determine the presence of the blind spot as follows.

1. Close your left eye and hold Figure 16.7 about 50 cm (20") from your face while staring at the cross with the right eye. Note that you can see both the cross and the dot.
2. While staring at the cross, slowly move the figure closer to the eye and watch for the dot to disappear.
3. At the point where the dot is not visible, have your partner measure and record the distance from your eye to the figure.
4. Repeat the process for the left eye but stare at the dot and watch for the cross to disappear. Record the distance for your left eye.

Note that when the dot or cross disappears, it is not replaced with a black blotch, as might be expected. This is because the brain uses the image of the surrounding area to fill the blind spot with a similar but an imaginary image.

Near Point Determination

The shortest distance from your eye that an object is in focus is known as the **near point.** To focus on close objects, the contraction of the ciliary body relaxes the suspensory ligaments allowing the lens to take a more spherical shape. The ability for the lens to do this depends upon its elasticity, which deteriorates with age. At age 20, the near point is about 3.5 in; at age 40, it is 6.8 in; at age 60, it is 33 in. After age 60, elasticity is minimal, a condition called **presbyopia.** Determine your near point as follows.

1. Close one eye and focus on this letter: T; gradually move the page closer to your face until the letter is blurred. Then move it away until you get a clear image. At this point, have your partner measure the distance from the page to your eye. Record this distance as the near point.

2. Determine and record the near point for the other eye.

Visual Acuity

The size of the cones and the distance between them in the fovea centralis determine the maximum acuity. Thus, it is possible for a "perfect" human eye to differentiate two points that are only 1 mm apart from a distance of 10 m. Usually, poor acuity is due to incorrect focusing of light rays on the retina.

The Snellen eye chart (Figure 16.8) was developed to measure visual acuity. It is printed with letters of various sizes. When you stand 20 ft from the chart and can read the letters designated to be read at 20 ft, you have 20/20 vision. If you can read only the letters designated to be read at 200 ft, you have 20/200 vision. Determine your visual acuity as follows.

1. Stand 20 ft from the Snellen eye chart posted in the laboratory while your partner stands by the chart, indicates the lines to be read, and evaluates your accuracy.

Figure 16.8 The Snellen eye chart.

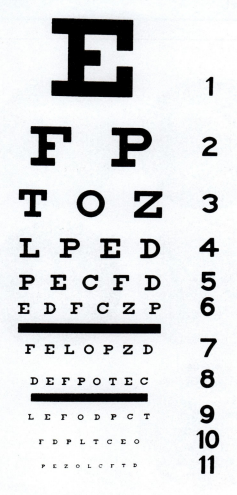

2. Test each eye independently while covering the other eye. If you use corrective lenses, test your acuity with and without them. Record your results.

Astigmatism

An uneven curvature of the lens or cornea prevents some light rays from being sharply focused on the retina. The result is **astigmatism,** a condition in which an image will be blurred in one axis and sharp in other axes. Proceed with the test as follows.

Using one eye at a time, stare at the center of Figure 16.9 and determine if all radiating lines are in focus and have the same intensity of blackness. If so, no astigmatism is present. If not, record the number(s) of the lines that are blurred or less dark.

Accommodation to Light Intensity

The diameter of the pupil may vary from 10 mm in darkness to only 1.5 mm in bright light. Sudden exposure of the retina to bright light causes an immediate and proportionate constriction of the pupil. In this reflex, action potentials from the retina are carried via the optic nerve to the brain and return to the iris to cause its contraction.

Use a subject with light-colored eyes to test this reflex and perform it in dim light. Hold an unlighted desk lamp about 6 in from the left eye. Have the subject look at the wall across the room, not at the lamp. While watching the pupil of the left eye, turn the light on for 1 second and then turn it off. Did the pupil constrict?

Now have the subject hold the edge of this manual against the forehead and extending down along the nose to exclude light from one side of the face. Repeat the test watching the right eye

Figure 16.9 Astigmatism test chart.

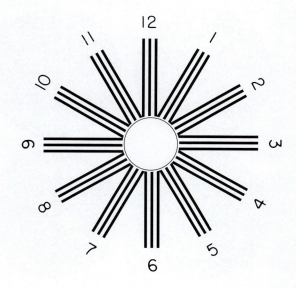

Senses

while shining light on the left eye. Record the results. What does this test tell you about the pathway of the action potentials?

Color Blindness

The perception of color by the brain results from its receiving action potentials from color-sensitive photoreceptors in the retina called **cones**. There are three types of cones, and each type responds maximally to a different color of light: red, blue, or green. The degree of stimulation of each type of cone determines the color that is perceived by the brain.

When the retina is exposed to *red* monochromatic light, the red-sensitive cones are stimulated 75%, the green-sensitive cones are stimulated 13%, and the blue-sensitive cones not at all. This ratio of stimulation, 75:13:0, is interpreted by the brain as red color.

When *blue* monochromatic light strikes the retina, red-sensitive cones are not stimulated, green-sensitive cones are activated at 14%, and blue-sensitive cones are stimulated 86%. The ratio of 0:14:86 is interpreted by the brain as blue light.

When *green* monochromatic light strikes the retina, the ratio of 50:85:15 is interpreted by the brain as green color. White light, composed of all colors of the spectrum, stimulates the cones equally.

Color blindness is a sex-linked hereditary trait affecting 8% of the male population and 0.5% of females. The most common type is red-green color blindness, in which either red-sensitive cones or green-sensitive cones are absent. People with these conditions have difficulty distinguishing reds and greens.

Test your color vision by viewing the Ishihara color-blindness test plates (*Ishihara's Tests for Colour Blindness,* Concise Edition). Start with plate 1 and progress through plate 14. Examine the samples of these test plates shown in Figure 16.10 so you will know what to expect. Your partner is to hold the test plates in good light about 30 in from your eyes. Your partner is to record your responses on *your* laboratory report. Then, reverse the roles and repeat the process. Note that you use the official color plates, not Figure 16.10 for the responses in your laboratory report.

Assignment

Complete Sections G through I of the laboratory report.

Figure 16.10 Ishihara color test plates. Alexander Kaludov/Alamy Stock Photo

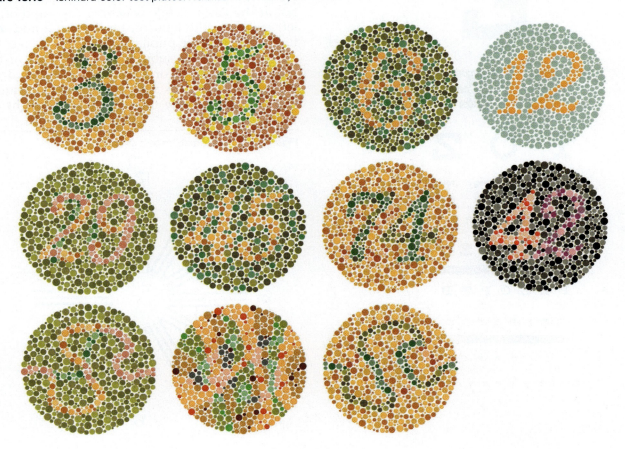

The External Ear

The external ear consists of two parts: the auricle and the external acoustic meatus. The **auricle,** sometimes called the **pinna,** is the outer shell-like structure of skin and elastic cartilage that is attached to the side of the head. The **external acoustic meatus** is the canal that extends from the auricle about an inch into the temporal bone and ends at the **tympanic membrane.** The skin lining the canal contains some small hairs and modified apocrine sweat glands that produce a waxy secretion called *cerumen.* The auricle collects sound waves and directs them through the external acoustic meatus to the tympanic membrane.

The Middle Ear

The middle ear consists of a small cavity, the **tympanic cavity,** within the temporal bone. This air-filled cavity lies between the tympanic membrane and the internal ear, and it contains three small bones, the ear **ossicles.** These tiny bones form a sensitive lever system that transmits the vibrations of the tympanic membrane to the fluids within the internal ear.

The first ossicle, the **malleus,** is a hammer-shaped or club-shaped bone that is attached to the tympanic membrane by the end of its "handle." The middle ossicle is an anvil-shaped bone called the **incus.** The third ossicle, the **stapes,** is a stirrup-shaped bone whose "footplate" fits into the **oval window** of the internal ear. Because of the linkage of the ossicles, vibrations of the tympanic membrane cause the stapes to move in and out of the oval window.

The **auditory tube** or eustachian tube leads downward from the middle ear to the nasopharynx. This slender duct allows the air pressure in the middle ear to be equalized with that of the outside atmosphere. A valve at the nasopharynx end of the tube usually keeps the tube closed. However, yawning or swallowing temporarily opens the valve to allow air pressure equalization.

The Internal Ear

The internal ear consists of a **bony labyrinth,** shown in illustration B, Figure 16.11, that is the hollowed-out portion of the temporal bone. Within the bony labyrinth is a **membranous labyrinth,** illustrated in Figure 16.12. The space between the bony and membranous labyrinths is filled with a fluid, the **perilymph.** The fluid within the membranous labyrinth is the **endolymph.** These fluids play important roles in hearing and equilibrium.

Note that the bony labyrinth consists of three semicircular canals, a vestibule, and the cochlea. The **vestibule** (label 13 in Figure 16.11) is the portion that contains the oval window. The **semicircular canals** branch off one end of the vestibule, and the **cochlea** is the coiled extension at the other end.

Two branches of the vestibulocochlear nerve (CN VIII) carry action potentials to the brain from the internal ear. The **vestibular branch** leads from the vestibule, and the **cochlear branch** emanates from the cochlea.

The Cochlea

The bony labyrinth of the cochlea contains three chambers extending along its full length. See the cross section in illustration C, Figure 16.11. The upper chamber, the **scala vestibuli** (label 17), is continuous with the vestibule. The lower chamber, the **scala tympani,** is named because it communicates with the tympanic cavity via the membranous **round window** (label 21). Both of these chambers are filled with perilymph. The middle chamber, the **cochlear duct,** is bounded by the **vestibular membrane** above and the **basilar membrane** below. The cochlear duct is filled with endolymph. On the upper surface of the basilar membrane is the **spiral organ,** which contains **hair cells** that act as receptors for sound stimuli.

The detail of the spiral organ is shown in illustration D. The hair cells sit on top of the basilar membrane with the tips of the hairs embedded in a gelatinous flap, the **tectorial membrane.** Axons leading from the hair cells become part of the cochlear branch of CN VIII.

Vestibular Apparatus

Figure 16.12 shows the membranous labyrinth and the details of the vestibular apparatus that consists of the saccule, utricle, and semicircular canals. The **saccule** and **utricle** are located within the vestibule. Sensory structures, the maculae, are located on the inner walls of these structures. Each **macula** contains hair cells that act as receptors for gravitational stimuli. The hairs are embedded in a gelatinous matrix that also contains crystals of calcium carbonate called **otoliths** (otoconia). Axons leading from the hair cells become part of the vestibular branch of CN VIII.

The hair cells that are receptors for dynamic equilibrium occur in the **ampullae,** the dilated basal portions of the semicircular canals. These hair cells are arranged in a crest, the **crista ampullaris,** within each ampulla. The hairs are embedded in a gelatinous mass called the **cupula.** Axons leading from the hair cells join the vestibular branch of CN VIII.

Assignment

1. Complete Section J of the laboratory report.
2. Examine prepared slides of cochlea, x.s. and compare your observations with Figure 16.11.

Figure 16.11 Anatomy of the ear. (C, D) Courtesy of Harold Benson

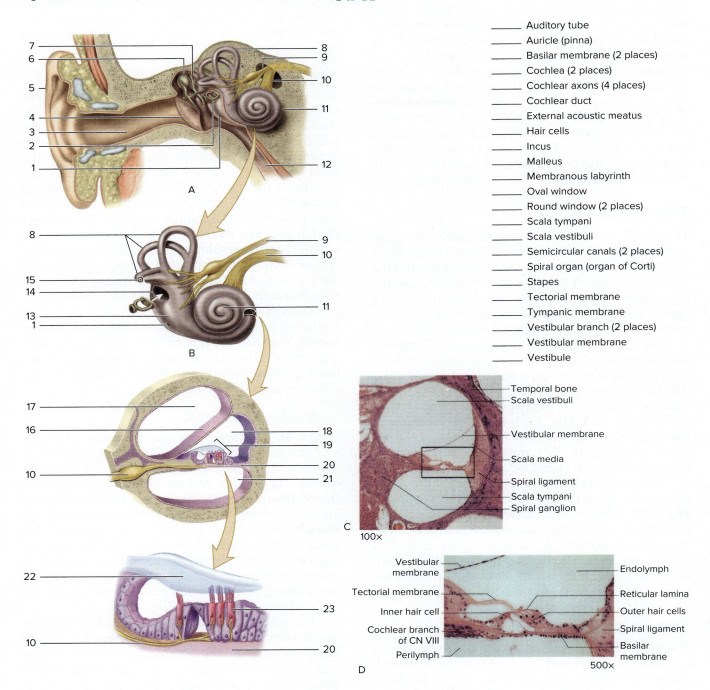

_____ Auditory tube
_____ Auricle (pinna)
_____ Basilar membrane (2 places)
_____ Cochlea (2 places)
_____ Cochlear axons (4 places)
_____ Cochlear duct
_____ External acoustic meatus
_____ Hair cells
_____ Incus
_____ Malleus
_____ Membranous labyrinth
_____ Oval window
_____ Round window (2 places)
_____ Scala tympani
_____ Scala vestibuli
_____ Semicircular canals (2 places)
_____ Spiral organ (organ of Corti)
_____ Stapes
_____ Tectorial membrane
_____ Tympanic membrane
_____ Vestibular branch (2 places)
_____ Vestibular membrane
_____ Vestibule

The Physiology of Hearing

Hearing is dependent upon (1) the conduction of sound waves from the tympanic membrane to the internal ear by the ear ossicles, (2) the stimulation of hair cells in the spiral organ, and (3) the formation of action potentials and their transmission to the auditory centers of the brain for interpretation. Figure 16.13 shows the cochlea uncoiled to reveal the relationships of the internal ear components involved in sound wave transmission.

Transmission to the Internal Ear

Sound waves striking the tympanic membrane cause it to vibrate. These vibrations are transmitted to the perilymph of the internal ear by the stapes moving in and out of the oval window.

Activation of the Basilar Membrane

Note in Figure 16.13 that the scala vestibuli and scala tympani are contiguous at the apex of the cochlea. Because the perilymph cannot be compressed, the

130 Control and Integration

Figure 16.12 The membranous labyrinth.

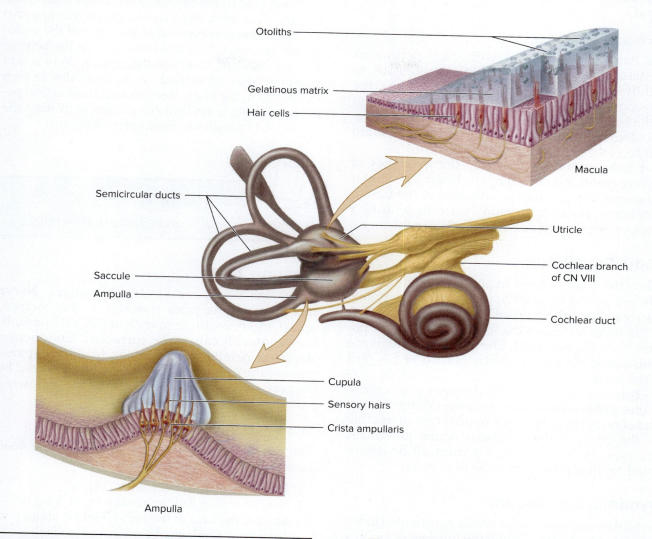

Figure 16.13 Pathway of sound wave transmission in the ear.

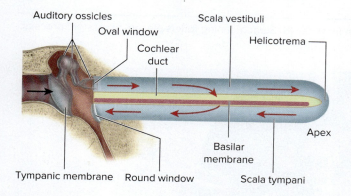

flexibility of the round window allows the fluid to move back and forth in these chambers in synchrony with the movements of the stapes. The pressure waves created by the sudden movements of the perilymph form bulges in the flexible cochlear duct, which activate the basilar membrane.

The basilar membrane contains about 20,000 transverse fibers of gradually increasing length, which are attached at one end to the bony center of the cochlea. Because they are unattached at the other end, they can vibrate like reeds of a harmonica to specific vibration frequencies. Fibers closest to the oval window are short (0.04 mm); they vibrate when stimulated by high-frequency sounds. Those at the apex of the cochlea are long (0.5 mm) and vibrate with low-frequency sounds. Because of the increasing lengths of these fibers, the human ear can detect sound frequencies from about 30 to 20,000 cycles per second.

Hair Cell Stimulation

When the fibers in a particular part of the basilar membrane vibrate in response to a specific sound frequency, the hairs of the hair cells above them are bent and stressed against the tectorial membrane. This action causes the hair cells to form action potentials that are carried via the cochlear branch of

Senses 131

CN VIII to the brain, where they are perceived as sound.

Pitch and Loudness

The hair cells associated with each portion of the basilar membrane send action potentials to slightly different portions of the brain. Thus, the detection of pitch is dependent upon the portion of the basilar membrane that is activated by a specific sound frequency and the part of the brain that receives the action potentials.

Loudness is determined by the frequency of action potentials received by the brain. The louder the sound, the greater the vibration of the basilar membrane, resulting in more hair cells sending a greater frequency of action potentials to the brain.

Static Equilibrium

When the head is tilted in any direction, the otoliths are pulled by gravity, causing the gelatinous matrix to bend the hairs of the hair cells. Bending of the hairs increases the formation of action potentials by the hair cells. The action potentials are transmitted via the vestibular branch of CN VIII to the brain.

Bending the head alone does not give a sense of disequilibrium because proprioceptors and the eyes send action potentials to the brain to neutralize the effect of the vestibular action potentials. The eyes, body, and maculae must all be disoriented for disequilibrium to be perceived.

Dynamic Equilibrium

Recall that the semicircular canals are oriented in the three planes of space. When the head moves in any direction, the semicircular canal in the direction of the movement moves in relation to the endolymph within it; that is, the canal moves but the fluid is stationary. This movement causes the cupula to move in either direction because of the force of the endolymph against it. This movement bends the hairs of the hair cells and results in the generation of action potentials that are carried via the vestibular branch of CN VIII to the brain. Reception of these action potentials by the brain initiates a reflex activating the appropriate muscles to maintain equilibrium.

Assignment

Complete Section K of the laboratory report.

Hearing Tests

There are two basic kinds of deafness. **Nerve deafness** results from damage to the cochlea, spiral organ, cochlear branch of CN VIII, or auditory center. It cannot be corrected. **Conduction deafness** results from damage to the tympanic membrane or ear ossicles. It may be corrected by surgery or hearing aids. Perform the following hearing tests in a quiet room.

The Rinne Test

The Rinne test distinguishes between nerve deafness and conduction deafness.

1. Plug one of your ears with cotton.
2. Set the tuning fork in motion by striking it against the heel of the hand. Hold it about 6 in from your unplugged ear with a tine facing the ear as shown in Figure 16.14. Persons with

Figure 16.14 The Rinne test. (A) Vibrating tuning fork is held 6 in from ear. (B) Stem of tuning fork is placed on mastoid process.
Jill Braaten/McGraw-Hill Education

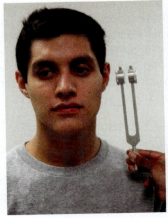

A

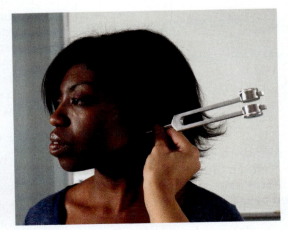

B

typical hearing and slight hearing loss will hear the sound, but those with a severe hearing loss will not hear the sound or will hear it only briefly.
3. As the sound fades to where it is no longer heard, place the stem of the tuning fork against the mastoid process behind your ear. If the sound reappears, conduction hearing loss is evident. If it does not, nerve hearing loss exists.
4. Test both ears and record the results.

The Weber Test

When the base of a vibrating tuning fork is placed against the center of the forehead of a person with healthy hearing, the sound is heard with equal loudness in both ears. See Figure 16.15. If conduction deafness exists in one ear, the sound will be louder in the defective ear than in the healthy ear. This difference occurs because the defective ear is more attuned to sound waves conducted through the bone. If nerve deafness exists in one ear, the sound will be louder in the healthy ear.

Place the base of a vibrating tuning fork against your forehead and compare the sounds in both ears. Record your results.

Static Balance Test

In addition to the vestibular apparatus, three other factors are important in maintaining equilibrium: (1) visual orientation, (2) proprioceptor sensations, and (3) cutaneous sensations. Sensory information from these four sources is subconsciously integrated to provide correct reflex responses in maintaining equilibrium.

A simple way to evaluate the effectiveness of the static balance mechanism is to have the subject stand perfectly still with the eyes closed. If there is damage to the system, the subject will waiver and tend to fall.

1. Use a meterstick to draw a number of vertical lines on the chalkboard about 5 cm apart. Cover an area about 1 m across. This chart will help detect body movements.
2. The subject is to remove his or her shoes and stand facing the examiner in front of the lined area.
3. Have the subject stand perfectly still for 30 seconds with arms at the sides and eyes open. Record the degree of swaying movement as small, moderate, or great.
4. Repeat the test with the subject's eyes closed. Record the results.
5. Repeat the test with the eyes closed while the subject stands on only one foot. What happens?

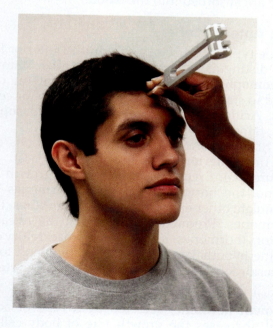

Figure 16.15 The Weber Test. Stem of tuning fork is placed on the middle of the forehead. Jill Braaten/McGraw-Hill Education

Assignment

Complete Sections L through O of the laboratory report.

Senses

Exercise 17

THE ENDOCRINE SYSTEM

Objectives

After completing this exercise, you should be able to

1. Identify the location of each endocrine gland, the hormones it produces, and their functions.
2. Indicate known effects of hyposecretion and hypersecretion.
3. Identify each endocrine gland when viewing its tissue microscopically.

Materials
Anatomical models containing endocrine glands
Prepared slides of endocrine glands

Endocrine glands are ductless glands whose secretions are absorbed into the blood and distributed throughout the body. The secretions of endocrine glands are **hormones,** chemical messengers that alter cellular metabolism. Although hormones are transported to all cells by blood, each hormone acts only on specific target cells that have receptors for that hormone. Other cells are not affected. In this way, the endocrine glands provide a chemical control of body functions.

In this exercise, you will study the location and histology of the endocrine glands and the action of their hormones. You will also learn of the disorders of the endocrine glands that result from either *hyposecretion* (deficient secretion) or *hypersecretion* (excessive secretion) of their hormones.

As you study the endocrine glands, locate each gland and label it in Figure 17.1. Then, correlate the description of its histology with the appropriate photomicrograph.

The Pituitary Gland

The **pituitary gland** (hypophysis) consists of **anterior** and **posterior lobes** that have different embryological origins and functions. It is attached to the hypothalamus by a slender stalk, the **infundibulum.** The anterior lobe has secretory cells; the posterior lobe does not. The histological section in Figure 17.2 shows the cleft separating the lobes and the two portions of the posterior lobe.

The hypothalamus plays a major role in controlling functions of the pituitary gland and, in this way, provides an important link between the nervous and endocrine systems. Special hypothalamic neurosecretory neurons secrete stimulatory (releasing) and inhibitory hormones that are carried by blood directly to the anterior lobe, where they control the release of anterior lobe hormones. The hypothalamic hormones and their actions are shown in Table 17.1. Posterior lobe hormones are produced by hypothalamic neurosecretory neurons whose axons release the hormones within the posterior lobe of the pituitary gland, where they are absorbed into the blood.

The Anterior Lobe

Three types of secretory cells in the anterior lobe are distinguished by their colors after staining. **Chromophobes** do not take up common dyes used to stain tissues, so they appear pale or colorless. **Acidophils** are stained pink, and **basophils** are stained blue to purple. See Figure 17.3.

Six hormones are produced by various cells in the **anterior lobe** (adenohypophysis). See Figure 17.4. Except for prolactin and growth hormone, the hormones are **tropic hormones** that stimulate other endocrine glands to produce their hormones. Production is regulated by releasing factors (hormones) formed by neurosecretory cells in the hypothalamus and carried by blood to the anterior lobe via the **hypophyseal portal system.**

Growth Hormone (GH)
Growth hormone, secreted by a type of acidophil, increases the growth rate of body cells by enhancing protein synthesis and utilization of carbohydrates and lipids. Hypersecretion of GH during growing years produces **gigantism** due to

Unit 4: Control and Integration

Figure 17.1 The endocrine glands. **APR**

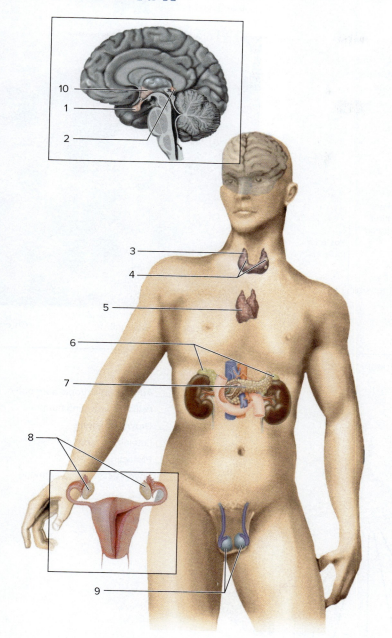

_____ Adrenal glands
_____ Hypothalamus
_____ Ovaries
_____ Pancreas
_____ Parathyroid glands
_____ Pineal gland
_____ Pituitary gland
_____ Testes
_____ Thymus
_____ Thyroid gland

TABLE 17.1
Regulation of the Anterior Lobe of the Pituitary Gland by Hypothalamus-Releasing and -Inhibiting Hormones

Hypothalamic Hormone	Action
GHRH: Growth-hormone-releasing hormone	Promotes GH secretion
Somatostatin (SST)	Inhibits GH and TSH secretion
PRH: Prolactin-releasing hormone	Promotes PRL secretion
PIH: Prolactin-inhibiting hormone	Inhibits PRL secretion
TRH: Thyrotropin-releasing hormone	Promotes TSH secretion
GnRH: Gonadotropin-releasing hormone	Promotes FSH and LH secretion
CRH: Corticotropin-releasing hormone	Promotes ACTH secretion

The Endocrine System

Figure 17.2 The pituitary gland. Courtesy of Harold Benson

Figure 17.3 Cells of the anterior lobe of the pituitary gland. Courtesy of Harold Benson

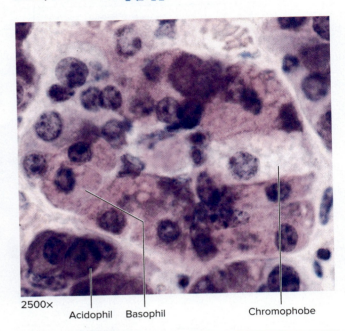

Figure 17.4 Target tissues of anterior pituitary hormones.

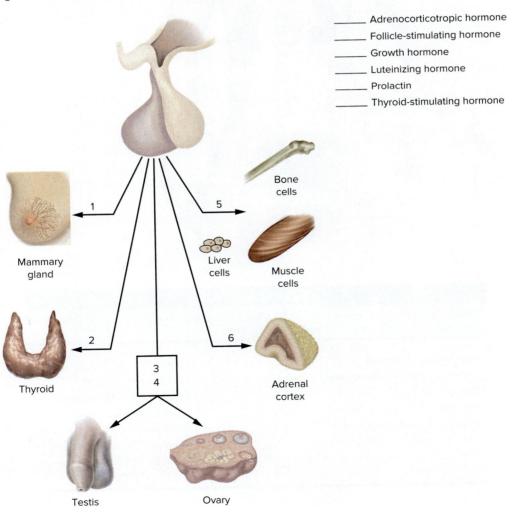

_____ Adrenocorticotropic hormone
_____ Follicle-stimulating hormone
_____ Growth hormone
_____ Luteinizing hormone
_____ Prolactin
_____ Thyroid-stimulating hormone

136 Control and Integration

excessive growth in the length of long bones. Hypersecretion in an adult causes **acromegaly,** which is characterized by an enlargement of the small bones in the hands and feet, the mandible, and frontal bones. Hyposecretion in the growing years causes **hypopituitary dwarfism.**

The secretion of GH is regulated by the antagonistic actions of two hypothalamic hormones: **growth-hormone-releasing hormone (GHRH)** stimulates GH secretion, and **somatostatin (SST)** inhibits GH secretion.

Prolactin (PRL)
PRL, secreted by another type of acidophil, stimulates the production of milk by mammary glands after childbirth. Its secretion is governed by two antagonistic hypothalamic hormones: prolactin-releasing hormone (PRH) and prolactin-inhibiting hormone (PIH). Usually, PIH predominates. Near the end of pregnancy, prolactin secretion rises sharply, due to an increase in PRH, and milk production becomes possible. Suckling by an infant inhibits PIH production, continuing prolactin secretion that, in turn, maintains milk production.

Hypersecretion of prolactin causes the absence of menstrual cycles in females and impotence in males.

Thyroid-Stimulating Hormone (TSH)
TSH, secreted by basophils, stimulates the production of thyroid hormone by the thyroid gland. TSH secretion is triggered by thyrotropin-releasing hormone (TRH) formed by the hypothalamus, which is inhibited by an increased level of thyroid hormone. An excessive production of TSH causes hyperthyroidism, and a deficient production results in hypothyroidism.

Adrenocorticotropic Hormone (ACTH)
ACTH, secreted by chromophobes, stimulates glucocorticoid production by the adrenal cortex. ACTH secretion is promoted by the corticotropin-releasing hormone (CRH) formed by the hypothalamus. ACTH regulation results from glucocorticoids exerting a negative-feedback control on CRH production.

Follicle-Stimulating Hormone (FSH)
FSH is secreted by basophilic cells that tend to be peripherally located. In females, FSH promotes development and maturation of ovarian follicles and the eggs they contain. In males, it promotes sperm formation by the testes. FSH production is promoted by gonadotropin-releasing hormone (GnRH) from the hypothalamus.

Luteinizing Hormone (LH)
LH is secreted by large basophilic cells scattered throughout the anterior lobe. In females, a rapid rise in LH, stimulated by GnRH, causes ovulation and development of a corpus luteum from the empty follicle. High levels of estrogen and progesterone exert a negative-feedback control on GnRH.

In males, LH is usually called **interstitial-cell-stimulating hormone (ICSH)** because it stimulates interstitial cells of the testes to secrete testosterone. Testosterone exerts a negative-feedback effect on GnRH production.

The Posterior Lobe

The two hormones from the **posterior lobe** are formed in the hypothalamus by neurosecretory neurons and released within the posterior lobe, which is also called the **neurohypophysis.** Their production is controlled by the hypothalamus.

Antidiuretic Hormone (ADH)
ADH promotes the reabsorption of water by the kidneys by increasing the permeability of the collecting tubules to water. Thus, ADH controls the osmotic balance of body fluids. Without ADH, excessive water loss occurs by the production of dilute urine. At times of severe blood loss, ADH also constricts arterioles to maintain adequate blood pressure. For this reason, ADH is sometimes called **vasopressin.**

Oxytocin
During childbirth, pressure on the cervix initiates a neural reflex that stimulates the secretion of oxytocin. In turn, **oxytocin** increases the strength of uterine contractions to facilitate childbirth.

The secretion of most hormones is controlled by a negative-feedback control mechanism, but oxytocin secretion is controlled by a positive-feedback control mechanism. Pressure on the cervix increases oxytocin secretion, which increases uterine contractions, increasing pressure on the cervix, which stimulates the secretion of more oxytocin. This cycle continually strengthens uterine contractions until childbirth is completed.

After childbirth, stimulation of a nipple by a nursing infant initiates a neural reflex that also releases oxytocin. In this case, oxytocin causes contraction of smooth muscles associated with the milk glands, which forces milk into the milk ducts, leading to release of milk from the breast.

The Thyroid Gland

The thyroid gland is located just below the larynx. It is formed of two lobes, one on each side of the trachea, that are joined by an isthmus of tissue across the front of the trachea. Histologically, the gland contains numerous secretory **follicles**

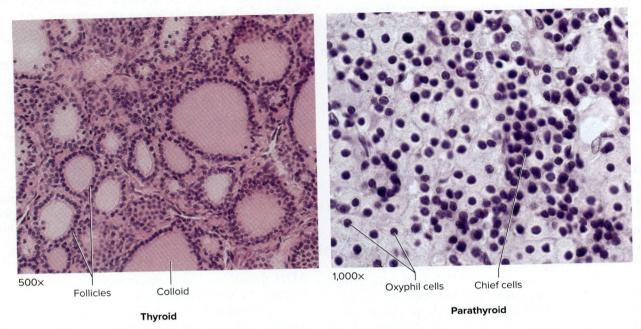

Figure 17.5 Thyroid and parathyroid glands. Courtesy of Harold Benson

that secrete the thyroid hormone. The follicles are filled with **thyroglobulin,** a colloidal material that holds excess thyroid hormone until it is released. The parafollicular cells, interstitial cells between the follicles, produce a different hormone, calcitonin. See Figure 17.5.

Thyroid Hormone

The **thyroid hormone (TH)** consists of several related compounds containing iodine in their molecules. Thyroxine (T_4) and triiodothyronine (T_3) are the most important, and they contain four iodine atoms and three iodine atoms, respectively, in each molecule. Iodine is essential for the physiological activity of thyroid hormone. Thyroid hormone increases the rate of metabolism of most body tissues. For example, it stimulates aerobic respiration, heart rate, protein synthesis, mental activity, and growth.

Like most hormones, the secretion of TH is regulated by a negative-feedback system. TRH from the hypothalamus stimulates the secretion of thyroid-stimulating hormone (TSH) by the anterior lobe of the pituitary gland. TSH promotes the secretion of TH, which, in turn, decreases the production of TRH by the hypothalamus and TSH by the anterior lobe of the pituitary gland.

Simple goiter is enlargement of the thyroid gland caused by a deficiency of iodine in the diet. Without sufficient iodine, inadequate amounts of T_4 and T_3 are produced, and the thyroid gland enlarges in an attempt to produce more TH. Severe hyposecretion (hypothyroidism) in children may result in **cretinism,** a condition characterized by physical and mental retardation. In adults, it may cause **myxedema,** which is characterized by obesity, depression, slow heart rate, and sluggishness.

Severe hypersecretion (hyperthyroidism) results in weight loss; nervousness; sleeplessness; rapid heart rate; and, sometimes, exophthalmos, the protrusion of the eyeballs.

Calcitonin

Calcitonin reduces the level of calcium in the blood by stimulating bone deposition by osteoblasts and inhibiting calcium removal by osteoclasts. A high level of blood calcium stimulates calcitonin production; a low level inhibits production. The action of calcitonin is antagonistic to the function of parathyroid hormone, which is discussed below.

The Parathyroid Glands

The four parathyroid glands are small glands embedded on the back of the thyroid gland. Figure 17.5 illustrates the histology of a parathyroid gland. The small, numerous **chief cells** produce the parathyroid hormone. The function of the larger **oxyphil cells** is unknown.

Parathyroid Hormone

The **parathyroid hormone (PTH)** raises the calcium concentration and lowers the phosphorus concentration of the blood. This is done by promoting reabsorption of bone tissue by osteoclasts and stimulating calcium reabsorption and phosphorus excretion by the kidneys. The opposing actions of PTH and calcitonin maintain a healthy concentration of calcium in the blood. A deficiency of PTH results in **hypocalcemia,** which causes hyperexcitability of the nervous system. When severe, it leads to muscle tremors, spasms, tetany, and even death. Excess PTH results in **hypercalcemia,** which depresses nervous system function leading to muscle weakness; sluggishness; emotional disturbances; and, sometimes, cardiac arrest.

Production of PTH is regulated by negative-feedback control. A high level of blood calcium inhibits production of PTH, while a low level of blood calcium promotes PTH production.

The Pancreas

The pancreas is a pennant-shaped gland that has both exocrine and endocrine functions. It is located between the stomach and duodenum. Each of the three pancreatic hormones (insulin, glucagon, and somatostatin) are produced by a different type of cell found in the **pancreatic islets.** Cells surrounding the pancreatic islets are acinar cells, which produce pancreatic juice. See Figure 17.6.

Insulin

Insulin (produced by beta cells) decreases blood glucose by (1) facilitating the passage of glucose through plasma membranes into body cells where it can be metabolized and (2) stimulating the conversion of glucose into glycogen in the liver. A deficiency of insulin results in **diabetes mellitus,** a condition in which **hyperglycemia,** an abnormally high blood glucose level, is present.

The concentration of blood glucose regulates insulin production. A high level stimulates production; a low level inhibits production.

Glucagon

Glucagon (produced by alpha cells) increases the blood glucose level by stimulating the liver to (1) convert glycogen into glucose and (2) synthesize glucose from noncarbohydrate sources. Thus, insulin and glucagon have opposite effects on the concentration of blood glucose. A deficiency of this hormone results in **hypoglycemia,** a low level of blood glucose.

Regulation is by negative-feedback control. A low level of blood glucose promotes glucagon production; a high level decreases glucagon production.

Somatostatin

Somatostatin (produced by delta cells) is produced in the hypothalamus as well as the pancreatic islets. It inhibits secretion of growth hormone by the anterior lobe of the pituitary gland and insulin and glucagon by islet cells.

The Adrenal Glands

An adrenal gland sits atop each kidney and consists of two parts: the inner **medulla** and the outer **cortex.** See Figure 17.7. These two parts arise from different embryological origins and have distinctive functions.

The Adrenal Cortex

The adrenal cortex secretes many different hormones that are collectively called *corticosteroids* (corticoids). These hormones may be subdivided into three different groups. *Mineralocorticoids* control the concentrations of water and electrolytes in the blood, specifically Na^+ and K^+. *Glucocorticoids* control reactions to stress and the metabolism of glucose and proteins. Sex hormones occur in such small concentrations that their effect is typically minimal. Of the 30 or more corticoids, the two most important are **aldosterone** and **cortisol.**

Figure 17.7 shows the three zones of the adrenal cortex. The outer **zona glomerulosa** lies just under the adrenal capsule. The **zona fasciculata** is the middle zone, and the **zona reticularis** is the innermost zone that is adjacent to the adrenal medulla.

Figure 17.6 Pancreas histology. Courtesy of Harold Benson

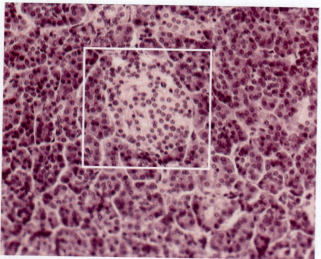

500×

Pancreatic islets (boxed area)

Figure 17.7 The adrenal gland. (A) Anatomy; (B) histology.

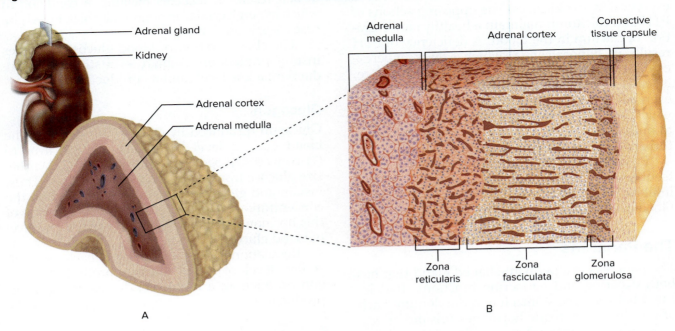

Aldosterone
Aldosterone is a mineralocorticoid secreted by cells in the zona glomerulosa. It increases the rate of Na$^+$ reabsorption and K$^+$ excretion by the kidneys, which, in turn, increases water reabsorption and decreases urine output. In this manner, the concentrations of Na$^+$ and K$^+$ in the blood are maintained within healthy ranges. A deficiency of aldosterone results in an increase of K$^+$ in body fluids, a decrease of Na$^+$, and a reduction in water concentration. When severe, these conditions cause reduced cardiac output, shock, and, possibly, death.

The control of aldosterone secretion is complex. The primary stimulus is an increase in K$^+$ concentration and a decrease in Na$^+$ concentration in body fluids.

Cortisol
Cortisol is a glucocorticoid produced primarily by the zona fasciculata. It serves to maintain metabolism and to enhance resistance to stress. Cortisol increases the quantity of amino acids, fatty acids, and glucose available to body cells by causing (1) a decrease in protein synthesis, (2) the release of fatty acids from adipose tissue, and (3) the formation of glucose from noncarbohydrates. At a therapeutic level, it also tends to block processes that produce inflammation.

A chronic excess of glucocorticoids may result in **Cushing syndrome,** a condition characterized by protein depletion, muscle and bone weakness, slow healing, hyperglycemia, hydration of tissues, scraggly hair, and possibly masculinization in females.

A chronic deficiency of glucocorticoids may result in **Addison disease,** a condition characterized by hypoglycemia, dehydration, and electrolyte imbalance, which may be fatal if untreated.

The production of cortisol and other glucocorticoids is stimulated by the secretion of ACTH from the anterior lobe of the pituitary gland, which is increased in times of stress.

Sex Steroids
Small amounts of **androgens** and **estrogens** are produced by the zona reticularis. The androgens are relatively unimportant in adult males because the testes produce so much more testosterone, but in females they are responsible for the development of pubic and axillary hair at puberty and maintain the libido (sex drive) in adult life. Adrenal estrogens are relatively unimportant in women of reproductive age, but after menopause they are the only source of estrogens.

Hypersecretion of androgens often accompanies Cushing syndrome producing **androgenital syndrome (AGS).** In children, it causes enlargement of the penis or clitoris and premature onset of puberty. In adult females, it produces masculinizing effects, such as increased body hair and deepening of the voice.

The Adrenal Medulla

Two related hormones are produced by the adrenal medulla: **epinephrine** and **norepinephrine.** The effects of these hormones resemble those of the sympathetic part of the autonomic division: increased heart and respiration rates, elevated blood pressure and blood glucose concentration, and decreased digestive action. The action of these hormones and the sympathetic part of the autonomic division prepare the body to meet energy-expending emergencies. The production of both hormones is stimulated by sympathetic axons.

The Thymus

The thymus lies in the mediastinum above the heart and consists of two long lobes. It is relatively large in young children but after puberty diminishes in size throughout life. Histologically, the gland consists of an outer **cortex** and an inner **medulla** as shown in Figure 17.8. The cortex is composed of lymphoid like tissue that contains massive numbers of lymphocytes. The medulla contains fewer lymphocytes and *thymic corpuscles* whose function is unclear. The thymus secretes a hormone, **thymosin,** which promotes the development and maturation of T lymphocytes (*T* for thymus) within the thymus. T lymphocytes develop within the cortex and migrate into the medulla where maturation is completed. Then, mature T cells enter the blood or lymph and exit the thymus.

The Pineal Gland

The pineal gland is located in the brain between the corpora quadrigemina and the corpus callosum. At puberty, the gland begins to form small calcium deposits called **pineal sand** whose significance is not known. See Figure 17.8.

The pineal gland's major secretion is the hormone **melatonin,** whose production varies with the day–night diurnal cycle and is regulated by light-generated action potentials reaching the pineal via a retina–hypothalamic pathway. Peak level occurs at night and lowest level occurs during midday. Further, seasonal variation in day length alters melatonin secretion.

In animals that reproduce seasonally, reproduction is associated with relative changes in day length, and changes in melatonin production regulate these events. The reproductive function of melatonin in humans is unclear, but it may inhibit precocious sexual development. Variations in melatonin secretion may be involved with circadian (roughly 24-hour) rhythms, such as sleep and body temperature. Variations also may play a role in seasonal mood changes experienced by some people.

Figure 17.8 The thymus and pineal glands. Courtesy of Harold Benson

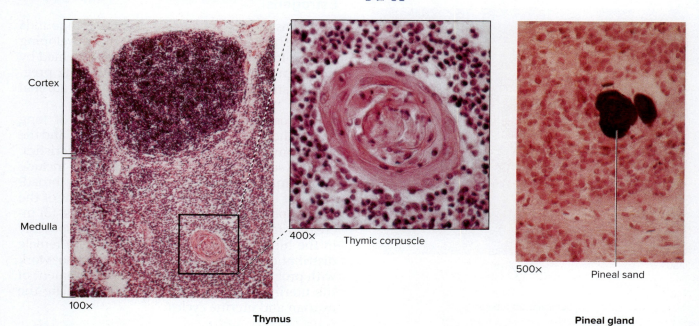

The Testes

The male gonads, the **testes,** contain numerous coiled seminiferous tubules that produce spermatozoa. Cells located in spaces between these tubules, the **interstitial cells,** secrete the male hormone, **testosterone.** Interstitial cells are shown between sectioned seminiferous tubules in Figure 17.9.

Human chorionic gonadotropin (hCG), secreted by the placenta, stimulates testosterone production during embryonic and fetal development. Testosterone secretion before birth is responsible for the development of the male sex organs and descent of the testes into the scrotum. From birth to puberty, little testosterone is produced.

At puberty, the secretion of interstitial-cell-stimulating hormone (ICSH) by the anterior lobe of the pituitary gland triggers increased testosterone secretion, which, in turn, brings about sexual maturation. Testosterone exerts a negative-feedback control on ICSH production.

In addition to stimulating the maturation of the male sex organs, increased testosterone level promotes a spurt in overall body growth, especially in height, and the development of secondary sexual characteristics. Male secondary sexual characteristics include (1) male distribution of body hair, (2) larynx enlargement that deepens the voice, (3) an increase in bone thickness and muscle development, and (4) development of broad shoulders.

Testosterone maintains the male sex organs, sex drive, and secondary sexual characteristics in adult life.

The Ovaries

The **germinal epithelium** of the ovary is formed early in embryological development. It produces numerous **primordial ovarian follicles** that reside in the ovarian cortex. Some of these primordial ovarian follicles migrate inward to become **primary ovarian follicles,** which consist of a primary oocyte enclosed in a sphere of follicular cells. Starting at puberty, follicle-stimulating hormone (FSH) from the anterior lobe of the pituitary gland stimulates some of the primary ovarian follicles to develop further. Usually, only one will become a **mature ovarian follicle** (Graafian follicle) containing a secondary oocyte that is released at ovulation in each ovarian cycle. See Figure 17.10. The empty follicle then becomes a **corpus luteum** under the stimulation of luteinizing hormone (LH) from the anterior lobe of the pituitary gland.

The ovaries produce the female sex hormones, estrogens and progesterone. FSH stimulates the production of **estrogens** by developing ovarian follicles. After ovulation, LH promotes the secretion of **progesterone** by the corpus luteum.

Estrogens

Estrogens consist of several related compounds and are produced primarily by the developing ovarian follicles. But estrogens are also formed by the corpus luteum; by the placenta during pregnancy; and, in minute quantities, by the adrenal cortex.

Estrogens promote the lengthening of long bones, the maturation of female sex organs, and the development of secondary sexual characteristics. Female secondary sexual characteristics include (1) feminization of the skeleton, such as broadening of the pelvic girdle and enlargement of the pelvic opening; (2) development of breasts; (3) increased fat deposition under the skin, especially in the hips, buttocks, and thighs; and (4) female distribution of body hair. Estrogens also work with progesterone to promote the development of the uterine lining, and they help to regulate the ovarian and uterine cycles.

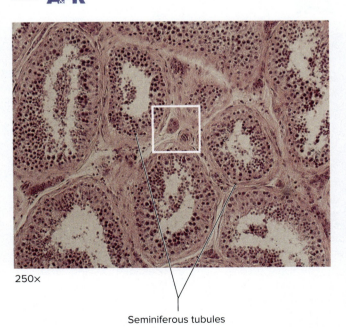

Figure 17.9 Histology of the testes. Courtesy of Harold Benson **APR**

250×

Seminiferous tubules

Figure 17.10 Ovarian tissue. Courtesy of Harold Benson

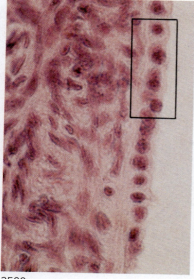

2500×
Germinal epithelium
(boxed cells)

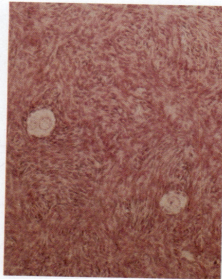

250×
Primordial ovarian follicles

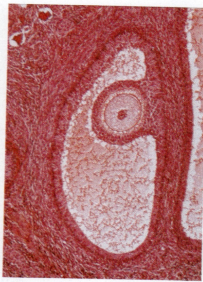

250×
Mature ovarian follicle

Progesterone

Progesterone is primarily secreted by a corpus luteum that is formed from an empty follicle after ovulation. Progesterone continues the development of the uterine lining that was begun under the stimulation of estrogens, preparing the uterus to receive an early embryo.

Without pregnancy, degeneration of the corpus luteum causes a decrease in progesterone production, which results in a breakdown of the endometrium, which results in menstruation. If pregnancy occurs, hCG stimulates the corpus luteum to form increased amounts of progesterone and some estrogen to maintain the uterine lining until this role is taken over by placental estrogens and progesterone.

Progesterone works with estrogen in promoting the development of secretory cells in the mammary glands of the breasts in preparation for milk production after childbirth.

Figure 17.11 summarizes the interactions of several endocrine glands, beginning with their overall regulation by the pituitary gland.

Assignment

1. Label Figures 17.1, 17.4, and 17.11.
2. Complete Sections A through E of the laboratory report.
3. Examine prepared slides of endocrine glands. Compare your observations with the photomicrographs. Make labeled drawings in Section F of the laboratory report that will help you identify the glands from microscope slides.
4. Complete the laboratory report.

The Endocrine System   143

Figure 17.11 The pituitary regulatory mechanism.

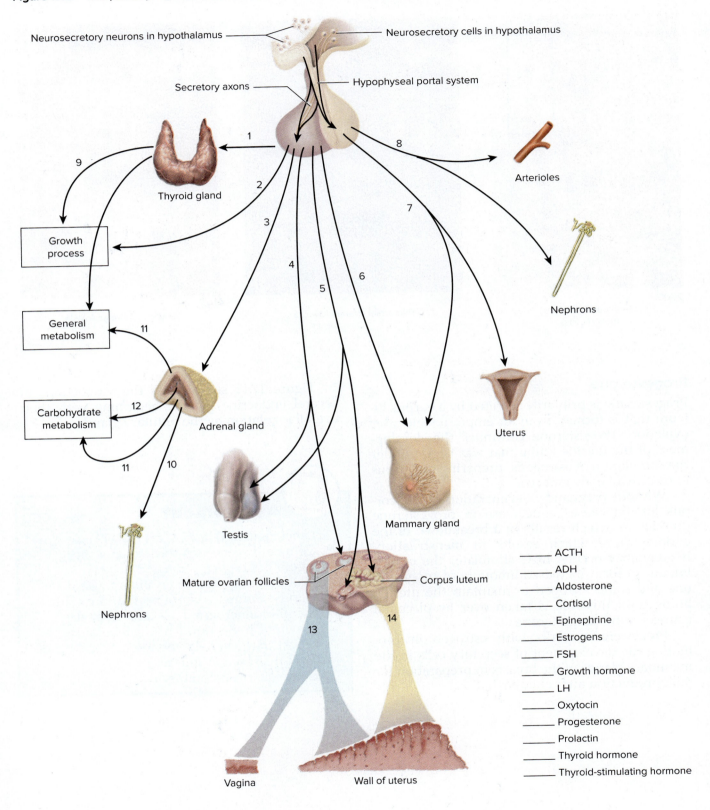

_____ ACTH
_____ ADH
_____ Aldosterone
_____ Cortisol
_____ Epinephrine
_____ Estrogens
_____ FSH
_____ Growth hormone
_____ LH
_____ Oxytocin
_____ Progesterone
_____ Prolactin
_____ Thyroid hormone
_____ Thyroid-stimulating hormone

144 Control and Integration

Exercise 18

BLOOD TESTS

A variety of blood tests are routinely performed as part of a physical examination because they provide a significant amount of information about the health of a patient. This exercise contains three of the simpler blood tests that can easily be performed in a teaching laboratory. Although these tests have diagnostic value, our purpose is not to diagnose but to understand the purpose and nature of the tests. Your instructor will determine whether you are to use your own blood, simulated blood, or sterilized blood for these tests. If you perform these tests using your own blood, follow the precautions below and your instructor's directions precisely. Normal range values vary with age, sex, and the laboratory completing the tests. In this lab general averages are used.

Precautions

In performing these tests, care must be taken to avoid (1) self-infection from environmental contaminants; and (2) transmission of bloodborne infections, such as hepatitis and AIDS, from one person to another. Your instructor will determine whether you are to use your own blood or a blood substitute for each test. **These tests can be performed safely using your own blood if the following safeguards are rigidly applied.**

1. Wash your tabletop at the beginning of the laboratory period with a suitable disinfectant.
2. Wash your hands with soap and water before and after doing the blood tests.
3. **Avoid contact with blood of other students.** *Wear protective disposable gloves whenever there is a chance of such contact—for example, when piercing your partner's finger with a lancet.*
4. Before piercing the finger for a drop of blood, disinfect the skin with 70% alcohol.
5. Use disposable lancets only once and deposit them immediately into an autoclavable biohazard sharps container or a receptacle containing disinfectant as specified by your instructor. **Never** *place a used lancet on a tabletop or toss it into a wastebasket.*
6. Disposable microscope slides, glassware, toothpicks, and anything that could puncture a biohazard bag should be disposed of in the biohazard sharps container.
7. Disposable paper towels, alcohol wipes, and similar items that may have been in contact with blood are to be placed *immediately* in the biohazard bag for autoclaving and disposal.
8. At the end of the laboratory period, wash the tabletop and equipment with a suitable disinfectant. Wash your hands with soap and water and rinse them with a disinfectant.

Unit 5: Internal Transport and Gas Exchange

Objectives

After completing this exercise, you should be able to

1. Identify the formed elements of blood when viewed microscopically or on charts.
2. Perform and interpret these tests: differential white cell count, hematocrit, and blood typing.

Materials

For each test:
- Alcohol pads
- Amphyl solution, 0.25%
- Biohazard bag
- Biohazard sharps container
- Household bleach solution, 10%
- Kimwipes
- Paper towels
- Sterile disposable lancets
- Protective disposable gloves

For white cell differential count:
- Bibulous paper
- Clean microscope slides
- Distilled water in dropping bottle
- Wax pencil
- Wright's blood stain in dropping bottle
- Prepared slides of a human blood smear

For hematocrit:
- Heparinized capillary tubes
- Microhematocrit centrifuge
- Microhematocrit tube reader
- Seal-Ease (Clay Adams)

For blood typing:
- Clean microscope slide
- Slide warming box with typing slide
- Toothpicks
- Typing sera: anti-A, anti-B, anti-D

Before You Proceed

Consult with your instructor about using protective disposable gloves when performing portions of this exercise.

Blood is the medium that transports substances to and from body cells. It consists of a fluid **plasma,** which forms 55% of the blood volume, and the **formed elements,** which constitute the remaining 45%. The formed elements consist of **erythrocytes** (red blood cells), **leukocytes** (white blood cells), and **platelets** (thrombocytes). Their relative abundance and functions are shown in Table 18.1. They are illustrated in Figure 18.1 and shown in photomicrographs in Figure 18.2 as they appear when stained with Wright's blood stain. Learn to recognize the characteristics of each type of formed element.

Caution: *Follow infection control procedures prescribed by your instructor when performing the blood tests.* Record your results of the tests on the laboratory report.

TABLE 18.1
Formed Elements of the Blood

Formed Element	Concentration	Function
Erythrocytes	4,000,000–6,000,000 per µl	Transport oxygen and carbon dioxide
Leukocytes	5,000–10,000 per µl	Destroy pathogens; neutralize toxins
Granulocytes		
Neutrophils	50%–70% of leukocytes	Phagocytosis
Eosinophils	1%–3% of leukocytes	Neutralize products of allergic reactions; destroy parasitic worms
Basophils	0.5%–1% of leukocytes	Release heparin and histamine; intensify inflammation and allergic reactions
Agranulocytes		
Lymphocytes	20%–30% of leukocytes	B lymphocytes form antibodies
		T lymphocytes form chemicals that destroy antigens and/or stimulate phagocytosis
Monocytes	2%–6% of leukocytes	Phagocytosis
Platelets	150,000–400,000 per µl	Initiate clotting process

Figure 18.1 Formed elements of blood. **APR**

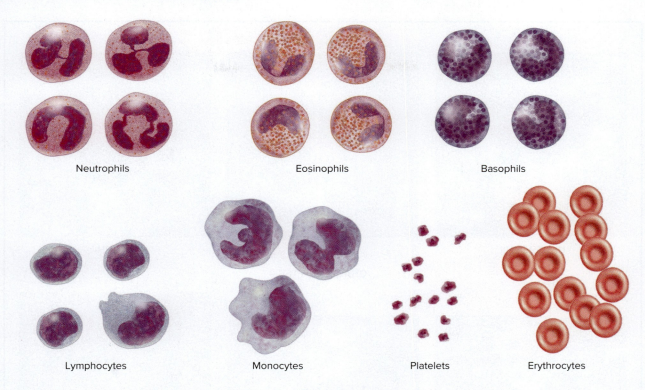

Differential White Blood Cell Count

The differential white blood cell (WBC) count is performed to determine the relative percentages of the five types of white blood cells shown in Figure 18.1. Three of them—**neutrophils, eosinophils,** and **basophils**—have cytoplasmic granules and are classified as **granulocytes. Monocytes** and **lymphocytes** lack cytoplasmic granules and are classified as **agranulocytes.** High or low counts for certain types of white blood cells may be related to pathological conditions. High neutrophil counts usually indicate bacterial infection. High eosinophil counts occur in allergic reactions or parasitic worm infestations. High monocyte counts usually indicate a chronic infection. High lymphocyte counts usually indicate antigen–antibody reactions.

A differential WBC count is performed by preparing a stained blood slide and counting a total of 100 white cells to determine the percentage of each type of cell.

Slide Preparation

If your instructor wants you to use a prepared slide of a human blood smear, obtain the slide and skip down to the next section.

Keys to success in preparing a good slide are (1) placing the correct size drop of blood on the slide and (2) spreading the blood evenly over the slide.

1. Clean three or four slides with soap and water, and keep them free of fingerprints.
2. Cleanse a fingertip with an alcohol pad and pierce it from the side with a lancet. Place the lancet *immediately* in the biohazard container. Place a drop of blood 2 cm from one end of a slide. *The drop should be 3–4 mm in diameter.*
3. Spread the blood on the slide using another clean slide as a spreader as shown in Figure 18.3. Note that the blood is *pulled* along by the slide. Hold the spreader slide at an angle of 50° to obtain a good smear. Place the spreader slide directly in the biohazard container.
4. After the blood has dried, use a wax pencil to draw a line at each end of the smear to confine the stain.
5. Place the slide on a folded paper towel and add Wright's blood stain, *counting the drops until the smear is covered.*
6. Wait *4 minutes,* then add an equal number of drops of distilled water. Let it stand for *10 minutes* while blowing on the slide every few minutes to keep the solutions mixed.

Figure 18.2 Photomicrographs of formed elements in blood (5,000×). (a–g) Courtesy of Harold Benson

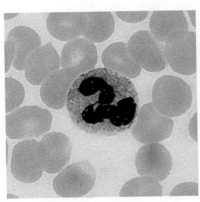
NEUTROPHIL   EOSINOPHIL   BASOPHIL
GRANULOCYTES

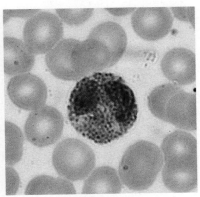

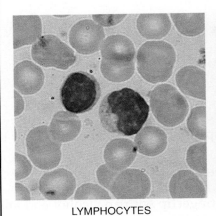
LYMPHOCYTES   MONOCYTE
AGRANULOCYTES

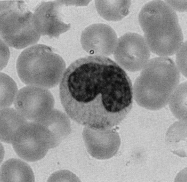

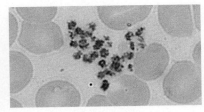
BLOOD PLATELETS

PLASMA CELL
(abnormal)

Neutrophils differ from the other two granulocytes in having small lavender granules in the cytoplasm. Nuclei of these leukocytes are characteristically lobulated with thin bridges between the lobules (see Figure 18.1). During infections, when many of these are being produced in the red bone marrow, the nuclei are horseshoe-shaped; these juvenile cells are often called "stab cells."

Eosinophils, generally, have bilobed nuclei as above; however, juveniles have nuclei that are horseshoe-shaped (Figure 18.1). Note that the red eosinophilic granules in the cytoplasm are large and pronounced.

Basophils, which are few in number, are distinguished from the other granulocytes by the presence of dark blue basophilic granules in the cytoplasm. The nuclei of these cells are large and varied in shape. The fact that heparin has been identified in these cells indicates that they probably secrete this substance into the blood.

Two sizes of lymphocytes—large and small—are shown above and in Figure 18.1. The small ones, which are more abundant than the large ones, have large, dense nuclei surrounded by a thin layer of basophilic cytoplasm. The large lymphocytes have indented nuclei and more cytoplasm than small lymphocytes.

Monocytes are the largest cells found in healthy blood. The nucleus may be ovoid, indented, or horseshoe-shaped, as illustrated in Figure 18.1. Monocytes have considerably more cytoplasm than lymphocytes and are voracious phagocytes.

Blood platelets, which form from megakaryocytes in the red bone marrow, function in the formation of blood clots.

Plasma cells are only rarely seen in the blood. The photomicrograph above was made from the blood of a patient with plasma cell leukemia. Although they resemble lymphocytes, the plasma cells have more cytoplasm that is very basophilic.

Figure 18.3 Smear preparation technique for differential WBC count.

1. A small drop of blood is placed about 3/4 in. away from one end of slide. The drop should not exceed 1/8 in. diameter.

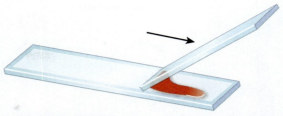

2. The spreader slide is moved in direction of arrow, allowing drop of blood to spread along slide's back edge.

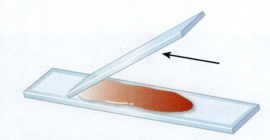

3. The spreader slide is moved along the slide, dragging the blood over the surface of the slide.

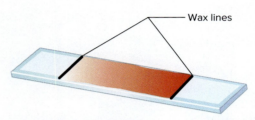

4. A china marking pencil is used to mark off both ends of the smear to retain the staining solution on the slide.

7. Gently rinse the slide in slowly running water. Blot the slide dry with bibulous paper. When the slide is dry, make the count as described in the next section.

Making the Count

Whether you are using a prepared slide or one that you have just stained, the counting procedure is the same. Ideally, one should use an oil immersion lens for this procedure; high-dry optics can be used, but they are not as reliable for best results. Proceed as follows:

1. Scan the slide with the low-power objective to find an area where cell distribution is best. A good area is one in which the cells are not jammed together or scattered too far apart.
2. Once an ideal area has been selected, place a drop of immersion oil on the slide near one edge and lower the oil immersion objective into the oil. If you are using a slide you have just prepared, you needn't worry about oil being placed directly on the stained smear.
   If the high-dry objective is to be used, omit placing oil on the slide.
3. Systematically scan the slide, following the pathway indicated in Figure 18.4. As each leukocyte is encountered, identify it, using Figures 18.1 and 18.2 for reference.
4. Record on the laboratory report the type of each white cell encountered until you have tabulated 100 cells.

Figure 18.4 Examination path for differential count.

5. When finished, either place the smear that you made in the biohazard container or return the prepared slide to its tray.

Hematocrit

Hemoglobin, the red blood pigment, is the oxygen-carrying component in erythrocytes. A person with a reduced concentration of hemoglobin is said to be anemic. One of the causes of anemia is a low erythrocyte concentration.

One test used to determine the concentration of erythrocytes in the blood is called the **hematocrit.** This test determines the percentage of the blood occupied by red blood cells. It is performed by using a microhematocrit centrifuge and heparinized microhematocrit capillary tubes.

In males, the healthy range is 40% to 54%; 47% is average. In females, the healthy range is 37% to 47%; 42% is average. Perform the test as described below and shown in Figure 18.5.

1. Cleanse and pierce the fingertip as before and place the lancet in the biohazard container. Wipe away the first drop of blood with the alcohol pad.

Figure 18.5 Determination of the hematocrit.

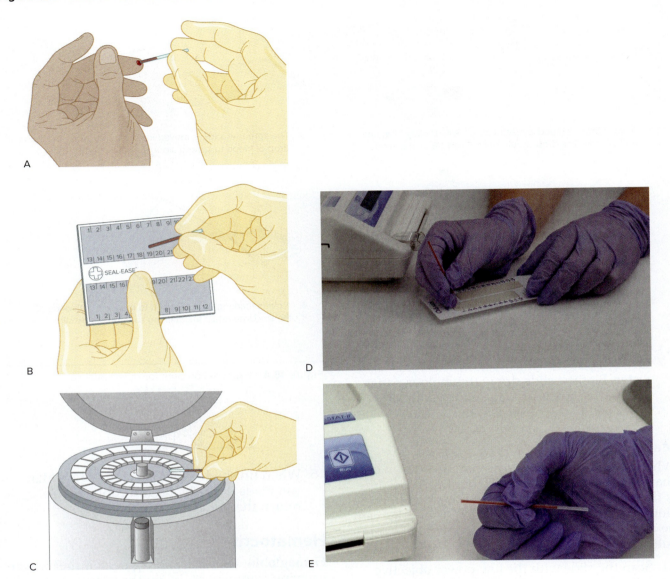

2. Place the red end of the heparinized capillary tube in the second drop of blood and lower the other end so the blood will fill about two-thirds of the tube.
3. Insert the blood end of the tube into Seal-Ease to plug it.
4. Place the tube in the centrifuge with the sealed end against the ring at the perimeter. Record your tube slot number. Load the centrifuge with an even number of tubes that are arranged opposite each other to balance the load.
5. Secure the inside cover and fasten the outside cover.
6. Turn on the centrifuge and set the timer for the time specified by your instructor.
7. Determine the hematocrit by using the mechanical tube reader. Instructions are on the instrument. The hematocrit may be read on the head of some centrifuges.
8. Place nonsharp blood-contaminated materials in the biohazard bag.

Blood Typing

Red blood cells possess a variety of antigens on their surfaces. The presence or absence of some of these antigens is used to determine one's blood type, which is genetically controlled and never changes throughout the lifetime of an individual.

An **antigen** is a chemical molecule that stimulates the production of a specific **antibody,** which,

Figure 18.6 Blood typing with a warming box.

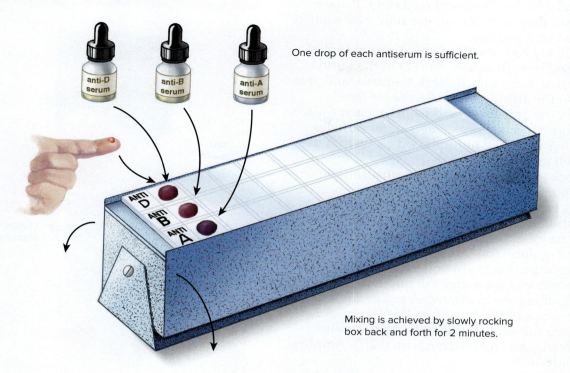

One drop of each antiserum is sufficient.

Mixing is achieved by slowly rocking box back and forth for 2 minutes.

in turn, will attach to the antigen. Although there are many antigens on the red blood cells, it is the **A, B,** and **D (Rh)** antigens that are most commonly used in blood typing. In this exercise, we will study blood typing using the A, B, and D antigens to determine your ABO and Rh blood type.

Transfusion typically involves a donor and a recipient with the same blood type. However, it may be possible for a recipient to receive a transfusion of a different blood type in emergency situations.

Not all blood types are compatible, and a transfusion of incompatible blood may be fatal. Thus, typing the blood of both the potential donor and the recipient is essential. Because of differences in concentration, it is the *antigens of the donor* and the *antibodies of the recipient* that must be considered. Table 18.2 shows the relationship of antigens and antibodies in blood. By understanding this table and the antigen–antibody reaction, you can predict the compatibility of various blood type combinations.

In the clinical setting, blood is usually diluted with saline when doing ABO blood typing. However, in this exercise, you will use whole blood because you will be determining the ABO group and Rh type simultaneously. Rh typing requires whole blood and a slide warming box to yield a temperature of about 50°C.

You will type your blood by mixing a drop of blood with a drop of **antiserum** that contains

TABLE 18.2
Antigen and Antibody Associations in Blood

Blood Type	Antigen	Antibody
O	none	anti-A, anti-B
A	A	anti-B
B	B	anti-A
AB	A,B	none
Rh+	D	none
Rh−	none	anti-D*

*Antibodies are formed by an Rh− person only after Rh+ erythrocytes enter his or her blood.

specific, known antibodies. If the corresponding antigen is present on the red blood cells, an antigen–antibody reaction occurs causing the erythrocytes to **agglutinate** (clump together). This clumping can be identified visually.

If you are to use simulated blood, follow the directions of your instructor. The procedure for determining your ABO and Rh blood type using your own blood is described below and shown in Figure 18.6. *Be sure to follow the safety precautions.*

1. Place a clean microscope slide across the three typing squares on the typing slide of the warming box. Allow 5 minutes for temperature equilibration.

Blood Tests

2. Cleanse and pierce a fingertip. Discard the lancet in the biohazard container. Place a small drop of blood on the slide over each of the three squares.
3. Quickly add one drop of antiserum to the blood in each square: anti-D to the anti-D square, anti-A to the anti-A square, and anti-B to the anti-B square. *Record the time.*
4. Mix the antiserum and blood in each square *using a separate clean toothpick for each one. Why?* Place the toothpicks directly into the biohazard container.
5. Rock the warming box for *2 minutes,* then examine the mixtures for agglutination, starting with the anti-D square. Clumps in the anti-D square must appear within 2 minutes to be valid. These clumps will be very small and require close scrutiny for detection. Agglutination in any square indicates the presence of the antigen in question. For example, clumping in the anti-A square indicates presence of the A antigen. See Figure 18.7.
6. Place the slide in the biohazard container. Clean your workstation with 0.25% Amphyl solution.

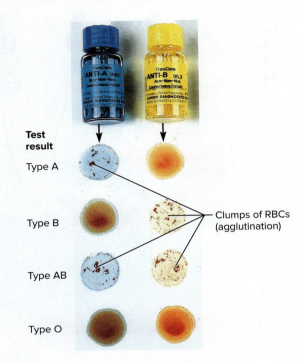

Figure 18.7 Interpretation of agglutination patterns.
DIOMEDIA/ISM/Jean-Claude RÉVY/Medical Images

Assignment

Complete the laboratory report.

Exercise 19

THE HEART

Objectives

After completing this exercise, you should be able to

1. Identify the parts of the heart on preserved specimens, charts, or models.
2. Describe the functions of each part of the heart.
3. Identify the heart sounds by stethoscope auscultation.

Materials
Sheep heart, fresh or preserved
Dissecting instruments and tray
Protective disposable gloves
Stethoscope
Alcohol wipes

Before You Proceed

Consult with your instructor about using protective disposable gloves when performing portions of this exercise.

The heart is a double pump that circulates blood via arteries and veins throughout the body. The two pumps are located side by side, and each consists of two chambers: an atrium (receiving chamber) and a ventricle (pumping chamber). The **right atrium** and **right ventricle** compose one pump. It receives deoxygenated blood from the body and pumps it to the lungs. The **left atrium** and **left ventricle** form the other pump. It receives oxygenated blood from the lungs and pumps it to all other parts of the body. Like all pumps, the heart has valves that prevent a backflow of blood and keep blood flowing in the correct direction.

External Anatomy

Locate the external structures of the heart in Figure 19.1 as you study this section.

Anterior Aspect

The heart depicted in this view is oriented in the same position as the frontal section of the heart shown in Figure 19.2. This will help you correlate interior and exterior relationships. Note on Figure 19.1 that the tilt of the heart in the **mediastinum** (label 13) places the **apex** (tip) on the left side of the midline and just above the diaphragm.

Locate the **superior vena cava** above the right atrium and the **inferior vena cava** where it projects below the **diaphragm**. Observe that the **aorta** forms an **aortic arch** over the heart and descends behind the heart to become the **descending aorta** (label 15). Find the **left** and **right pulmonary arteries** projecting from under the aortic arch, and locate the short **ligamentum arteriosum** (label 8).

Several arteries and veins of the coronary circulation are shown in the anterior view in Figure 19.1. The **right coronary artery** and a **cardiac vein** lie in the **coronary sulcus** (not labeled), a shallow groove between the right atrium and right ventricle. The **circumflex artery** (not labeled) and a cardiac vein are located in the coronary sulcus (not labeled) between the left atrium and left ventricle. The **anterior interventricular artery** and the **great cardiac vein** are found in the **anterior interventricular sulcus** (not labeled), which marks the location of the interventricular septum. Fat deposits envelop all of the major vessels of the coronary circulation.

Posterior Aspect

Locate the **aortic arch,** which carries blood from the left ventricle to the body, and the **superior vena cava** and **inferior vena cava,** which return blood to the **right atrium.** Just below the aortic arch are the **left** and **right pulmonary arteries** (blue), which carry deoxygenated blood from the pulmonary trunk to the lungs. The two **right pulmonary veins** (label 19) are seen near the base of the superior vena cava. The two **left pulmonary veins** are visible below the

Unit 5: Internal Transport and Gas Exchange

Figure 19.1 External anatomy of the heart.

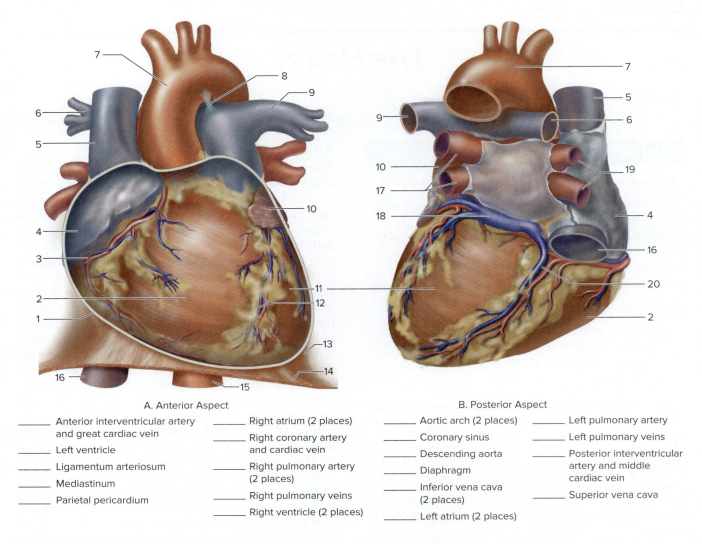

A. Anterior Aspect

____ Anterior interventricular artery and great cardiac vein
____ Left ventricle
____ Ligamentum arteriosum
____ Mediastinum
____ Parietal pericardium
____ Right atrium (2 places)
____ Right coronary artery and cardiac vein
____ Right pulmonary artery (2 places)
____ Right pulmonary veins
____ Right ventricle (2 places)

B. Posterior Aspect

____ Aortic arch (2 places)
____ Coronary sinus
____ Descending aorta
____ Diaphragm
____ Inferior vena cava (2 places)
____ Left atrium (2 places)
____ Left pulmonary artery
____ Left pulmonary veins
____ Posterior interventricular artery and middle cardiac vein
____ Superior vena cava

pulmonary arteries. Pulmonary veins return oxygenated blood from the lungs to the **left atrium.**

The posterior view shows a few major vessels of the coronary circulation. Locate the **posterior interventricular artery** and **middle cardiac vein** in the **posterior interventricular sulcus** (not labeled). This vein empties into the **coronary sinus** (label 18).

Assignment

1. Label Figure 19.1.
2. Complete Sections A, B, and C of the laboratory report.

Internal Anatomy

With the basic function of the heart in mind, locate in Figure 19.2 the heart structures described below and correlate each structure with its function. Vessels colored red carry oxygenated blood; those colored blue carry deoxygenated blood. Note the red does NOT always mean a vessel is an artery.

Wall of the Heart

The wall of the heart consists of three layers. The thick **myocardium** is composed of cardiac muscle tissue. It is covered by two thin serous membranes attached to the muscle tissue: an inner **endocardium** and an outer **epicardium,** or **visceral pericardium.** External to the epicardium is the double-layered **pericardial sac,** composed of an inner **parietal pericardium** and an outer **fibrous pericardium.** The **pericardial cavity** (label 20) lies between the epicardium and parietal pericardium. Fluid secreted by the serous membranes reduces friction, enabling the heart to move freely within the pericardial sac.

154   Internal Transport and Gas Exchange

Figure 19.2 Internal anatomy of the heart.

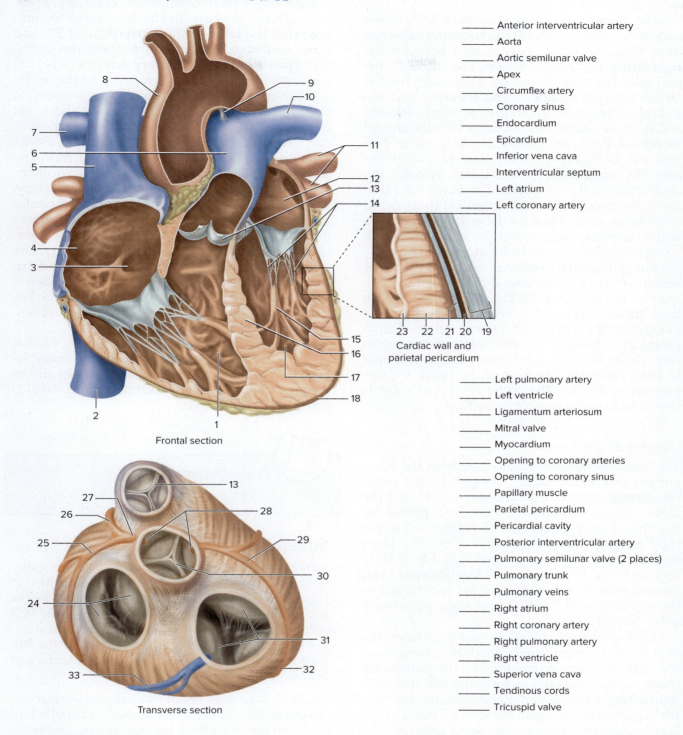

_____ Anterior interventricular artery
_____ Aorta
_____ Aortic semilunar valve
_____ Apex
_____ Circumflex artery
_____ Coronary sinus
_____ Endocardium
_____ Epicardium
_____ Inferior vena cava
_____ Interventricular septum
_____ Left atrium
_____ Left coronary artery

Cardiac wall and parietal pericardium

_____ Left pulmonary artery
_____ Left ventricle
_____ Ligamentum arteriosum
_____ Mitral valve
_____ Myocardium
_____ Opening to coronary arteries
_____ Opening to coronary sinus
_____ Papillary muscle
_____ Parietal pericardium
_____ Pericardial cavity
_____ Posterior interventricular artery
_____ Pulmonary semilunar valve (2 places)
_____ Pulmonary trunk
_____ Pulmonary veins
_____ Right atrium
_____ Right coronary artery
_____ Right pulmonary artery
_____ Right ventricle
_____ Superior vena cava
_____ Tendinous cords
_____ Tricuspid valve

Frontal section

Transverse section

Chambers of the Heart

The **left** and **right atria** are the upper, thin-walled chambers, which receive blood returning to the heart. There is no connection between the atria. A wall of tissue, the **interatrial septum** (not labeled) separates the two atria. During fetal life, an opening in the interatrial septum allows blood to flow between the atria. This opening is closed and sealed after birth, leaving only a slight depression, the **fossa ovalis** (not labeled) to mark the site. The lower, thick-walled chambers are the ventricles, which pump blood into the arteries leaving the heart. Note the **interventricular septum** separating the **left ventricle**

The Heart   155

from the **right ventricle,** and the thicker myocardium of the left ventricle.

The heart has two atrioventricular (AV) valves and two semilunar valves. The **tricuspid valve** has three flaps, or cusps; it is located between the right atrium and right ventricle. The **mitral (bicuspid) valve** has only two cusps; it separates the left atrium and left ventricle. The tricuspid and mitral valves are indicated in the transverse section, and they are shown but not labeled in the frontal section of the heart. Atrioventricular valves prevent the backflow of blood from the ventricles into the atria during **ventricular systole,** the contraction of the ventricles. The cusps have thin cords, the **tendinous cords** (label 14) that anchor them to **papillary muscles** (label 15) on the walls of the ventricles and prevent the valve cusps from being forced into the atria during contraction.

The **pulmonary semilunar valve** and the **aortic semilunar valve** are located at the base of the pulmonary trunk and aorta, respectively. They each have three cusps and prevent the backflow of blood from the arteries into the ventricles during **ventricular diastole,** the relaxation of the ventricles. The pulmonary semilunar valve is indicated in both the frontal section of the heart and the transverse section. The aortic semilunar valve is shown only in the transverse section.

Vessels of the Heart

The major vessels leading to and from the heart are the two venae cavae, four pulmonary veins, the coronary sinus, a pulmonary trunk, and the aorta. During diastole, blood is returned to the atria. Deoxygenated blood is returned to the right atrium from body regions above the heart by the **superior vena cava** (label 5) and from body regions below the heart by the **inferior vena cava.** Simultaneously, oxygenated blood is returned to the left atrium by the **left** and **right pulmonary veins** (label 11).

During systole, blood is pumped from the ventricles. Deoxygenated blood is carried from the right ventricle to the lungs via the **pulmonary trunk,** which branches to form the **left** and **right pulmonary arteries.** Simultaneously, the left ventricle pumps oxygenated blood to all parts of the body except the lungs via the **aorta** (label 8). The emergence of the aorta from the left ventricle is not visible in the figure.

The **ligamentum arteriosum** (label 9) is a nonfunctional remnant of the **ductus arteriosus** that allows blood to flow from the pulmonary artery to the aorta during fetal life.

Blood that flows through the heart does not provide oxygen and nutrients to the heart itself. This function is provided by the coronary circulation, which consists of coronary arteries and cardiac veins. In the transverse section, note that the **left coronary artery** (label 27) and **right coronary artery** branch from the aorta. Two **openings to the coronary arteries** (label 28) are located in the aorta just distal to the aortic semilunar valve. Each artery produces several branch arteries to serve the heart tissues. As mentioned in the external anatomy section of this exercise, the coronary blood vessels are located in sulci (grooves) in the heart surface. The transverse section shows that the left coronary artery branches to form the **circumflex artery,** which encircles the heart in the atrioventricular sulcus, and the **anterior interventricular artery** (label 26), which runs down the anterior interventricular sulcus (as shown in Figure 19.1). The right coronary artery encircles the right ventricle giving off several branches. The last branch is the **posterior interventricular artery** (label 32), which continues down the posterior interventricular sulcus (as shown in Figure 19.1).

Deoxygenated blood is returned in cardiac veins that carry blood to the **coronary sinus** (label 33), which empties blood into the right atrium through an **opening to the coronary sinus** (label 3) near the tricuspid valve.

Assignment
Label Figure 19.2.

Sheep Heart Dissection

While dissecting the sheep heart, locate as many of the structures as possible that are shown in Figures 19.1 and 19.2. Follow the directions below and compare to the labeled images in Figures 19.3 and 19.4. Because sheep walk on four legs and humans are bipedal, the anterior surface of the human heart is comparable to the ventral surface of the sheep heart.

1. Look for remnants of the **parietal pericardium** around the great vessels. It usually has been removed from laboratory specimens.
2. Note that the **epicardium** (visceral pericardium) is firmly attached to the heart.
3. Locate the **anterior interventricular sulcus** and the **left** and **right ventricles** as shown in Figure 19.3. Remove some fat to expose the **coronary vessels** in this sulcus.
4. Locate the **left** and **right atria.** The right atrium is seen on the left side of Figure 19.3.

Figure 19.3 Anterior view of sheep heart. J and J Photography

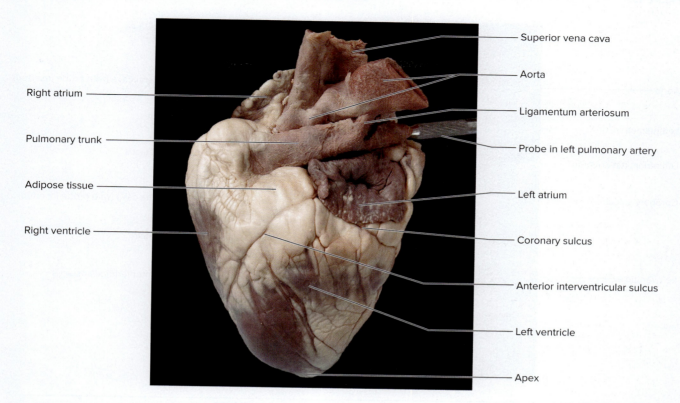

5. Identify the **aorta**. It is the large vessel just to the left of the right atrium.
6. Locate the **pulmonary trunk** between the aorta and left atrium. Trace it to where it branches into the **left** and **right pulmonary arteries**. Try to find the **ligamentum arteriosum**.
7. Examine the posterior surface of the heart as in Figure 19.4. Locate the thin-walled **pulmonary veins,** usually embedded in fat. Insert a probe to see that they enter the left atrium.
8. To view the internal view of the heart, as shown in Figure 19.5, you need to make a frontal section of the heart. Using a scalpel or a long knife, place the base of the heart down in the dissecting tray with the apex up. Face the anterior surface of the heart toward your body and make a cut in the frontal plane all the way through the heart dividing it into equal anterior and posterior portions.
9. Observe the **tricuspid valve** between the right atrium and ventricle. Wash under the tap to remove any blood clots present.
10. Observe the inner wall of the right atrium and locate the opening of the coronary sinus.
11. Observe the inner ventricular surface. Note the atrioventricular valve cusps, **tendinous cords, papillary muscles,** and supporting muscle band (moderator band) extending from the ventricular septum to the ventricular wall.
12. Look into the pulmonary trunk to view the **pulmonary semilunar valve.** Note the thickness of the cusps.
13. Next focus on the left ventricle. Spread the heart wall to observe the **mitral valve.** Compare the thickness of the left and right ventricular walls.
14. If your cut was in the correct location you should be able to see the **aortic semilunar valve.** Locate the openings of the **coronary arteries** in the aortic wall just above the valve. Insert a probe to verify that the openings lead into the coronary arteries.
15. Dispose of the heart as directed by your instructor. Wash and dry your tray and instruments.

Assignment

Complete Section D of the laboratory report.

Heart Sounds

There are three heart sounds that result from contraction and valvular movements. The **first sound** (lub) results from ventricular contraction and the simultaneous closure of the atrioventricular valves.

The Heart 157

Figure 19.4 Posterior view of sheep heart. J and J Photography

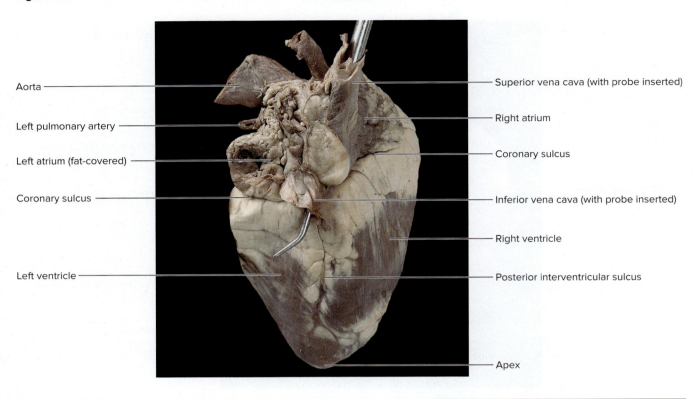

Figure 19.5 Frontal section of heart. Same view as Figure 19.2. J and J Photography

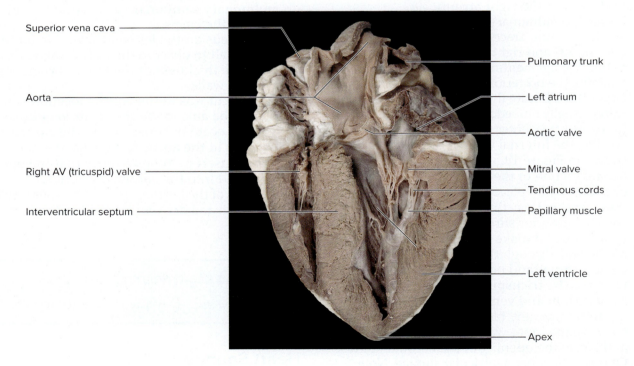

The **second sound** (dub) results from the simultaneous closure of the aortic and pulmonary semilunar valves. The **third sound** seems to be caused by the vibration of the ventricular walls and the atrioventricular valve cusps during systole. It is difficult to detect unless the subject is lying down. Abnormal heart sounds are called **murmurs** and usually result from a damaged or defective atrioventricular valve that allows blood to leak back into an atrium. Many murmurs have no clinical significance.

Auscultation of (listening to) the heart sounds provides important information for a physician and aids diagnosis of many heart conditions, including heart murmurs. For example, abnormal splitting of the heart sounds occurs in heart block, septal defects, and hypertension. For best results, you must use the **auscultatory areas** shown in Figure 19.6. These areas do not coincide with the anatomical locations of the valves, because heart sounds are projected to different locations on the thoracic cage.

Your instructor will demonstrate the heart sounds using an audio monitor. After you can distinguish the heart sounds, work in pairs to auscultate each other's heart with a stethoscope.

Figure 19.6 Auscultatory areas.

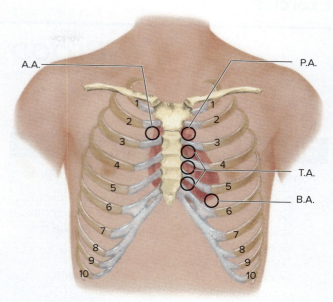

A.A. Aortic Area
B.A. Mitral Area
P.A. Pulmonic Area
T.A. Tricuspid Area

Assignment

1. Clean the earpieces of the stethoscope with an alcohol pad. Fit them into your ears, directing them inward and forward.
2. With the subject in a sitting position, try to maximize the first sound by placing the stethoscope on the tricuspid and mitral auscultatory areas. Note that the mitral valve auscultatory area is at the apex of the heart (fifth rib) and that the tricuspid area is 2- to 3-in medial from this site.
3. Now, move the stethoscope to the aortic and pulmonic areas to maximize the second sound. The aortic auscultatory area is on the right edge of the sternum at the level of the second rib, and the pulmonic area is on the left side of the sternum between the second and third ribs. Note that the aortic auscultatory area is the only one on the right side of the sternum.
4. Try to detect a **splitting of the second sound** while the subject is inspiring (inhaling). This is a typical phenomenon that occurs during inspiration because more venous blood is forced into the right side of the heart and causes a delayed closure of the pulmonary semilunar valve. Use hand signals, not verbalization, to inform the subject when to inspire. Have the subject inspire when you raise your hand and expire (exhale) when you lower it.
5. Now, have the subject lie down, face up, and repeat steps 2 to 4. Try to detect the **third sound** by placing the stethoscope over the apex of the heart. Do you detect any murmurs?
6. If time permits, compare the sounds before and after exercise.
7. Complete Sections E and F of the laboratory report.

The Heart

Exercise 20

Blood Vessels

Objectives

After completing this exercise, you should be able to

1. Diagram and describe the basic circulatory plan.
2. Identify the major arteries, veins, and lymphatic vessels on charts or models and describe their locations.
3. Distinguish healthy arteries and veins and atherosclerotic arteries when viewed microscopically.
4. Compare circulation in a fetus and an adult.
5. Determine the blood pressure and pulse of a subject.

Materials
    Colored pencils
    Prepared slides of
        Artery and vein, x.s.
        Atherosclerotic artery, x.s.
    Stethoscope
    Alcohol pads
    Sphygmomanometer

The Circulatory Plan

Blood is the carrier of materials to and from body cells. It is propelled by heart contractions through **arteries** to all parts of the body and returned to the heart by the **veins**. At the tissue level, materials diffuse from blood into cells and from cells into blood as it passes slowly through **capillaries,** the tiny vessels connecting **arterioles** and **venules.**

Figure 20.1 is an incomplete flow diagram of the basic circulatory plan. You are to complete and label this figure by adding and labeling the arteries and veins described below. The purpose of this activity is to understand the general plan of circulation rather than learn the specific arteries and veins.

Starting at the heart, the **right atrium** receives deoxygenated blood from all parts of the body via the superior and inferior venae cavae. The **superior vena cava** returns blood from the head, neck, upper limbs, and other regions above the heart. The **inferior vena cava** returns blood from regions below the heart. From the right atrium, blood flows into the **right ventricle.** During contraction, blood is pumped from the right ventricle through the **pulmonary trunk** and **pulmonary arteries** to the lungs. Oxygenated blood is returned from the lungs into the **left atrium** by the **pulmonary veins** and flows into the **left ventricle.** During contraction, blood is pumped to all parts of the body *except the lungs* via the **aorta.**

Oxygenated blood is carried to the head and upper limbs by arteries branching from the aorta, but the branches are represented here by a single vessel for simplicity. As the aorta continues downward it is called the **descending aorta** (draw it descending on the anatomical left of the heart). Branches from the aorta serve organs below the heart. The artery supplying the liver is the **hepatic artery.** Intestines are supplied by the **superior mesenteric artery.** Kidneys receive blood via **renal arteries.** The pelvis and lower limbs are supplied by several arteries, but only one is shown for simplicity.

Veins from the pelvis and lower limbs combine to form the **inferior vena cava,** which collects blood from regions below the heart and returns it to the right atrium. (Show the inferior vena cava ascending on the anatomical right side of the figure.) The **renal** and **hepatic veins** carry blood from the kidneys and liver, respectively, directly to the inferior vena cava. However, blood from the intestines is first carried to the liver by the **hepatic portal vein** before entering the vena cava via the **hepatic vein.**

Assignment

1. Complete and label Figure 20.1.
2. Color vessels carrying oxygenated blood *red* and those carrying deoxygenated blood *blue*. Indicate the direction of blood flow by arrows.
3. Complete Section A of the laboratory report.

Unit 5: Internal Transport and Gas Exchange

Figure 20.1 The circulatory plan. A&PR

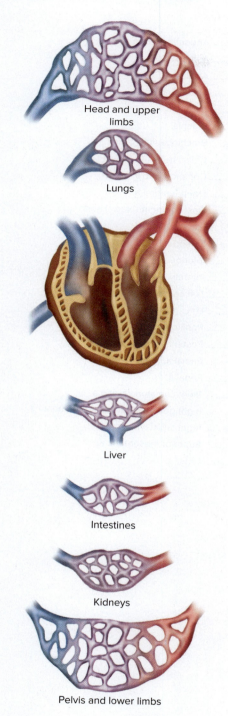

Major Arteries

Locate the major arteries in Figure 20.2 as you read this section. Blood leaves the left ventricle via the **ascending aorta,** which arches over the heart forming the **aortic arch** and continues downward as the **descending (thoracic) aorta.** In the abdominal area, it is called the **abdominal aorta.**

The short **brachiocephalic trunk,** the first artery to branch from the aorta, emerges from the right side of the aortic arch and quickly branches into (1) the **right common carotid artery,** which goes to the right side of the neck and head, and (2) the **right subclavian artery,** which passes behind the clavicle. The right subclavian subsequently becomes the **right axillary artery** and then the **right brachial artery** of the arm.

The middle artery from the aortic arch is the **left common carotid artery,** which supplies the left side of the neck and head.

The **left subclavian artery,** the third artery branching from the aorta, emerges from the aortic arch and enters the left shoulder behind the clavicle. In the armpit area, it becomes the **left axillary artery** (label 16), which, in turn, becomes the **left brachial artery** in the arm. The brachial artery branches into the **radial** and **ulnar arteries,** which follow the forearm bones.

Several major arteries branch from the **abdominal aorta.** The **celiac trunk** (label 5) emerges from the aorta just below the diaphragm and forms three branches (not shown). The **left gastric artery** supplies the stomach and esophagus; the **splenic artery** supplies the spleen; and the **hepatic artery** supplies the liver. Just below the celiac is the **superior mesenteric artery,** which serves most of the small intestine and part of the large intestine. The kidneys receive blood from the **renal arteries.** Below the renal arteries, a pair of **gonadal arteries** (label 7) emerge from the front of the abdominal aorta. They serve the ovaries in females and testes in males. Just below the gonadal arteries, a single **inferior mesenteric artery** exits the abdominal aorta to carry blood to the large intestine and rectum.

In the lumbar region, the aorta divides to form the **left** and **right common iliac arteries.** Each common iliac then divides into a small **internal iliac artery** and a large **external iliac artery,** which continues downward to become the **femoral artery** in the thigh. In the upper part of the thigh, the **deep femoral artery** branches from the femoral artery. In the knee region, the femoral artery becomes the **popliteal artery,** which branches to form the **anterior** and **posterior tibial arteries.**

Major Veins

Refer to Figure 20.3. The **superior vena cava** receives blood from major veins above the heart and returns it to the right atrium. It begins at

Figure 20.2 Major arteries of the body. (Note a. = artery and aa. = arteries.) Right and left are only indicated if the arteries are not the same on both sides.

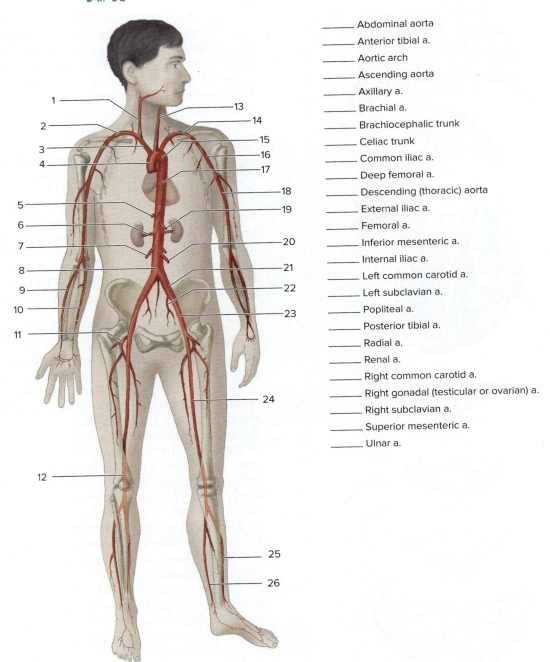

_____ Abdominal aorta
_____ Anterior tibial a.
_____ Aortic arch
_____ Ascending aorta
_____ Axillary a.
_____ Brachial a.
_____ Brachiocephalic trunk
_____ Celiac trunk
_____ Common iliac a.
_____ Deep femoral a.
_____ Descending (thoracic) aorta
_____ External iliac a.
_____ Femoral a.
_____ Inferior mesenteric a.
_____ Internal iliac a.
_____ Left common carotid a.
_____ Left subclavian a.
_____ Popliteal a.
_____ Posterior tibial a.
_____ Radial a.
_____ Renal a.
_____ Right common carotid a.
_____ Right gonadal (testicular or ovarian) a.
_____ Right subclavian a.
_____ Superior mesenteric a.
_____ Ulnar a.

the base of the neck by the junction of the short left and right **brachiocephalic veins** (label 13), which receive blood from the head and upper limbs.

In the neck are four veins: two medial **internal jugular veins** that empty into the brachiocephalic veins and two **external jugular veins** that empty into the **subclavian veins**, which are located deep to the clavicles. Each subclavian vein is continuous with a short **axillary vein** (label 3). Near its proximal end, the axillary receives blood from a **cephalic vein** (label 4) that drains the lateral surface of the upper limb. Distally, the axillary vein is formed by the merging of two veins. The **basilic vein** drains the medial surface of the upper limb, and the **brachial vein** (label 16) drains

162 Internal Transport and Gas Exchange

Figure 20.3 Major veins of the body. (Note v. = vein and vv. = veins.) Right and left are only indicated if the veins are not the same on both sides.

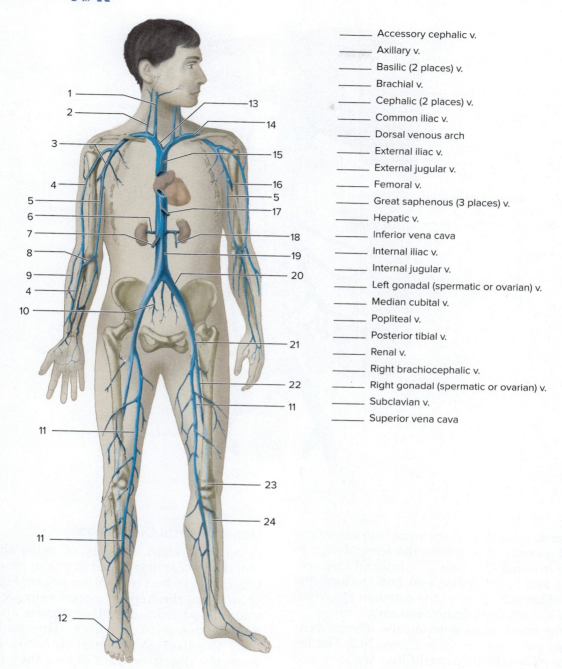

_____ Accessory cephalic v.
_____ Axillary v.
_____ Basilic (2 places) v.
_____ Brachial v.
_____ Cephalic (2 places) v.
_____ Common iliac v.
_____ Dorsal venous arch
_____ External iliac v.
_____ External jugular v.
_____ Femoral v.
_____ Great saphenous (3 places) v.
_____ Hepatic v.
_____ Inferior vena cava
_____ Internal iliac v.
_____ Internal jugular v.
_____ Left gonadal (spermatic or ovarian) v.
_____ Median cubital v.
_____ Popliteal v.
_____ Posterior tibial v.
_____ Renal v.
_____ Right brachiocephalic v.
_____ Right gonadal (spermatic or ovarian) v.
_____ Subclavian v.
_____ Superior vena cava

the back of the arm. The short **median cubital vein** (label 8), joining the basilic and cephalic veins, is a common site for venipuncture. The **accessory cephalic vein** (label 9) drains the lateral forearm and enters the cephalic vein at the elbow a short distance above the junction of the median cubital and cephalic veins.

Blood is returned from the lower limbs by both deep and superficial veins. The **posterior tibial vein** (label 24) is a deep vein that collects blood from the foot and leg. It becomes the **popliteal vein** in the knee region and continues as the **femoral vein** of the thigh. The surface vein of the lower limb is the **great saphenous vein,** which originates at the **dorsal**

Blood Vessels   163

Figure 20.4 Hepatic portal circulation.

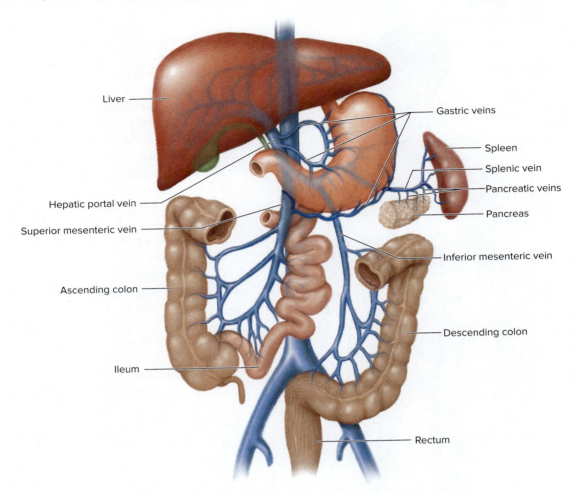

venous arch (label 12) on the upper surface of the foot and ascends to join with the femoral vein to form the **external iliac vein**. The **internal iliac vein** (label 10) and external iliac vein unite to form the **common iliac vein**. Both of the common iliac veins combine to form the **inferior vena cava**.

Of the many veins entering the inferior vena cava, only three are shown in Figure 20.3. The **hepatic vein** returns blood from the liver and is located just below the heart. The two **renal veins** return blood from the kidneys. The branch connecting to the left renal vein is the **left gonadal vein,** which carries blood from the left testis or ovary. The **right gonadal vein** enters the inferior vena cava just below the right renal vein.

The larger veins contain valves that prevent a backflow of blood and keep blood moving toward the heart. If the valves break down and the veins dilate, blood flow is reduced. This condition is known as *varicose veins,* and it is promoted by aging and inactivity.

Hepatic Portal Circulation

A portal system consists of veins that carry blood from capillary beds in one or more organs to a capillary bed in another organ. Figure 20.4 shows that the **hepatic portal vein** receives venous blood from digestive organs and spleen and carries it to the liver. This pathway allows the liver to process nutrients absorbed from the digestive tract before they enter the general circulation. The hepatic portal vein receives blood from four major veins. Blood from the intestines is drained by the **superior mesenteric** and the **inferior mesenteric veins.** The **splenic vein** brings blood from the spleen, pancreas, and part of the stomach to join with the inferior mesenteric vein. The **gastric veins** serve most of the stomach. After flowing through capillary beds in the liver, blood from the liver enters the inferior vena cava via the **hepatic vein.**

Assignment

1. Label Figures 20.2 and 20.3.
2. Demonstrate the presence of valves in veins by applying pressure with your fingertip to a prominent vein on the back of your hand near your wrist. Then, continue to apply pressure as you trace the vein toward your knuckles. Note that blood does not flow back to fill the vein until you release the pressure.
3. Examine a prepared slide of artery and vein, x.s. Note the thicker, more muscular wall of the artery.
4. Examine a prepared slide of an atherosclerotic artery, x.s. Locate the fatty plaque that partially plugs the vessel. A high level of blood cholesterol and genetic tendencies lead to the development of atherosclerosis. Make drawings of healthy and atherosclerotic arteries and a healthy vein in Section D of the laboratory report.
5. Complete Section B of the laboratory report.

Fetal Circulation

Circulation during fetal development is necessarily different from that of an adult. During fetal development, the placenta is the site of the exchange of nutrients and oxygen from maternal blood to fetal blood and, simultaneously, the transfer of wastes and carbon dioxide from fetal blood to maternal blood. Also, the lungs are nonfunctional and receive a reduced supply of blood.

Fetal circulation through the heart is shown in Figure 20.5. Vessels colored purple are carrying a mixture of oxygenated and deoxygenated blood. Locate the red-colored **umbilical vein** (label 3), which brings oxygenated blood from the placenta to the fetus via the **umbilical cord** (label 16). The umbilical vein continues to the liver. Near the liver, the umbilical vein gives off a branch, the **ductus venosus,** that bypasses the liver to join a vein from the liver and merges with the inferior vena cava. This arrangement allows oxygen-rich blood to quickly enter the general circulation. Note that there is a small constriction at the beginning of the ductus venosus where a **sphincter muscle** is present in the vessel wall. The sphincter muscle remains open before birth to allow most of the blood to flow directly into the inferior vena cava, but it closes soon after birth.

Two **umbilical arteries** carrying blood to the placenta are wrapped around the umbilical vein in the umbilical cord. They are extensions of the **hypogastric arteries** that pass along each side of the urinary bladder.

Two additional fetal circulatory adaptations enable blood with relatively more oxygen to be delivered quickly to body cells. In the fetal heart, the **foramen ovale** is an opening between the right and left atria. It allows much of the blood entering the right atrium to flow into the left atrium and on into the left ventricle, which pumps blood to all parts of the body except the lungs. A smaller amount of blood from the right atrium enters the right ventricle, which pumps blood into the pulmonary artery. However, much of the blood in the pulmonary artery does not go to the lungs but instead passes through the **ductus arteriosus** (label 10) into the aorta, increasing the amount of blood carried to body cells. Both of these fetal adaptations allow much of the blood entering the right ventricle to bypass the nonfunctional lungs, which aids the rapid transport of oxygen and nutrients to body cells.

When the fetus is separated from the placenta after birth, circulatory changes occur. The constriction and closure of the ductus venosus helps to decrease the loss of blood via the umbilical vein. It subsequently becomes the **ligamentum venosum** of the liver. The umbilical vein becomes nonfunctional and is converted into the **round ligament** extending from the umbilicus to the liver. The distal portions of the hypogastric arteries become fibrous cords, and the proximal portions remain functional and supply the urinary bladder.

As soon as the infant begins to breathe, more blood is channeled to the lungs. The decreased use of the ductus arteriosus leads to its closure and conversion into the **ligamentum arteriosum.** The resulting increase in blood volume and pressure in the left atrium closes the flap on the foramen ovale. Subsequently, connective tissue seals this opening to permanently separate the atria.

Failure of the foramen ovale or ductus arteriosus to close after birth results in a circulatory defect that allows continued mixing of oxygenated and deoxygenated blood. Such defects may deprive body tissues of adequate oxygen, seriously impairing the newborn. These defects are usually repaired surgically.

Assignment

Label Figure 20.5 and add arrows to show the direction of blood flow.

Blood Vessels

Figure 20.5 Fetal circulation.

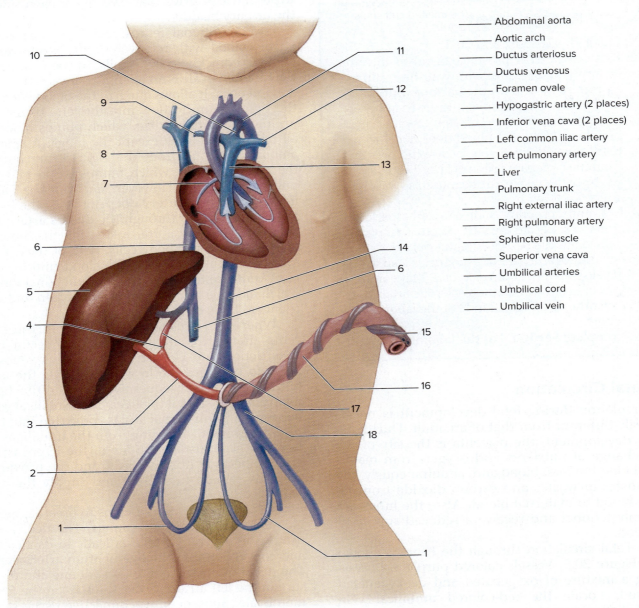

_____ Abdominal aorta
_____ Aortic arch
_____ Ductus arteriosus
_____ Ductus venosus
_____ Foramen ovale
_____ Hypogastric artery (2 places)
_____ Inferior vena cava (2 places)
_____ Left common iliac artery
_____ Left pulmonary artery
_____ Liver
_____ Pulmonary trunk
_____ Right external iliac artery
_____ Right pulmonary artery
_____ Sphincter muscle
_____ Superior vena cava
_____ Umbilical arteries
_____ Umbilical cord
_____ Umbilical vein

Blood Pressure Determination

Blood pressure is usually taken when a person is at rest, and average healthy blood pressure values apply to a person at rest. However, blood pressure is not constant, and many factors affect it. For example, blood pressure is increased by an increase in heart rate, blood volume, and blood viscosity or a decrease in blood vessel diameter. These factors, in turn, are affected by a person's health, physical activity, and emotional state. The contractions of the ventricles of the heart pump blood into the pulmonary artery and aorta in pulses about 70 times a minute. This pumping action causes hydrostatic pressure in the arteries to fluctuate during the heart cycle. Blood pressure is greatest during **ventricular systole**, the contraction phase, and least during **ventricular diastole**, the relaxation phase.

There are several ways to measure blood pressure, but we will use a standard **sphygmomanometer** and a **stethoscope** to measure blood pressure indirectly. In this method, a stethoscope is used to listen for the **Korotkoff sounds** that are produced by blood rushing through a partially occluded brachial artery.

166 Internal Transport and Gas Exchange

Figure 20.6 Sphygmomanometer cuff is wrapped around the arm, keeping lower margin of cuff above a line through the epicondyles of the humerus. The inflated cuff shuts off the flow of blood in the brachial artery. Korotkoff sounds are listened for as pressure is gradually lowered by releasing air from the cuff. Iryna Kurhan/123RF

Figure 20.6 shows how to apply the cuff of the sphygmomanometer to the arm with its lower margin above the epicondyles of the humerus. When the cuff is properly positioned, it is inflated to shut off the flow of blood in the brachial artery. As shown, the bell of the stethoscope is positioned over the brachial artery. Then, air is released from the cuff through the control valve on the hand pump while listening for the Korotkoff sounds.

Korotkoff sounds are caused by turbulent blood flow through a partially constricted artery. They are not produced by blood flow in a completely open artery. Korotkoff sounds change as air pressure in the cuff is decreased. First, they sound like a clear tapping (at each heartbeat), and they get louder and deeper in tone as the pressure in the cuff is decreased. Then, they become muffled and suddenly cease.

When the first Korotkoff sound is heard, the pressure on the gauge is read. This is the **systolic pressure.** At this point, the air pressure in the cuff is equal to the blood pressure that is just sufficient to force blood through the partially occluded artery. As air continues to be released, the Korotkoff sounds change in tone and then cease. The pressure at which the sounds suddenly cease is the **diastolic pressure.** At this point, the artery is no longer partially occluded during ventricular diastole.

Average systolic pressure in adults is 110 mm Hg; average diastolic pressure is 70 mm Hg. Such blood pressures are usually designated as 110/70. Systolic and diastolic pressures may vary by ±10 mm Hg and still be considered healthy.

Your instructor will demonstrate the procedure and Korotkoff sounds using a microphone and audio monitor. After you understand the procedure, work with a partner to determine each other's blood pressure. Record your results on the laboratory report.

Assignment

1. Have the subject's forearm resting on the table with the palm upward. Wrap the cuff of the sphygmomanometer around the arm. The method of attachment will vary with the type of instrument. Your instructor will indicate how to secure it.
2. Close the air valve on the neck of the rubber bulb *but not so tightly that you can't open it.* Attach the pressure gauge on the cuff so that you can easily read it.
3. Pump air into the cuff by squeezing the bulb. Stop when the pressure gauge reads 180 mm Hg. *Do not leave the cuff inflated for more than 30 seconds.*
4. Slip the diaphragm of the stethoscope under the cuff over the brachial artery—about midway between the epicondyles of the humerus.
5. Loosen the valve control screw *slightly* to slowly release air from the cuff while listening for the Korotkoff sounds. When the first sound is heard, read the pressure on the gauge. This is the systolic pressure.
6. Continue listening to the Korotkoff sounds as air continues to escape from the cuff. At the point at which the sounds cease, read the pressure on the gauge. This is the diastolic pressure.
7. Repeat several times until you can get consistent results. Wait several minutes between repetitions.
8. Measure the blood pressure after the subject has done some exercise.
9. Complete Section F of the laboratory report.

The Pulse

Ventricular contraction produces a surge of blood into the arterial tree causing a sudden pressure wave that produces expansion of the arteries. During ventricular relaxation, the arterial walls recoil to their prior state. This alternating expansion and recoil of the elastic walls of arteries during each cardiac cycle is called the **pulse.** The pulse results from the difference between systolic and diastolic blood pressures—the **pulse pressure**.

The pulse is commonly taken by *palpation* (feeling of organs with the hands) of any artery that lies close to the body surface. The most common sites for taking the pulse are (1) the radial artery, which runs over the distal, anterior end of the radius at the wrist; and (2) a common carotid artery lying between the larynx and sternocleidomastoid muscle on either side of the neck. Figure 20.7 shows locations (pressure points) where the pulse may be felt.

The pulse rate reflects the heart rate, and the pulse strength reflects the difference between systolic and diastolic blood pressure. The pulse rate varies with body position, exercise, physical condition, and emotional state. In this section, you will palpate the pulse at various pressure points and note any rate changes.

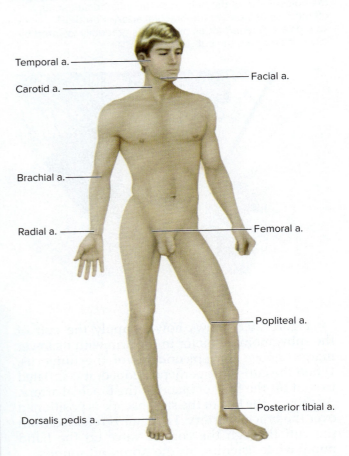

Figure 20.7 Pressure points where the pulse may be detected. (a. = artery)

Assignment

1. Work in pairs during this assignment.
2. Place the tips of your index and middle fingers over your partner's radial artery to locate the pulse. Determine your partner's pulse rate (beats/minute) when lying supine, sitting, and standing. Allow 3 minutes at each position for equilibration before taking the pulse rate. Count the beats for 15 seconds and multiply by 4 to obtain the beats per minute. Record the results on your partner's laboratory report.
3. Have your partner jog in place for 3 minutes. Have them sit down and immediately take their pulse rate. Repeat measuring the pulse rate at 1-minute intervals while they remain seated and still. The more quickly their pulse rate returns to the pre-exercise seated value from step 2 above, the better physical conditioning they have.
4. Locate your partner's pulse in the brachial artery, common carotid artery, temporal artery, and posterior tibial artery.
5. Complete Section G of the laboratory report.

Exercise 21

THE LYMPHOID SYSTEM

Objectives

After completing this exercise, you should be able to

1. Identify the major lymphatic vessels on charts or models and describe their locations.
2. Describe lymphocyte formation.

Materials
Prepared slide of lymph node, l.s.

The lymphoid system collects excess interstitial fluid (tissue fluid) from interstitial spaces in body tissues and returns it to the blood. On the way back to the blood, the fluid is cleansed of bacteria and cellular debris as it passes through lymph nodes. In addition, lymph nodes contain masses of lymphocytes that play a major role in immunity.

The Lymph Pathway

The smallest lymphatic vessels are the **lymphatic capillaries,** tiny closed-end vessels that are abundant among cells of body tissues. After entering a lymphatic capillary, interstitial fluid is called **lymph.** Lymphatic capillaries combine to form **lymphatic vessels**, which are shown (label 11) in Figure 21.1. Like veins, lymphatic vessels contain numerous valves that prevent a backflow of lymph, which is propelled by contractions of skeletal muscles depressing the walls of the vessels.

As lymph moves through the lymphatic vessels it will pass through one or more **lymph nodes.** Lymph nodes are especially abundant in cervical, axillary, abdominal, pelvic, and inguinal areas.

The lymphatic vessels from the right upper limb, right side of the thorax, and right side of the neck and head join to form the **right lymphatic duct** (label 5), which drains into the right subclavian vein.

Lymphatic vessels from the lower limbs, abdomen, left side of the thorax, left upper limb, and left side of the neck and head empty into the **thoracic duct.** At its abdominal origin is a saclike enlargement, the **cisterna chyli,** that receives lymph from the intestines. The thoracic duct empties into the left subclavian vein.

Lymph Node Structure

The enlarged lymph node in Figure 21.1 reveals its structure. The small indentation at its upper end is the **hilum,** the site where blood vessels and an **efferent lymphatic vessel** exit the node. Each node has only one hilum and one efferent lymphatic vessel, but it may have two or more **afferent lymphatic vessels** that carry lymph into the node.

A **capsule** of dense connective tissue envelops the node. Just inside the capsule is the **cortex** (label 18) that contains many **lymphoid nodules.** Lymphoid nodules are masses of lymphocytes. Active lymphoid nodules have a characteristic **germinal center** that stains lightly in a histological preparation. The central portion of the node is called the **medulla.**

Assignment

Label Figure 21.1 and add arrows to show the direction of lymph flow.

Lymphocytes and Immunity

Although all types of white blood cells are involved in an immune response, lymphocytes play a predominant role in immunity. They are responsible for *active acquired immunity*, immunity that develops only after exposure to specific **antigens,** certain proteins on the surface of pathogens.

There are two basic types of lymphocytes, **B lymphocytes (B cells)** and **T lymphocytes**

Unit 5: Internal Transport and Gas Exchange

Figure 21.1 The lymphoid system.

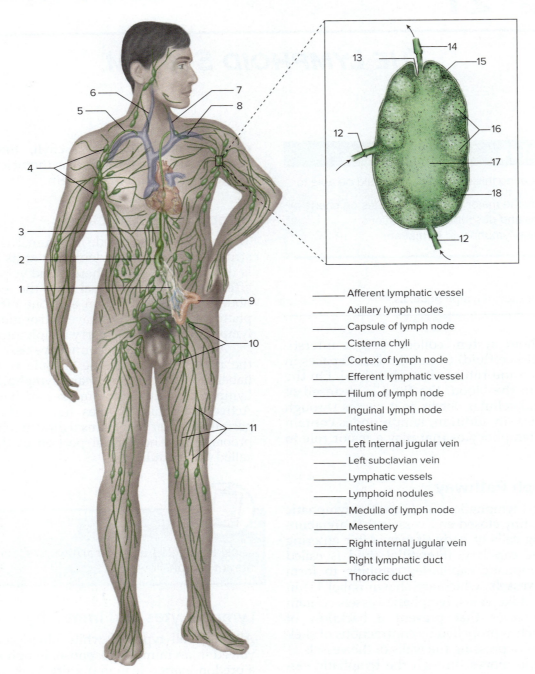

_____ Afferent lymphatic vessel
_____ Axillary lymph nodes
_____ Capsule of lymph node
_____ Cisterna chyli
_____ Cortex of lymph node
_____ Efferent lymphatic vessel
_____ Hilum of lymph node
_____ Inguinal lymph node
_____ Intestine
_____ Left internal jugular vein
_____ Left subclavian vein
_____ Lymphatic vessels
_____ Lymphoid nodules
_____ Medulla of lymph node
_____ Mesentery
_____ Right internal jugular vein
_____ Right lymphatic duct
_____ Thoracic duct

(T cells), although they appear morphologically similar. There are three basic types of T cells: *helper T cells* (T_H), *cytotoxic T cells* (T_C), and *memory T cells* (T_M).

Like all blood cells, both types of lymphocytes are derived from **stem cells** in red bone marrow before and for a short time after birth. Lymphocytes go through a maturation process to become immunocompetent, which means capable of responding to exposure to an antigen (Figure 21.2). The maturation of B lymphocytes occurs in red bone marrow. T lymphocytes mature in the thymus. The maturation process yields thousands of different varieties of B cells and T cells. Each variety is capable of recognizing a different specific antigen. Thus, any pathogen that may enter the body will be recognized by some (but not all) B cells and T cells. After maturation, both B cells and T cells are carried to lymph nodes, spleen, and other lymphoid tissue throughout the body, and some circulate in the blood.

Figure 21.2 Activation of a helper T cell or a cytotoxic T cell and cell-mediated immunity. (APC = antigen presenting cell; T_H = helper T cell; T_C = cytotoxic T cell; T_M = T memory cell) **A&P**

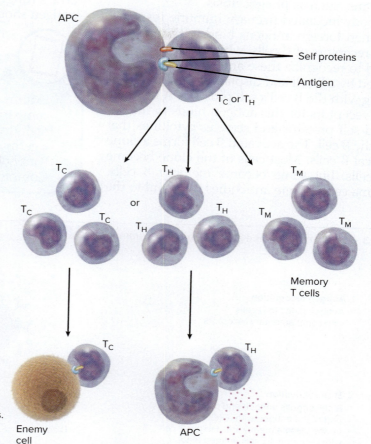

1. **Activation of a T cell**
   T cell binds to nonself antigen and self proteins of APC and becomes activated.

2. **Clone formation**
   Activated T cell undergoes repeated mitotic division producing a clone of identical T cells.

3. **Action of clone cells**
   T_C cells bind to cells displaying the targeted antigen and inject chemicals to destroy them.

   T_H cells bind to the targeted antigen and self proteins on APC, become activated, and release cytokines, which stimulate activity of T_C and B cells, inflammation, and phagocytosis by neutrophils and macrophages.

   T_M cells launch a secondary response if the targeted antigen later reappears.

There are two basic forms of immunity: T cells provide cell-mediated immunity, and B cells provide antibody-mediated immunity.

Cell-Mediated Immunity

In **cell-mediated immunity,** T cells directly attack foreign antigens or diseased body cells, such as cancer cells. It also destroys *intracellular* pathogens and the invaded cells.

Cell-mediated *primary immune response* begins when an *antigen-presenting cell (APC),* often a macrophage, engulfs a foreign antigen. A part of the antigen is presented on the cell surface along with the APC's self antigens. When a T cell, which has been programmed in the maturation process to recognize the antigen, binds to both the antigen and the APC's self proteins, it becomes activated. The activated T cell undergoes repeated cell divisions forming a clone of identical T cells (Figure 21.2).

If the activated T cell is a helper T cell (T_H), it forms a clone of mostly active T_H cells and fewer dormant memory T cells (T_M). When active T_H cells bind with the antigen, they secrete *cytokines* (chemical) that promote phagocytosis by neutrophils and macrophages and stimulate activity of T_C cells and B cells. After the foreign antigen has been destroyed, the dormant T_M cells will launch a quicker and stronger *secondary immune response* if the antigen should reappear.

If the activated T cell is a cytotoxic T cell (T_C), it forms a clone of mostly active T_C cells and fewer memory T cells (T_M). When an active T_C cell binds to the antigen, it injects lethal chemicals that kill the cell and any pathogen within. It then detaches and searches for more antigens to destroy. After the foreign antigen has been destroyed, the dormant T_M cells will launch a quicker and stronger secondary immune response if the antigen should reappear.

Antibody-Mediated Immunity

Antibody-mediated immunity involves both B cells and helper T cells, and it provides a defense against *extracellular* pathogens. Instead of directly attacking

The Lymphoid System

a foreign antigen, it involves B cells producing **antibodies** that bind to the targeted antigens, which makes it easier for them to be destroyed by other means, such as phagocytosis.

Antibody-mediated primary immune response begins when foreign antigens bind to the complementary receptors of B cells, which have been programmed to recognize these antigens. The antigen is engulfed by a B cell and displayed on the cell surface along with the B cell's self proteins. A helper T cell with receptors for this antigen binds to the antigen and self proteins and secretes cytokines that activate the B cell. The activated B cell forms a clone of identical B cells. Most cells of the clone become plasma cells, but some become memory B cells. The plasma cells secrete antibodies that bind to the targeted antigens, tagging them for easier destruction. After the foreign antigen has been destroyed, the dormant memory B cells will launch a quicker and stronger secondary immune response if the antigen should reappear (Figure 21.3).

Assignment

1. Examine a prepared slide of lymph node, comparing it with Figure 21.1. Using the high-dry objective, observe lymphocytes in a germinal center. Can you see any dividing lymphocytes?
2. Complete the laboratory report.

Figure 21.3 B cell activation and antibody-mediated immunity.

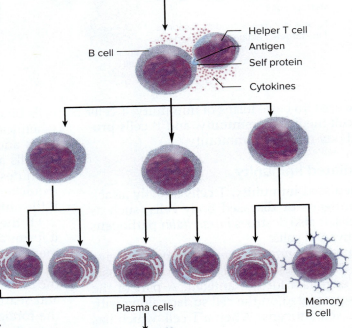

1. **Antigen recognition**
   Antigen binds to B cells with complementary receptors.

2. **B cell activation**
   B cell engulfs antigen and displays part of the antigen with a self protein on the plasma membrane. Helper T cell with receptors for this antigen binds to the antigen and self protein and secretes cytokines, which activate the B cell.

3. **Clone formation**
   Activated B cell divides repeatedly producing a clone of identical B cells programmed for this antigen.

4. **Differentiation**
   Most clone cells become plasma cells. Some become memory B cells.

5. **Action**
   Plasma cells produce and release antibodies that can bind to this antigen, tagging it for destruction by phagocytosis and other means. Memory B cells remain to launch an attack against the antigen if it should later reappear.

Exercise 22

THE RESPIRATORY SYSTEM

Objectives

After completing this exercise, you should be able to

1. Identify the components of the respiratory system on charts, models, and a sheep pluck and describe their functions.
2. Trace the flow of air into and out of the lungs.
3. Identify lung and tracheal tissues when viewed microscopically.
4. Describe the factors that stimulate breathing.
5. Describe the mechanics of breathing.
6. Determine the various lung volumes that can be measured using spirometry.

Materials

Model of a median section of the head
Torso model with removable organs
Corrosion preparation of the bronchial tree
Sheep plucks without livers
Dissecting instruments and trays
Prepared slides of human trachea, healthy lung tissue, emphysematous lung tissue, and "smoker's lung" tissue
Protective disposable gloves
Paper bag
Paper cup
Drinking straw

Before You Proceed

Consult with your instructor about using protective disposable gloves when performing portions of this exercise.

Body cells must be continuously supplied with oxygen in order to carry out their metabolic activities. The metabolic activities of cells, in turn, produce carbon dioxide that must be removed from the body. The primary function of the respiratory system is to carry out the exchange of these respiratory gases between the body and the atmosphere. Secondary functions include warming and filtering the air that is inspired.

The exchange of respiratory gases involves four related processes: (1) breathing air into and out of the lungs, (2) the exchange of oxygen and carbon dioxide between the air in the lungs and the blood, (3) the transport of oxygen and carbon dioxide by the blood, and (4) the exchange of oxygen and carbon dioxide between the blood and the tissues.

The respiratory organs include not only the lungs, the primary organs of gas exchange, but also the series of tubes and passageways that enable air to enter the lungs during inspiration (inhalation) and exit the lungs during expiration (exhalation). The respiratory organs are often subdivided into upper and lower respiratory tracts. Those respiratory organs located outside the thorax constitute the upper respiratory tract. Respiratory organs within the thorax form the lower respiratory tract.

In this exercise, you will study both macroscopic and microscopic aspects of the respiratory organs as well as the basic functions of each organ. The respiratory organs include the nose, pharynx, larynx, trachea, bronchi, and lungs. After you complete the anatomical portion of the lab you will (1) investigate the control of breathing and (2) determine your lung volumes using either nonrecording spirometers or a computer-based spirometer.

The Upper Respiratory Tract

The median section of the head in Figure 22.1 shows the parts of the upper respiratory tract. The **nasal cavity** is divided into left and right chambers by the **nasal septum** (not shown), which is composed of the perpendicular plate of the ethmoid, vomer bone, and associated cartilages. The surface area of the nasal cavity is increased by the **superior, middle,** and **inferior nasal conchae**, which project from the lateral walls. Air enters the nasal cavity via the **nares** (nostrils).

Unit 5: Internal Transport and Gas Exchange

Figure 22.1 Upper respiratory system. (A) Median section; (B) divisions of the pharynx.

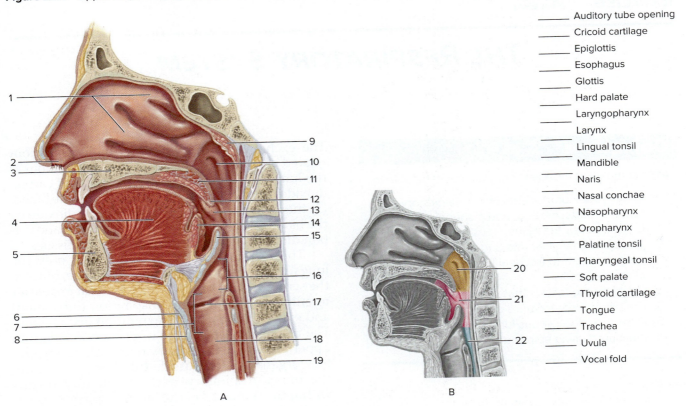

_____ Auditory tube opening
_____ Cricoid cartilage
_____ Epiglottis
_____ Esophagus
_____ Glottis
_____ Hard palate
_____ Laryngopharynx
_____ Larynx
_____ Lingual tonsil
_____ Mandible
_____ Naris
_____ Nasal conchae
_____ Nasopharynx
_____ Oropharynx
_____ Palatine tonsil
_____ Pharyngeal tonsil
_____ Soft palate
_____ Thyroid cartilage
_____ Tongue
_____ Trachea
_____ Uvula
_____ Vocal fold

The **palate** separates the nasal cavity from the oral cavity. The palate consists of a front portion supported by bone, the **hard palate,** and a back **soft palate,** which terminates in a median finger-like projection, the **uvula.**

The **pharynx** consists of three parts. The **oropharynx** lies behind the oral cavity, and the **nasopharynx** is just above it and behind the nasal cavity. Below the oropharynx is the **laryngopharynx,** where air enters the larynx and food enters the esophagus. The openings of the **auditory tubes** (label 10) are visible on the lateral walls of the nasopharynx.

As air passes through the nasal cavity, it is warmed, filtered, and humidified by the mucous membrane lining. Airborne particles are trapped in mucus secreted by goblet cells. The mucus and entrapped particles are moved by beating cilia to the pharynx and swallowed.

From the nasal cavity, air passes through the pharynx, larynx, and trachea on its way to the lungs. The **larynx** has walls formed of an upper **thyroid cartilage** (label 6) and a lower **cricoid cartilage.** Within the larynx are the **vocal folds** (label 17), which vibrate to produce sounds when activated by passing air. The **epiglottis** is a flaplike structure diagonally situated over the **glottis** (label 16), the upper opening to the larynx.

It prevents food from entering the larynx when swallowing. The opening into the **esophagus** (label 19), which carries food to the stomach, lies just behind the larynx.

Clumps of lymphoid tissue compose the tonsils. The **pharyngeal tonsils** (adenoids) (label 9) are located in the roof of the nasopharynx; the **palatine tonsils** (label 13) occur on either side of the oropharynx; and the **lingual tonsils** are on the base of the tongue.

The Lower Respiratory Passages

Figure 22.2 illustrates the lower respiratory structures. The **trachea** extends below the larynx to the center of the thorax, where it branches into the **main** (primary) **bronchi,** which enter the lungs. Within the lungs, the main bronchi divide to form the **lobar** (secondary) **bronchi,** one for each lobe. Continued branching forms the **segmental** (tertiary) **bronchi** and **bronchioles,** which end in a cluster of tiny sacs called **pulmonary alveoli.** The exchange of gases occurs between the air in the pulmonary alveoli and blood in capillaries surrounding the pulmonary alveoli. Cartilaginous "rings" support the walls of both the trachea and bronchi. Bronchioles lack cartilaginous rings but contain smooth muscle in their walls. In an **asthma** attack, contraction of the

Figure 22.2 The respiratory system. (A) Gross anatomy; (B) pulmonary alveoli.

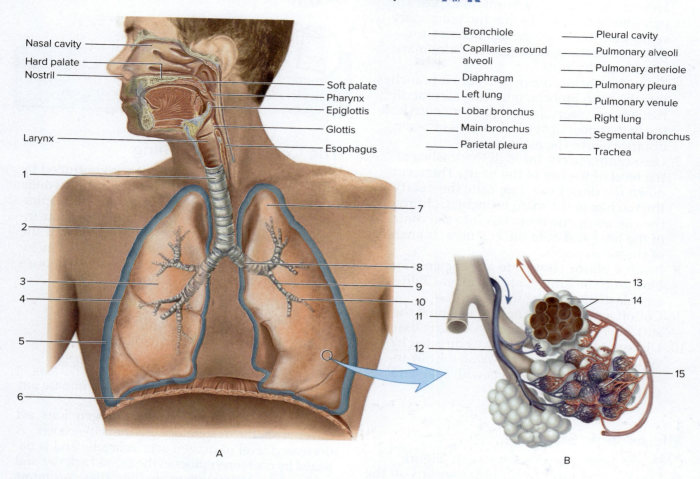

_____ Bronchiole
_____ Capillaries around alveoli
_____ Diaphragm
_____ Left lung
_____ Lobar bronchus
_____ Main bronchus
_____ Parietal pleura
_____ Pleural cavity
_____ Pulmonary alveoli
_____ Pulmonary arteriole
_____ Pulmonary pleura
_____ Pulmonary venule
_____ Right lung
_____ Segmental bronchus
_____ Trachea

muscles and an excessive production of mucus in the air passages restrict the air flow.

Each lung is covered by a tightly adhering membrane, the **visceral pleura** (label 4), and the **parietal pleura** lines the inner wall of the thoracic cavity. Fluid secreted into the **pleural cavity,** the space between these membranes, reduces friction during lung movements. The **diaphragm,** the primary muscle of respiration, separates the thoracic and abdominal cavities.

Assignment

1. Label Figures 22.1 and 22.2.
2. Complete Sections A, B, and C of the laboratory report.
3. Locate the parts of the respiratory system on models available in the laboratory.
4. Examine a corrosion preparation of the bronchial tree and identify the air passages composing it.

Sheep Pluck Dissection

A sheep "pluck" consists of the larynx, trachea, lungs, and heart removed during routine slaughter. The study of a fresh sheep pluck is valuable because its components are similar to those in humans. Although it is as safe to handle as any material from a meat market, use universal precautions. Record your observations on the laboratory report.

1. Lay out the pluck on the dissecting tray. Locate the heart, lungs, trachea, larynx, and diaphragm.
2. Identify the epiglottis and the thyroid and cricoid cartilages of the larynx. Look into the larynx to find the vocal folds.
3. Use scissors to cut through the trachea below the larynx and examine the cut end. Note the shape of the cartilaginous rings and the smooth, slimy inside of the trachea.
4. Note that the lungs are divided into lobes. Are the same number of lobes found in both sheep and human lungs? Feel the surface of a lung. What membrane makes it so smooth?

The Respiratory System

5. Separate or remove the connective tissue around the pulmonary trunk to expose the pulmonary arteries. Locate the pulmonary veins. Can you locate the ligamentum arteriosum extending between the pulmonary trunk and the aorta?
6. Trace the trachea down to where it branches into the main bronchi, removing surrounding tissue as necessary. Note that a bronchus leading to the upper right lobe branches off some distance above the main bronchi.
7. Use scissors to cut through the trachea at the level of the top of the heart. Then, cut down the dorsal side (opposite the heart) of the trachea to the main bronchi. Continue the cut along one bronchus into the center of the lung and note the extensive branching of the air passages.
8. Insert a plastic straw into a bronchiole and blow into the lung. Note the extensibility and elasticity of lung tissue.
9. Cut off a piece of a lung and examine the cut surface. Note the sponginess of lung tissue.
10. Dispose of the sheep pluck as directed by your instructor. **Important:** Wash the tray, instruments, and your hands with soap and water.

Microscopic Study

Compare your observations with Figure 22.3 and make drawings of your observations on the laboratory report.

1. Examine the photomicrographs of the nasal cavity in Figure 22.3. Note the nasal septum, concha, and pseudostratified ciliated epithelium.
2. Examine a prepared slide of trachea. Locate the hyaline cartilage composing cartilaginous rings and the ciliated cells and goblet cells of the epithelial lining.
3. Examine a prepared slide of healthy lung tissue. Note the porous nature of lung tissue. Look for the thin-walled pulmonary alveoli, a bronchiole, and blood vessels. Why are thin pulmonary alveolar walls advantageous?
4. Examine a prepared slide of emphysematous lung tissue. Note that many walls of the pulmonary alveoli have been destroyed, decreasing the respiratory surface of the lung. Emphysema also reduces the elasticity of lung tissue so that it is difficult to force air out of the lungs.
5. Examine a prepared slide of "smoker's lung." Note the breakdown of pulmonary alveoli and the deposits of carbon and tar particles, which can cause cancer.

Assignment
Complete Section E of the laboratory report.

The Control of Breathing

The rhythmic cycle of respiration is controlled by the **respiratory rhythmicity center** in the medulla oblongata. The primary stimulus on the respiratory rhythmicity center for inspiration is an increase in the hydrogen ion concentration of the blood and cerebrospinal fluid. Recall that an increase in carbon dioxide concentration increases the hydrogen ion concentration.

$$CO_2 \qquad H_2O \qquad H_2CO_3$$
carbon dioxide + water $\rightleftharpoons$ carbonic acid

$$H_2CO_3 \qquad H+ \qquad HCO_3^-$$
carbonic acid $\rightleftharpoons$ hydrogen ion + bicarbonate ion

Both carbon dioxide and hydrogen ions act directly on the respiratory rhythmicity center. A decreased level of oxygen acts indirectly and is detected by chemoreceptors in the carotid arteries and aortic arch. Action potentials from these receptors are then transmitted to the respiratory rhythmicity center. Usually, oxygen concentration is of little significance in stimulating breathing.

Hyperventilation

Hyperventilation reduces the concentrations of carbon dioxide and hydrogen ions in the blood so that the respiratory rhythmicity center is depressed and, in turn, the desire to breathe is reduced. Demonstrate this effect as follows. Record your results on the laboratory report.

1. Breathe deeply at the rate of 15 cycles per minute for 1 to 2 minutes. It will be increasingly difficult to breathe. Stop if you *start* to get dizzy.
2. Place a paper bag over your nose and mouth, and breathe into it for about 3 minutes. Note how much easier it is to breathe into the bag. Why?
3. After 5 minutes of resting breathing, measure how long you can hold your breath.
4. Repeat step 1 and determine how long you can hold your breath.

Figure 22.3 Histology of respiratory structures. (A-D): Courtesy of Harold Benson

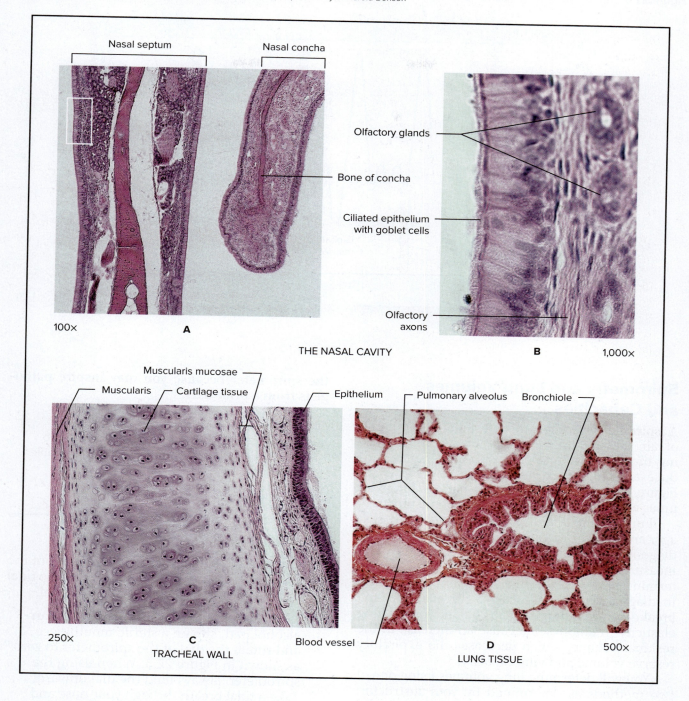

Deglutition Apnea

Obviously, breathing and swallowing food or drink cannot occur simultaneously. A reflex called **deglutition apnea** prevents one from wanting or attempting to breathe while swallowing. Demonstrate this reflex and record your results on the laboratory report.

1. Hold your breath until you have a strong need to breathe.

2. Then, sip water through a straw and note whether the desire to breathe decreases.

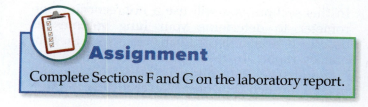

Assignment

Complete Sections F and G on the laboratory report.

The Respiratory System

Figure 22.4 Spirogram of lung volumes and capacities.

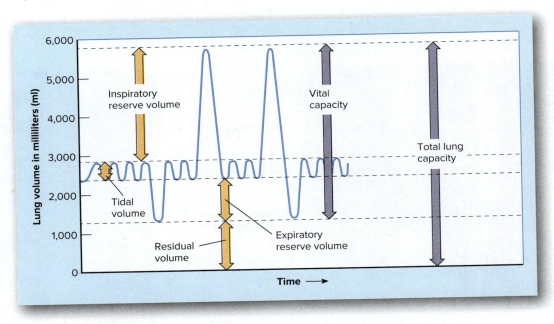

Spirometry and Lung Volumes and Capacities

A **spirometer** is used to determine the volume of air moving into and out of the lungs during breathing. A recording spirometer may produce a **spirogram** similar to the one shown in Figure 22.4, which indicates the various lung volumes and capacities. Note their relationships.

Lung volumes and capacities vary with sex, age, and weight of the subject and are useful in diagnosing certain pulmonary disorders. The measurement of lung volumes is essential to determine how well the lungs are functioning and in diagnosing lung disorders. For example, the breakdown of pulmonary alveoli and loss of lung elasticity in emphysema cause an increase in the residual volume, which decreases the expiratory reserve volume and vital capacity.

You will determine lung volumes using one of two methods as determined by your instructor: Nonrecording spirometers or a computer-based spirometer.

Using Spirometers

In this section, you will use a nonrecording spirometer to determine your lung volumes as described below. Record your results on the laboratory report. In these experiments, you are to *expire* (exhale) *only* through the spirometer and inspire through your nose. **Never inspire through the spirometer because you may inspire pathogens from a prior user.**

Materials
  Spirometers (older models could be referred to as Propper spirometers) with sterile, disposable mouthpieces
  Alcohol pads
  Biohazard bag

Tidal Volume

The amount of air that moves into and out of the lungs during a resting respiratory cycle is the **tidal volume.** Tidal volume averages about 500 ml.

1. Cleanse the stem of a spirometer with an alcohol pad, slip on a sterile mouthpiece, and rotate the dial of the spirometer to zero as shown in Figure 22.5. When using the spirometer, always hold the dial upward.
2. Take a tidal breath through your nose and expire through the spirometer. Repeat for a total of three expirations. *Do not inspire or expire forcibly.*
3. Record the total volume from the spirometer dial. Divide this number by 3 to determine your tidal volume.

Minute Ventilation (MV)

The amount of tidal air that moves into and out of your lungs in 1 minute is your **minute ventilation.**

Figure 22.5 Dial face of spirometer is rotated to zero prior to measuring expirations. J and J Photography

To determine this, count the number of respirations in 1 minute and multiply this number by your tidal volume.

Expiratory Reserve Volume (ERV)

The amount of air that you can expire beyond the tidal volume is your **expiratory reserve volume**. It is usually about 1,100 ml.

1. Set the spirometer dial on 1,000.
2. After two to three tidal expirations through your nose, expire forcibly all of the *additional* air that you can through the spirometer.
3. Subtract 1,000 from the dial reading to determine your expiratory reserve volume.

Vital Capacity (VC)

As shown in Figure 22.4, the **vital capacity** of the lungs is the total of the tidal, expiratory reserve, and inspiratory reserve volumes. Vital capacity varies with age, sex, and size, so you must obtain your predicted normal vital capacity from one of the tables in Appendix A. Vital capacity averages about 4,300 ml but may vary as much as 20% from established norms and still be considered normal.

1. Set the spirometer dial on zero.
2. After inspiring as deeply as possible and expiring completely, take a second deep breath and expire as much air as possible through the spirometer. An even, forced expiration is best.
3. Repeat three times and record the value for each expiration. The values should not vary by more than 100 ml. Divide the total by 3 to determine your vital capacity. Compare your vital capacity with the predicted normal values in Appendix A.
4. Place the mouthpiece in the biohazard bag. Wipe the spirometer with an alcohol pad.

Inspiratory Capacity (IC)

The **inspiratory capacity** is the maximum volume of a deep inspiration after expiring the tidal air. It is usually about 3,000 ml. This volume must be calculated because the spirometer cannot measure inspirations. Determine your inspiratory capacity as follows:

$$IC = VC - ERV$$

Inspiratory Reserve Volume (IRV)

The **inspiratory reserve volume** is the volume of air that can be inspired after filling the lungs with tidal air. Calculate your IRV as follows:

$$IRV = IC - TV$$

Residual Volume (RV)

The **residual volume** is air that cannot be expelled from the lungs—about 1,200 ml. It cannot be determined by usual spirometric methods.

Using Computer-Based Spirometry

The principal advantage of using a computer based spirometer setup is that all of the calculations are made *automatically* from data that are fed into the program. For whichever platform you are using, there will be separate instructions. It is assumed here that you have a basic knowledge of the platform you are using.

In performing these experiments, you will be working with a laboratory partner. While one acts as the subject, breathing into the spirometer, the other person can run the computer and prompt the subject as to the procedures. In general, you will measure the same values in a similar way as in the previous activity. In addition, a computer-based program will allow you to measure various FEV_T values.

To determine the *forced expiratory volume* (FEV_T) of an individual, the subject takes a deep breath and expels it as fast as possible. An individual with no respiratory impairment should be able to expire 95% of his or her vital capacity within 3 seconds; however, it is the percentage of vital capacity that is expelled within the first second that is of paramount importance. **An individual with no pulmonary impairment should be able to expel 75% of his or her vital capacity within 1 second.** Individuals with emphysema and asthma will have a much lower percentage due to air entrapment.

The Respiratory System

In doing these two tests, both the **volume** and **rate** of air flow will be measured. It is for this reason that the subject must exhale as completely and rapidly as possible. The subject must keep trying to squeeze out every little bit of air until the software stops and graphs the values. Refer to the directions for your specific equipment to complete the FEV_T.

Assignment

Complete the laboratory report and attach your spirograms to the laboratory report.

Exercise 23

THE DIGESTIVE SYSTEM

Objectives

After completing this exercise, you should be able to

1. Identify the parts of the digestive system on charts and models and describe the function of each.
2. Identify tissues studied microscopically.
3. Describe the role of digestive enzymes in the process of digestion.
4. Describe the effect of temperature and pH on the action of salivary amylase.

Materials
Torso model with removable organs
Prepared slides of
    Vallate papillae with taste buds
    Stomach wall, x.s.
    Ileum wall, x.s.
    Jejunum wall, x.s.
    Liver

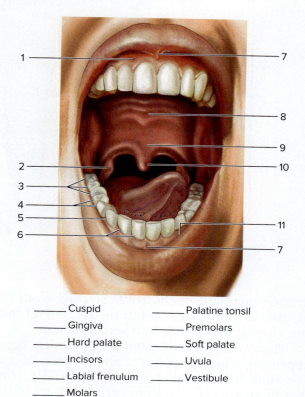

Figure 23.1 The oral cavity.

_____ Cuspid
_____ Gingiva
_____ Hard palate
_____ Incisors
_____ Labial frenulum
_____ Molars
_____ Palatine tonsil
_____ Premolars
_____ Soft palate
_____ Uvula
_____ Vestibule

The digestive system consists of the alimentary canal and the associated glands. The **alimentary canal** is the tube through which food passes, and it consists of the mouth, pharynx, esophagus, stomach, small intestine, and large intestine. The associated glands are the salivary glands, liver, gallbladder, and pancreas. As you study the descriptions that follow, locate the structures on Figures 23.1 through 23.5.

The Oral Cavity

The teeth, tongue, and salivary glands play significant roles in the digestive functions of the oral cavity. When food is taken into the mouth and chewed, it is mixed with saliva secreted by the salivary glands. The food mass is then pushed backward by the tongue into the **oropharynx,** where the swallowing reflex is initiated. The oropharynx is the portion of the pharynx that is behind the oral cavity.

The basic structure of the oral cavity is shown in Figure 23.1. Note that the cheeks have been cut and the lips retracted to expose relevant structures. A mucous membrane lines the lips and cheeks. It forms thickened folds, the **labial frenula,** at the midline on the inner surface of the lips. The **oral vestibule** (label 11) is the space between the cheeks or lips and the teeth of each jaw.

The roof of the mouth is formed by the **hard palate,** which is reinforced by bone, and the **soft palate.** The **uvula** is the fingerlike downward projection from the back of the soft palate. Note that portions of the **palatine tonsils** are visible on the sides of the oropharynx.

The Teeth

All 20 deciduous teeth usually have erupted by the time a child is two years old. Five teeth are present

Unit 6: Nutrient Absorption and Water Balance

Figure 23.2 Tooth anatomy. **APR**

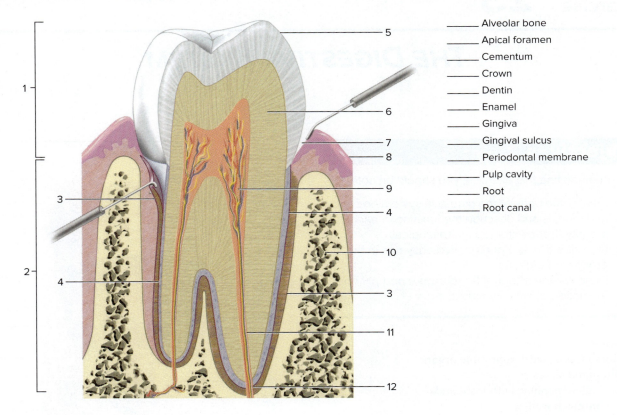

_____ Alveolar bone
_____ Apical foramen
_____ Cementum
_____ Crown
_____ Dentin
_____ Enamel
_____ Gingiva
_____ Gingival sulcus
_____ Periodontal membrane
_____ Pulp cavity
_____ Root
_____ Root canal

in each quadrant of the jaws. Starting from the median line, they are **central incisor, lateral incisor, cuspid, first molar,** and **second molar.**

There are 32 permanent teeth, which begin appearing when a child is about six years of age. The adult dentition is shown in Figure 23.1. Note that there are eight teeth in each quadrant of the jaws. Starting at the median line, they are **central incisor, lateral incisor, cuspid** (canine), **first premolar** (bicuspid), **second premolar** (bicuspid), **first molar, second molar,** and **third molar.** The chisel-like incisors are effective in biting off bits of food. The pointed cuspids are used to puncture and tear food. Premolars and molars have low, rounded cusps suitable for grinding food.

Figure 23.2 illustrates the basic anatomy of a tooth. Note the two major subdivisions of a tooth: a **root** that is embedded in alveolar bone, and a **crown** that is not embedded in bone. The **neck** of the tooth is at the junction of the crown and root.

Four substances compose a tooth. **Enamel** is the hardest substance in the body and forms an ideal chewing surface for the crown. **Dentin** forms most of a tooth and lies under the enamel in the crown. It is similar to bone tissue in composition and hardness. **Cementum** (label 4) is modified bone tissue that covers the root and attaches the tooth to the **periodontal membrane.** Fibers of the periodontal membrane penetrate into the bone, holding the tooth firmly in place.

The **pulp cavity** occupies the central part of a tooth. It extends down into the roots in the **root canals.** Pulp consists of loose connective tissues and contains blood vessels, lymphatic vessels, and nerves that enter via the **apical foramina** (label 12) of the roots.

The mucosa covering the maxillae, mandible, and the neck of each tooth is the **gingiva.** The upper margin of the gingiva is pulled away from the tooth in Figure 23.2 to reveal the **gingival sulcus,** a common site for bacterial putrefaction that may cause periodontal disease.

The Tongue

The tongue consists of skeletal muscle covered by a mucous membrane. The **root** (label 1 in Figure 23.3) of the tongue is covered by the **lingual tonsils.** Note in Figure 23.3 the **epiglottis** extending upward near the root and the **palatine tonsils** located laterally in the oropharynx. The **body of the tongue** extends from the root to the tip. The **lingual frenulum** (not shown) extends between the lower surface of the tongue and the floor of the oral cavity.

The upper surface of the tongue is covered by tiny projections called **papillae.** The enlarged cutout in Figure 23.3 shows the different types. **Filiform papillae** have tapered points, are most numerous, and provide a rough texture for handling the food. **Fungiform papillae** (label 7) are low, rounded projections scattered over the surface. Eight to twelve large **vallate papillae** occur in a V-shaped pattern on the back of the body. Both fungiform and vallate papillae contain **taste buds,** which are shown in the circular furrow of a vallate papilla.

The Salivary Glands

Saliva is secreted from three pairs of salivary glands as shown in Figure 23.4. The **parotid gland**

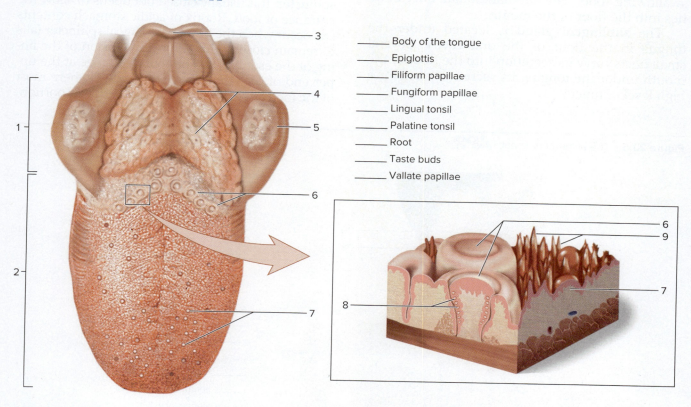

Figure 23.3 Tongue anatomy.

_____ Body of the tongue
_____ Epiglottis
_____ Filiform papillae
_____ Fungiform papillae
_____ Lingual tonsil
_____ Palatine tonsil
_____ Root
_____ Taste buds
_____ Vallate papillae

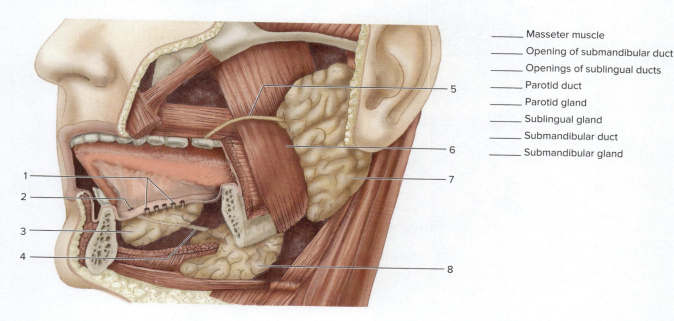

Figure 23.4 Major salivary glands.

_____ Masseter muscle
_____ Opening of submandibular duct
_____ Openings of sublingual ducts
_____ Parotid duct
_____ Parotid gland
_____ Sublingual gland
_____ Submandibular duct
_____ Submandibular gland

The Digestive System

is located in front of the ear and on top of the masseter muscle. Its watery secretion contains salivary amylase, which catalyzes the digestion of starch, and enters the mouth through a duct that opens near the upper second molar.

The **submandibular gland** lies just beneath the ramus of the mandible. Its secretion contains *mucin,* which aids in holding food together and imparts a slippery nature to the saliva, facilitating swallowing food. The submandibular duct empties into the floor of the mouth.

The **sublingual gland** is located under the tongue in the floor of the oral cavity. Several small ducts carry its secretion into the floor of the mouth under the tongue. Its secretion contains a high level of mucin.

The Esophagus and Stomach

Refer to Figure 23.5 as you study the discussion of the alimentary canal. When swallowed, food passes into the **esophagus,** a long tube extending from the pharynx to the stomach. Wavelike **peristaltic contractions** begin at the top of the esophagus and carry the food to the **stomach.**

The upper opening of the stomach is guarded by a sphincter muscle, the **lower esophageal (cardiac) sphincter.** It is usually closed but opens to allow the entrance of food. Reflux of acidic stomach contents can occur when the lower esophageal sphincter fails to remain closed. This can cause erosion of the lining of the esophagus. The rounded bulge at the upper end of the stomach is the **fundus,** where most of the food to be digested is held. The lower portion,

Figure 23.5 The alimentary canal.

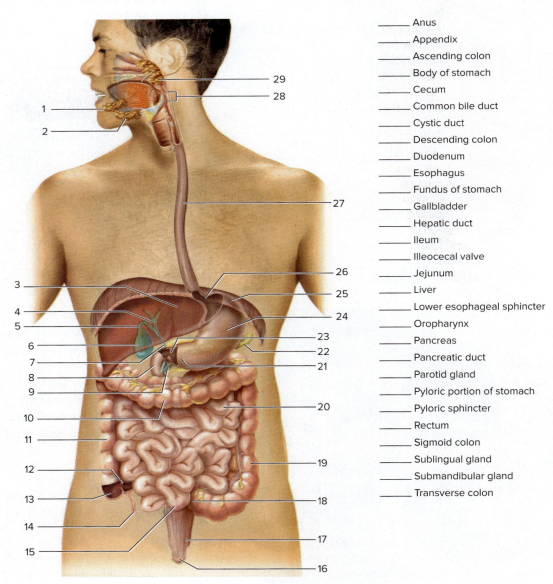

_____ Anus
_____ Appendix
_____ Ascending colon
_____ Body of stomach
_____ Cecum
_____ Common bile duct
_____ Cystic duct
_____ Descending colon
_____ Duodenum
_____ Esophagus
_____ Fundus of stomach
_____ Gallbladder
_____ Hepatic duct
_____ Ileum
_____ Illeocecal valve
_____ Jejunum
_____ Liver
_____ Lower esophageal sphincter
_____ Oropharynx
_____ Pancreas
_____ Pancreatic duct
_____ Parotid gland
_____ Pyloric portion of stomach
_____ Pyloric sphincter
_____ Rectum
_____ Sigmoid colon
_____ Sublingual gland
_____ Submandibular gland
_____ Transverse colon

the **pyloric region,** is smaller in diameter and is the primary site of digestion in the stomach. The stomach region between the pylorus and the fundus is called the **body.**

Gastric juice, secreted by the stomach mucosa, is highly acidic and contains enzymes to catalyze the digestion of proteins and milk fats. Food is mixed with gastric juice and is converted into a semiliquid substance called **chyme.** The chyme enters the small intestine through another sphincter, the **pyloric sphincter** (label 23) of the stomach.

The Small Intestine

The small intestine is about 23-ft long and consists of three parts: the duodenum, jejunum, and ileum. In Figure 23.5, the small intestine has been shortened for clarity. The first 10 to 12 in is the **duodenum** (label 8), which receives bile that emulsifies fats from the **liver** and **gallbladder** and digestive secretions from the **pancreas.**

The **hepatic duct** from the liver joins the **cystic duct** from the gallbladder to form the **common bile duct.** Pancreatic juice is carried from the pancreas by the **pancreatic duct** (label 9), which empties into the common bile duct near the duodenum. This union forms a short, dilated duct, the **hepatopancreatic ampulla** (not labeled), that opens into the duodenum. The *hepatopancreatic sphincter,* located at the distal end of the ampulla, controls the passage of bile and pancreatic juice into the duodenum. When no food is in the duodenum, the sphincter constricts so that bile from the liver is forced up the cystic duct to the gallbladder for temporary storage. When food is present in the duodenum, the sphincter opens, allowing the passage of bile, and the gallbladder contracts, expelling stored bile into the cystic and common bile ducts. **Gallstones** may form in the gallbladder if the bile contains too much cholesterol or is too concentrated.

The middle section of the small intestine is the **jejunum** (label 20), which is 7- to 8-ft long; the remainder is the **ileum.**

In the small intestine, chyme is mixed with bile and pancreatic juice, which enter via the common bile duct, and with intestinal juice that is secreted by the intestinal mucosa. Digestion of food and absorption of the end products of digestion are completed in the small intestine. Absorption occurs through **villi,** tiny, fingerlike projections of the intestinal mucosa. The villi are extremely numerous and greatly increase the absorptive surface of the small intestine.

The Large Intestine

Indigestible food and water pass from the small intestine into the **large intestine** through the **ileocecal valve** (label 12 in Figure 23.5). The large intestine is about 3 1/2 to 4 feet in length. There are three continuous sections of the large intestine: cecum, colon, and rectum. The ileum opens into the proximal portion of the large intestine, the **ascending colon,** near its junction with the pouch-like **cecum** from which the narrow **appendix** extends. The ascending colon runs up the right side of the abdominal cavity and is continuous with the **transverse colon,** which extends across the abdominal cavity below the diaphragm. The **descending colon** runs downward on the left side of the abdomen and continues as the S-shaped **sigmoid colon.** The final 5 to 6 in is the **rectum,** which terminates with an opening, the **anus.**

The principal functions of the large intestine are the decomposition of undigested food by resident bacteria, absorption of certain vitamins formed by the bacteria, and reabsorption of water. These actions bring about the formation of the feces, which are passed from the body by defecation.

Assignment

1. Label Figures 23.1 through 23.5.
2. Complete Sections A through C of the laboratory report.
3. Locate the digestive organs on the human torso model. Note their relative positions.

Microscopic Study

Compare your observations of the following prepared slides with Figures 23.6 through 23.9.

1. *Vallate papillae with taste buds.* Note the structure of the papillae and the location of the taste buds. Only substances in solution can reach the taste buds.
2. *Stomach wall.* Locate the columnar epithelium and gastric pits. Try to find chief and parietal cells near the base of a gastric pit. Chief cells produce pepsinogen, a precursor of pepsin, and parietal cells produce hydrochloric acid.
3. *Ileum wall.* Note how the villi increase the surface area of the mucosa. Try to find a **lacteal,** a small lymph capillary, in a villus. Colloidal fats are absorbed into lacteals.
4. *Jejunum wall.* Locate the four layers of the intestinal wall: serosa (visceral peritoneum), muscularis, submucosa, and mucosa. Observe the columnar epithelium with goblet cells on the surface of the villi. Nutrients are absorbed in the villi.
5. *Liver.* Locate a lobule with its **central vein** and **sinusoids** radiating from it between

The Digestive System

Figure 23.6 Vallate papillae and taste buds. (A–B): Courtesy of Harold Benson

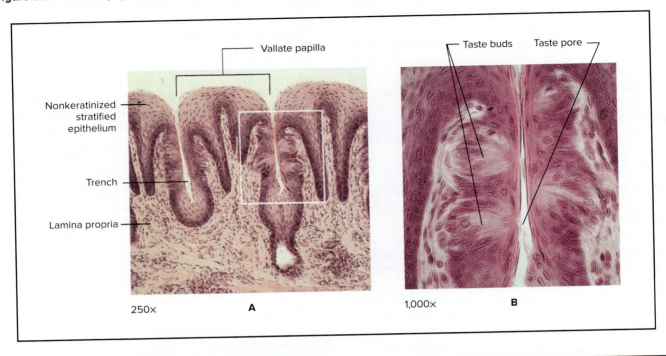

Figure 23.7 Stomach mucosa. (A–B): Courtesy of Harold Benson

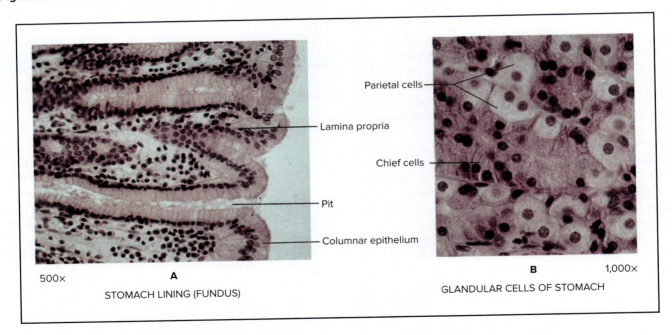

rows of liver cells. Find branches of the **hepatic portal vein, hepatic artery,** and **bile duct** at a "corner" of the lobule. On its way to the central vein, blood from the portal vein and hepatic artery flows through the sinusoids, bathing the liver cells. All central veins merge to form the hepatic vein. Bile flows from the liver cells through small bile channels (canaliculi) into the bile duct.

Assignment

Make drawings of your observations in Section D of the laboratory report.

Figure 23.8 The jejunum and ileum. (A–B): Courtesy of Harold Benson

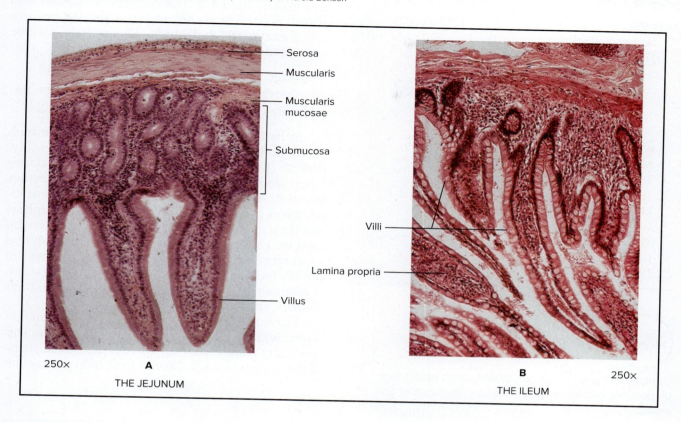

Figure 23.9 Liver tissue. (A–B): Courtesy of Harold Benson

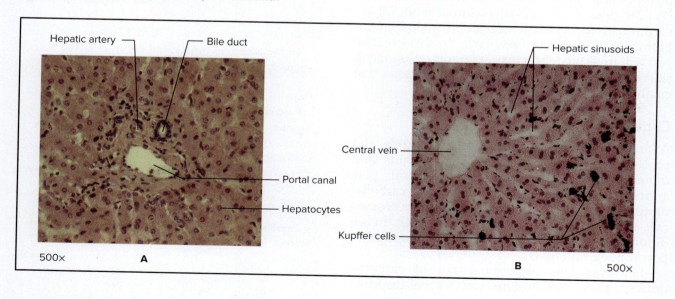

The Digestive System

Digestion

Materials
- Beaker, 250 ml
- Dropping bottles of
  - Iodine (IKI) solution
  - 10% saliva solution
  - pH 5 buffer solution
  - pH 7 buffer solution
  - pH 9 buffer solution
- 0.1% soluble starch solution in beaker
- Medicine droppers
- Test tubes, 13 mm diam. × 100 mm long
- Test tube rack, Wasserman type
- Electric hot plate
- Water baths at 20°C and 37°C
- Thermometers, Celsius
- Glass-marking pen
- Protective disposable gloves

Before You Proceed

Consult with your instructor about using protective disposable gloves when performing portions of this exercise.

Organic nutrients are required for the metabolic activities of body cells. The processes involved in providing organic nutrients to body cells are (1) ingestion of food, (2) digestion of food, (3) absorption of nutrients into the blood, and (4) distribution of nutrients to body cells. In this exercise you will investigate the process of digestion using starch as a food and amylase as the enzyme that catalyzes starch digestion.

Digestive Enzymes

A healthy diet contains carbohydrates, lipids, and proteins plus small quantities of vitamins and minerals. Although vitamins and minerals play a vital role in metabolic processes, the carbohydrates, lipids, and proteins provide the necessary raw materials for tissue repair and growth. In addition, they are the sources of energy required to power body functions.

However, the large molecules of carbohydrates, lipids, and proteins cannot be absorbed by the mucosa of the digestive tract and enter the blood. They must be reduced to absorbable **nutrients** by digestion.

Digestion is a complex process that involves both mechanical and chemical components. Food is mechanically broken down into smaller pieces by mastication (chewing), which increases the surface area of the food. The conversion of food into absorbable nutrients occurs by a chemical process called **enzymatic hydrolysis**. This process may be expressed by the following:

$$\text{large nonabsorbable molecules} \xrightarrow[(+H_2O)]{\textit{(digestive enzymes)}} \text{small absorbable molecules}$$

In hydrolysis, water combines with large molecules and splits them into smaller molecules. Hydrolysis without the catalytic action of enzymes is too slow to sustain life. The **digestive enzymes** found in saliva and gastric, pancreatic, and intestinal juices greatly increase the rate of the hydrolysis reactions and enable us to derive nutrients from the food we eat. Relatively few enzymes can catalyze a great number of reactions because the enzymes are not altered in the reaction and are simply recycled. A summary of the major digestive enzymes, their sources, and their actions are shown in Table 23.1.

Enzymes are proteins. Each type of enzyme interacts with a specific type of substrate molecule. This specificity results from the three-dimensional shape of the enzyme, which allows it to fit onto a particular

TABLE 23.1
Summary of the Major Digestive Enzymes

Enzyme	Digestive Action
Saliva	
Amylase	Starch to maltose
Gastric Juice	
Lipase	Milk fat to fatty acids** and monoglycerides**
Pepsin	Proteins to peptides
Pancreatic Juice	
Amylase	Starch to maltose
Trypsin	Proteins to peptides
Chymotrypsin	Proteins to peptides
Carboxypeptidase	Peptides to peptides and amino acids
Lipase	Lipids to fatty acids** and monoglycerides**
Intestinal Juice	
Maltase	Maltose to glucose*
Sucrase	Sucrose to glucose* and fructose*
Lactase	Lactose to glucose* and galactose*
Lipase	Lipids to fatty acids** and monoglycerides**
Peptidase	Peptides to amino acids***

*End products of carbohydrate digestion
**End products of lipid digestion
***End products of protein digestion

type of substrate molecule. The shape of an enzyme is largely determined by hydrogen bonds that form between certain amino acids in the molecule. Temperature and pH changes may alter the hydrogen bonding and change the shape of the enzyme so that it is no longer able to fit onto the substrate molecule. When this happens, the enzyme is inactivated. Most enzymes function optimally within rather narrow temperature and pH ranges.

Assignment

Complete Sections E, F, and G of the laboratory report.

Starch Digestion Experiments

The enzyme **amylase** catalyzes the hydrolysis of starch, a polysaccharide, into maltose, a disaccharide. Starch digestion begins in the mouth through the action of salivary amylase and is completed in the small intestine by the action of pancreatic amylase.

In the experiments that follow, you will investigate the effect of temperature and pH on the digestive action of amylase on starch dissolved in an aqueous solution.

The experiments are best performed by students working in pairs. If time is limited, it may be necessary to divide the class into three groups, with each performing part of the experiments as shown in Table 23.2. Results are to be shared with the entire class.

Solutions will be provided in dropping bottles. You will dispense the solutions with medicine droppers in quantities of "drops" or "droppers." A dropper means *one dropper full of solution,* approximately 1 ml. You must maintain cleanliness and use good laboratory procedures in dispensing the solutions in order to achieve good results. Contamination must be avoided. Do not allow a dropper to touch any solution except the one it is used to dispense. You can avoid problems in performing the experiments if you understand the procedures thoroughly and establish a division of labor among your partners before you start.

Amylase Preparation

Prior to the laboratory session, your instructor prepared either a 10% saliva solution or a 1% commercial amylase solution. The directions that follow assume the use of a 10% saliva solution. Glassware in contact with saliva is to be placed in a 0.25% Amphyl solution after use.

Controls

In the experiments, the activity of the amylase will be determined by using a color test: the iodine (IKI) test for starch. Prepare a set of control tubes as follows to compare with your experimental results:

1. Label two clean test tubes 1 and 2, and add to them the ingredients as follows:
   Tube 1: one dropper starch, two drops iodine solution
   Tube 2: one dropper distilled water, two drops iodine solution
2. Your results are interpreted as follows:
   Tube 1: A blue-black coloration is positive for starch.
   Tube 2: An amber coloration is negative for starch.

Effect of Temperature

In this section, you will compare the activity of amylase at 20°C (room temperature) and 37°C (body temperature). A water bath is provided at each temperature. The general protocol for the experiment is shown in Figure 23.10. Proceed as follows for each temperature to be tested:

Figure 23.10 Amylase activity at 20°C and 37°C.

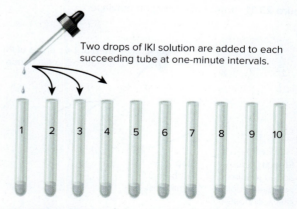

Ten Tubes: Each contains 1 ml of starch solution and 1 drop of 10% saliva. Keep in water baths at 20°C and 37°C.

TABLE 23.2
Division of Labor

Group	Assignment
A (1/3 of class)	Controls, 20°C, boiling
B (1/3 of class)	Controls, 37°C, boiling
C (1/3 of class)	Controls, pH effects

The Digestive System

1. Label 10 test tubes 1 through 10 and arrange them sequentially in the test tube rack.
2. Add one dropper of starch solution to each of the 10 test tubes.
3. Place the rack of test tubes and the dropping bottle of saliva in the water bath. Allow 5 minutes for temperature equilibration.
4. *Record the time,* and place one drop of saliva solution into each tube starting with tube 1. Be sure that each drop lands directly in the starch solution and does not run down the side of the tube. Mix by agitation.
5. *Exactly 1 minute* after tube 1 received the saliva solution, add two drops of IKI to tube 1.
6. *One minute later,* add two drops of IKI to tube 2. Continue to add IKI sequentially to the tubes at 1-minute intervals until all 10 tubes have received IKI solution.
7. Remove the tubes from the water bath after adding the iodine solution and compare the color of the solutions with the controls to determine the time required to digest the starch. Record your results on the laboratory report.
8. Discard the tube contents in the sink and place the tubes in the pan containing Amphyl solution.

Effect of Boiling

In this section, you will compare the activity of amylase that has been heated to boiling with unheated amylase. Figure 23.11 depicts the general procedure.

1. Label three test tubes 1, 2, and 3.
2. Add five drops of saliva solution to tubes 1 and 2, and five drops of distilled water to tube 3.
3. Place about 100 ml of water in a beaker and heat it to boiling on a hot plate. Place tube 1 in the boiling water bath for 3 minutes.
4. Add one dropper of starch solution to each tube and mix by agitation. Place the tubes in a water bath at 37°C for 5 minutes.
5. Remove the tubes from the water bath and add two drops of IKI to each tube. Record the results on the laboratory report.
6. Discard the tube contents in the sink and place the tubes in the pan of Amphyl solution.

Effect of pH

In this section, you will determine the effect of three different hydrogen ion concentrations on amylase activity. The general procedure is shown in Figure 23.12.

1. Label nine test tubes 1 through 9. Arrange them in sequence in the test tube rack and write the pH designation on each tube as shown in Figure 23.12. Note that there are three tubes at each pH.
2. Add one dropper of buffer solution as follows:
   pH 5 buffer to tubes 1, 4, and 7
   pH 7 buffer to tubes 2, 5, and 8
   pH 9 buffer to tubes 3, 6, and 9
3. Add two drops of saliva solution to each tube, mix by agitation, and place the rack with the tubes in the water bath at 37°C. Also, place the dropping bottle of starch in the water bath. Allow 5 minutes for temperature equilibration.
4. *Record the time* and, starting with tube 1, add one dropper of starch solution to each tube. Mix by agitation.

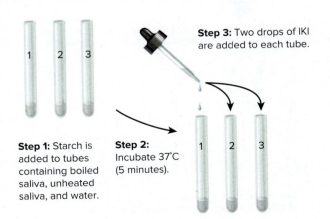

Figure 23.11 Effect of boiling on amylase.

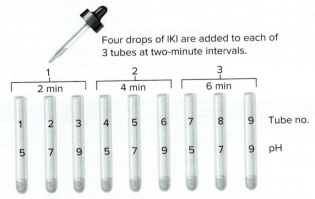

Figure 23.12 Effect of pH on amylase activity.

Nine Tubes: Each contains 1-ml buffered solution, two drops of 10% saliva, and 1 ml of starch solution in 37°C water bath.

5. *Two minutes* after the recorded time, add four drops of IKI to tubes 1, 2, and 3. *Four minutes* after the recorded time, add four drops of IKI to tubes 4, 5, and 6. *Six minutes* after the recorded time, add four drops of IKI to tubes 7, 8, and 9.
6. Remove the tubes from the water bath after adding the iodine solution and compare them with the controls. Record the results on the laboratory report.
7. Discard the contents and place the tubes in the Amphyl solution. Wash your workstation with Amphyl solution.

Assignment

Complete the laboratory report.

Exercise 24

THE URINARY SYSTEM

Objectives

After completing this exercise, you should be able to

1. Identify the components of the urinary system on charts or models and describe their functions.
2. Identify the parts of the kidney on charts, models, or specimens and describe their functions.
3. Identify a glomerulus and glomerular capsule when viewed microscopically.
4. Identify the typical components and characteristics of urine.
5. List the common abnormal components and characteristics of urine, and identify possible clinical conditions associated with them.
6. Perform a simple urinalysis.

Materials

Models of the urinary system and a kidney
Sheep kidneys, fresh or preserved
Sheep kidneys, triple injected, sectioned
Dissecting kits and trays
Long, sharp knife
Prepared slides of renal cortex and renal medulla
Protective disposable gloves
Urine "unknowns"
10SG-Multistix reagent strips and analysis charts (Ames)
Urine collection cups, plastic
Test tubes and racks
Amphyl, 0.25%
Biohazard bag
Glass-marking pen
Protective disposable gloves

Before You Proceed

Consult with your instructor about using protective disposable gloves when performing portions of this exercise.

Organs of the Urinary System

The components of the urinary system are shown in Figure 24.1: two kidneys, two ureters, the urinary bladder, and the urethra. The **kidneys** are

Figure 24.1 The urinary system.

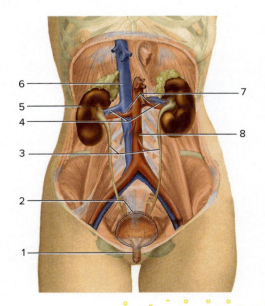

_____ Abdominal aorta
_____ Inferior vena cava
_____ Kidney
_____ Renal arteries
_____ Renal veins
_____ Ureters
_____ Urethra
_____ Urinary bladder

Unit 6: Nutrient Absorption and Water Balance

bean-shaped, reddish brown organs located on either side of the vertebral column and behind the parietal peritoneum (retroperitoneal). Each kidney receives blood from a **renal artery,** which branches from the **abdominal aorta.** Blood leaves each kidney through a **renal vein,** which empties into the **inferior vena cava.**

Urine, formed by the kidneys, is carried from each kidney to the **urinary bladder** through a slender tube, a **ureter.** Peristaltic contractions of the ureter wall propel the urine to the bladder. Each ureter originates as a funnel-like **renal pelvis** in the kidney and descends parallel to the vertebral column and retroperitoneally. The lower end enters the back of the urinary bladder.

Urine is temporarily stored in the distensible urinary bladder and then voided from the bladder via a short tube, the **urethra.** The male urethra is about 20 cm in length; the female urethra is approximately 4 cm in length. *Cystitis,* inflammation of the urinary bladder, is more common in females than males because the shorter length of the female urethra provides an easier entrance for pathogens.

The passage of urine from the bladder is called **micturition** and is controlled by two sphincter muscles in males. The **internal urethral sphincter** is located at the junction of the bladder and urethra, before the prostate. It is formed of smooth muscle and is under autonomic control. The **external urethral sphincter** is located surrounding the urethra about 2 cm from the bladder, after the prostate. It consists of skeletal muscle and is under voluntary control. In females, only the external urethral sphincter exists surrounding the urethra.

When about 300 ml of urine has accumulated in the urinary bladder, the stretching of the bladder walls initiates an urge to urinate and a subconscious reflex, which causes the walls to contract. This contraction begins to move urine into the urethra creating a sensation of urgency. When the external urethral sphincter is consciously relaxed, micturition occurs.

Assignment

1. Label Figure 24.1.
2. Locate the parts of the urinary system on the model and note their relationships to the adjacent structures.

Kidney Anatomy

Figure 24.2 shows the structure of a kidney in frontal section. The kidney is enveloped in a thin **fibrous capsule** (label 3) and encased in a thick

Figure 24.2 Anatomy of the kidney.

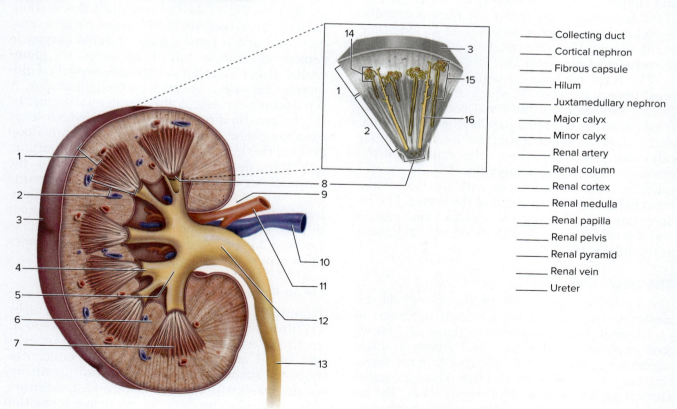

_____ Collecting duct
_____ Cortical nephron
_____ Fibrous capsule
_____ Hilum
_____ Juxtamedullary nephron
_____ Major calyx
_____ Minor calyx
_____ Renal artery
_____ Renal column
_____ Renal cortex
_____ Renal medulla
_____ Renal papilla
_____ Renal pelvis
_____ Renal pyramid
_____ Renal vein
_____ Ureter

Figure 24.3 The nephron.

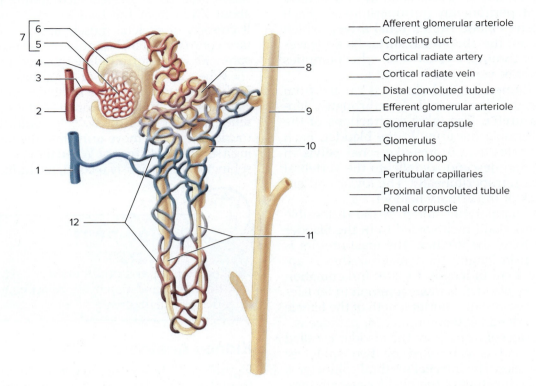

_____ Afferent glomerular arteriole
_____ Collecting duct
_____ Cortical radiate artery
_____ Cortical radiate vein
_____ Distal convoluted tubule
_____ Efferent glomerular arteriole
_____ Glomerular capsule
_____ Glomerulus
_____ Nephron loop
_____ Peritubular capillaries
_____ Proximal convoluted tubule
_____ Renal corpuscle

layer of fat (not shown), which provides protection and support. The **renal cortex,** the outer portion of the kidney, lies just under the fibrous capsule. Its reddish brown color results from an extensive blood supply. The inner portion of the kidney is the **renal medulla** (label 2).

The renal medulla contains the cone-shaped **renal pyramids** (label 7), which are separated by inward extensions of the renal cortex known as **renal columns.** The **renal papilla,** the tip of a renal pyramid, projects into a short tube called a **minor calyx** (label 4). Each minor calyx opens into the centrally located **major calyx** (label 5). The major calyx, in turn, is continuous with the funnel-like **renal pelvis** at the upper end of the **ureter.** The indentation in the kidney where renal blood vessels and the ureter are joined to the kidney is known as the **hilum** (label 9).

The Nephrons

The basic functional unit of the kidney is the **nephron.** There are about a million nephrons in each kidney. The enlargement in Figure 24.2 shows four nephrons, and Figure 24.3 shows a single nephron in greater detail. About 80% of the nephrons are located in the renal cortex. These are called **cortical nephrons** (label 14 in Figure 24.2). The remaining nephrons, the **juxtamedullary nephrons,** are located partially in the renal cortex and partially in the renal medulla.

Refer to Figure 24.3 as you study this section. Each nephron consists of two major parts: a renal corpuscle and a renal tubule. A **renal corpuscle** consists of an inner tuft of capillaries, the **glomerulus** (label 5), and an outer, double-walled **glomerular capsule** that envelops the glomerulus. A **renal tubule** consists of three sequential segments: (1) the **proximal convoluted tubule,** which leads from the glomerular capsule; (2) the **nephron loop** (loop of Henle)—the downward U-shaped portion; and (3) the **distal convoluted tubule** (label 10)—the terminal segment that empties into a collecting duct. Several tubules empty into a single **collecting duct.** There are thousands of collecting ducts in each renal pyramid. Their presence produces the faint lines extending from the base to the apex of each pyramid.

Urine Formation

Nephrons form urine by three processes: (1) filtration, (2) reabsorption, and (3) secretion. In this way, water and essential substances in the blood are conserved while the concentrations of surplus substances, including **nitrogenous wastes** (urea, uric acid, and creatinine), are reduced. The process of urine formation

Nutrient Absorption and Water Balance

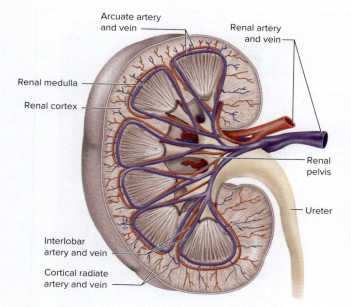

Figure 24.4 Renal circulation.

TABLE 24.1
Quantity of Selected Substances in a 24-Hour Period in (1) Blood Flowing Through the Kidneys, (2) Glomerular Filtrate Passing Through the Glomerular Capsule, and (3) Urine

Substance	Blood	Glomerular Filtrate	Urine
Total volume	806 l	180 l	1 l
Proteins	7,500 g	10 g	0
Chloride ions	3,180 g	667 g	7 g
Potassium ions	170 g	36 g	3 g
Sodium ions	2,924 g	612 g	5 g
Glucose	860 g	180 g	0
Creatinine	8.6 g	1.5 g	1.5 g
Urea	215 g	45 g	25 g
Uric acid	43 g	9 g	0.8 g

maintains the healthy concentration of substances in blood.

The blood supply to the kidney is shown in Figure 24.4. Note that the arteries and veins are essentially parallel to each other. The **renal artery** branches to form **interlobar arteries** that extend through the renal columns to form **arcuate arteries** that run along the base of the renal pyramids. The arcuate arteries give off small and numerous branches, the **cortical radiate arteries** (label 2 in Figure 24.3), which permeate the renal cortex and give off extremely numerous branches, the afferent glomerular arterioles. Each **afferent glomerular arteriole** (label 3 in Figure 24.3) carries blood to a **glomerulus**. An **efferent glomerular arteriole** (label 4 in Figure 24.3) carries blood from a glomerulus to the **peritubular capillaries** (label 12 in Figure 24.3) that enmesh the renal tubule. From the capillaries, blood drains into an **cortical radiate vein,** which opens into an **arcuate vein.** Arcuate veins empty into **interlobar** veins, which carry blood to the **renal vein.**

The efferent glomerular arteriole has a smaller diameter than the afferent glomerular arteriole. This difference elevates the blood pressure within the glomerulus and forces the **glomerular filtrate,** a dilute fluid derived from blood, into the glomerular capsule. About 600 ml of blood flows through the glomeruli each minute: 860 l in 24 hours. The kidneys form about 125 ml of glomerular filtrate per minute: 180 l per day. The filtrate consists of all substances present in the blood except the formed elements. However, very few plasma proteins enter the glomerular filtrate because their molecules are too large.

As the filtrate passes through the renal tubule, various substances are reabsorbed into the peritubular capillaries. A few substances are secreted from the capillaries into the glomerular filtrate. This selective reabsorption and secretion of substances by the renal tubule plays a major role in maintaining the constancy of the body fluids.

The remaining glomerular filtrate passes into a collecting duct, where it may be further concentrated or diluted by the reabsorption of water or mineral ions. The glomerular filtrate flows from the renal papilla into a minor calyx and on into the major calyx, renal pelvis, and ureter. When it reaches the renal pelvis, it is called urine. Table 24.1 compares the quantity of selected substances in blood, tubular fluid, and urine.

Assignment

1. Label Figures 24.2 and 24.3 and complete Section A of the laboratory report.
2. Examine a sectioned, triple-injected kidney under a demonstration dissecting microscope. Note the many renal corpuscles. Can you distinguish the glomerulus, glomerular capsule, and tubules?
3. Examine a prepared slide of renal cortex and renal medulla. Compare your observations with Figure 24.5 and locate the labeled structures. Note that the walls of the glomerular capsule and the renal tubule are only one cell thick. Make diagrams of your observations in Section B of the laboratory report.
4. Complete Section C of the laboratory report.

Figure 24.5   Kidney histology. (A–E): Courtesy of Harold Benson

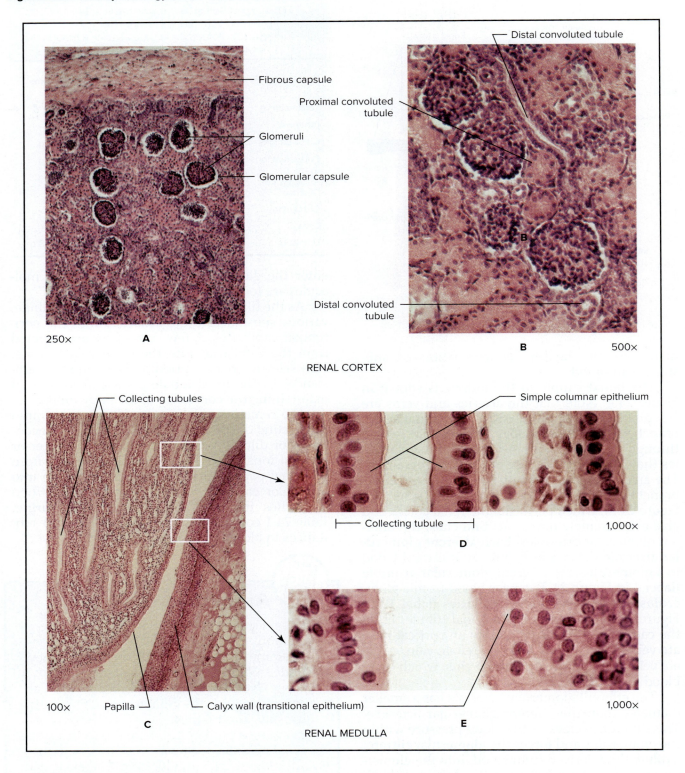

Sheep Kidney Dissection

1. Obtain a sheep kidney for study. If it is still encased in fat, remove the fat carefully with your hands. Look for the adrenal gland embedded in the fat near one end of the kidney. Cut the gland in half and note that it has a distinct outer renal cortex and an inner renal medulla.
2. Insert a dissecting needle into the kidney to distinguish the tougher fibrous capsule from the softer underlying tissue.

Nutrient Absorption and Water Balance

3. With a long, sharp knife, cut the kidney longitudinally to make a frontal section similar to Figure 24.2. Wash the cut surfaces.
4. Locate all of the macroscopic structures shown in Figure 24.2.

Assignment
Complete Section D of the laboratory report.

Typical Components of Urine

Urine is a complex aqueous solution of organic and inorganic substances. Most of the solutes are either waste products of cellular metabolism or products formed from certain foods.

The volume, pH, and solute concentration of urine may vary within a considerable range in a healthy person as the constancy of the internal environment is maintained. In contrast, the physical and chemical characteristics of urine may change dramatically in certain pathological conditions. Thus, a **urinalysis** is a commonly used clinical test because it provides much information about the condition of the body.

The physical characteristics of healthy urine are shown in Table 24.2, and the major organic and inorganic solutes in healthy urine are found in Table 24.3. Some variations from these values are considered healthy, but marked variations are considered abnormal.

Abnormal Characteristics of Urine

In this section, you will study the abnormal characteristics of urine, which may have clinical implications.

Physical Characteristics

Volume
Typical volume ranges from 0.7 to 2.5 l per day. An excessive production of urine, **polyuria,** occurs in diabetes insipidus and diabetes mellitus; very low or no urine production, **anuria,** occurs in renal failure.

Color
Urochrome gives the color to healthy urine. It is the end product of hemoglobin breakdown. Certain foods may cause a color change without having pathological significance. Carrots may produce a yellow color, beets a reddish color, and rhubarb a brownish color.

TABLE 24.2
Physical Characteristics of Healthy Urine

Characteristic	Description*
Volume	0.7–2.5 l in 24 hour. Varies with diet, water intake, concentration of plasma solutes, temperature, physical activity, emotional state, and other factors.
Color	Pale yellow to amber due to urochrome. The greater the solute concentration, the darker the color. Certain foods and drugs affect the color.
Turbidity	Clear when fresh but may become turbid after standing.
Odor	Characteristic aromatic odor, but becomes ammonia-like after standing due to bacterial breakdown of urea.
pH	4.8–7.5, average of 6.0. Acidic in high-protein diets; alkaline in vegetable diets.
Specific gravity	1.015–1.025 in 24-hour specimens. Single specimens from 1.002 to 1.030. The greater the concentration of solutes, the higher the specific gravity.

*Based on 24-hour urine specimen.

TABLE 24.3
Principal Solutes of a Healthy 24-Hour Urine Specimen

Solute	Grams	Comments
Organic		
Urea	25.0	Forms up to 90% of nitrogenous wastes. Produced by liver after deamination of amino acids.
Creatinine	1.5	By-product of muscle metabolism.
Uric acid	0.8	Product of nucleic acid catabolism. Tends to form crystals; a common constituent of kidney stones (renal calculi).
Inorganic		
NaCl	15.0	Most abundant mineral salt. Amount varies with intake.
K^+	3.3	As chloride, phosphate, and sulfate salts.
Sulfates	2.5	As sodium, potassium, magnesium, or calcium compounds.
Phosphates	2.5	As sodium, potassium, magnesium, or calcium compounds.

Certain colors of urine may indicate pathological conditions.

Red to smoky brown: presence of hemoglobin or red blood cells. One or both are found in urine in kidney or urinary tract injury, certain kidney or urinary tract infections, and hemolytic anemia.

The Urinary System

Brownish yellow or green: presence of bile pigments. Bile pigments are found in urine in jaundice.

Red-amber: may indicate the presence of porphyrin. Porphyrin occurs in urine in liver damage, jaundice, Addison's disease, and other conditions.

Turbidity
Fresh urine usually is clear, but it may become cloudy after standing. Abnormal turbidity usually is caused by the presence of phosphates, urates, blood, pus, or bacteria.

Odor
Variations in odor may be due to diet or drugs, but only odors related to diseases are considered abnormal. For example, the urine may have a fruity odor in diabetes mellitus.

pH
The average pH of urine is 6.0, but the healthy range is from 4.8 to 7.5. Excessive acid reactions occur with fever, **nephritis** (inflammation of the kidney), **glomerulonephritis** (inflammation of the kidney involving glomeruli), and acute **cystitis** (inflammation of the urinary bladder). Excessive alkaline reactions occur in chronic cystitis and prostatic obstruction.

Specific Gravity
Specific gravity is a measure of the concentration of solids in the urine. Pure water has a specific gravity of 1.000. The specific gravity of a 24-hour specimen of healthy urine ranges from 1.015 to 1.025. Single specimens may range from 1.002 to 1.030. The greater the concentration of solutes, the higher is the specific gravity. A low specific gravity (dilute urine) may result from excessive water intake, diabetes insipidus, or chronic nephritis. A high specific gravity (concentrated urine) may result from low water intake, diabetes mellitus, fever, or acute nephritis.

Contents

Proteins
Proteins are typically absent in the urine because their large size usually prevents their passage into the glomerular capsule. However, under certain conditions, albumin, the smallest and most abundant of the plasma proteins, is present in the urine—a condition known as **albuminuria.** Temporary albuminuria may be caused by excessive physical exertion. Chronic albuminuria results from excessive permeability of the glomerular membrane and may be caused by bacterial toxins and glomerulonephritis.

Traces of mucin, a glycoprotein, may be present due to secretion from the urinary epithelium. Excessive mucin suggests irritation of the urinary tract.

Glucose
Only trace amounts of glucose are present in healthy urine. The presence of larger amounts of glucose in the urine, **glycosuria,** occurs when the glucose level of the blood exceeds healthy limits (hyperglycemia), and glucose is not completely reabsorbed from the tubular fluid. Persistent glycosuria is usually diagnostic of diabetes mellitus. Temporary glycosuria may result from other causes such as the excessive ingestion of carbohydrates.

Ketones
When the body uses an excessive amount of fatty acids in aerobic respiration, the incomplete metabolism of fatty acids produces ketones, which are released into the blood and excreted in urine. The presence of ketones in urine, **ketonuria,** may result from an excessively low carbohydrate diet or diabetes mellitus. Diabetic ketosis may lead to acidosis, which can culminate in coma and death.

Hemoglobin
The presence of hemoglobin in urine, **hemoglobinuria,** occurs when hemoglobin is released into blood by disintegrating red blood cells. This condition may result from hemolytic anemia, hepatitis, transfusion reactions, burns, malaria, and other causes.

Erythrocytes
The presence of red blood cells in urine, **hematuria,** indicates bleeding in the urinary tract, resulting from disease or trauma. It is often the first symptom of malignant kidney tumors.

Leukocytes
The presence of leukocytes or other components of pus in the urine, **pyuria,** indicates inflammation in the urinary tract, such as glomerulonephritis, cystitis, or urethritis (inflammation of the urethra).

Bilirubin
Bilirubin, a bile pigment, is formed by the breakdown of hemoglobin. In healthy individuals, only trace amounts are present in urine. The presence of larger concentrations, **bilirubinuria,** usually indicates liver pathology, such as hepatitis or cirrhosis, in which bilirubin is not removed from the blood by liver cells.

Urobilinogen
Bacteria in the large intestine convert bilirubin to urobilinogen. Some of this compound is absorbed into the blood and removed by liver cells in the formation of bile. Excessive amounts in the urine, **urobilinogenuria,** usually indicate an excessive destruction of hemoglobin or liver pathology.

Casts

Casts are concentrations of cells or cellular debris that have hardened in the tubules before being forced from the tubules by the buildup of tubular fluid behind them. See Figure 24.6. Casts always indicate kidney pathology, such as glomerulonephritis.

Renal Calculi

Calculi (kidney stones) are usually formed of crystals of uric acid, calcium oxalate, or calcium phosphate. They usually form in the pelvis of the kidney. They may lodge in the ureters, bladder, or urethra, where they may cause pain, hematuria, and pyuria.

Microbes

Urine typically contains low concentrations of bacteria (<1,000 bacteria/ml) derived from the urethral flora. An excessive concentration of bacteria in the urine, **bacteriuria,** or the presence of yeasts or protozoans indicates a urinary infection.

Assignment

1. Study Tables 24.2 and 24.3.
2. Complete Sections E and F of the laboratory report.

Urinalysis

A routine urinalysis includes a description of the physical characteristics and a determination of the presence or absence of abnormal components. Microscopic examination of urine sediments and tests for the presence, identification, and concentration of microbes may also be performed; in some cases, these tests are the most important parts of the analysis. Figure 24.6 illustrates the types of crystals, cells, and casts that may occur in urine. However, microscopic examination of urine

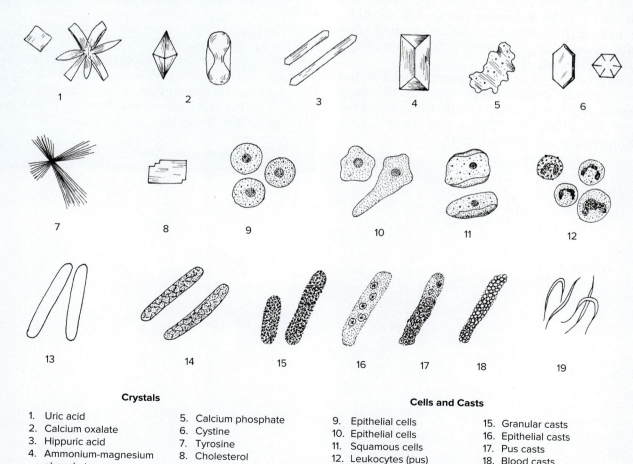

Figure 24.6 Microscopic elements in urine.

Crystals
1. Uric acid
2. Calcium oxalate
3. Hippuric acid
4. Ammonium-magnesium phosphate
5. Calcium phosphate
6. Cystine
7. Tyrosine
8. Cholesterol

Cells and Casts
9. Epithelial cells
10. Epithelial cells
11. Squamous cells
12. Leukocytes (pus)
13. Hyaline casts
14. Granular casts
15. Granular casts
16. Epithelial casts
17. Pus casts
18. Blood casts
19. Mucus threads

The Urinary System

sediments is beyond the scope of this course. As an alternative, your instructor may set up slides of urine sediments under demonstration microscopes for your examination.

Your instructor has prepared several simulated "unknown" urine specimens. You will be assigned two "unknowns" to analyze in addition to your own urine specimen. You are to determine the physical characteristics of the urine specimens by visual examination and test for abnormal urine constituents using 10SG-Multistix reagent strips. After the analysis, you are to determine if any of the specimens indicate pathological conditions.

Each Multistix dipstick contains squares of chemical reagents that react with specific urine components. The reagent squares change color to indicate the presence of or the degree of concentration of these urine components. From top to bottom, a 10SG-Multistix dipstick has reagent squares that test for leukocytes (pus), nitrite, urobilinogen, protein (albumin), pH, blood (erythrocytes or hemoglobin), specific gravity, ketone, bilirubin, and glucose.

Each Multistix container has a color chart illustrating positive and negative results. After dipping a Multistix into a urine specimen, you must wait for the minimum number of seconds, as noted on the color chart, before reading your results. *Be sure that you know how to read the results before using a Multistix.*

Collect a sample of your own urine as directed by your instructor using the plastic collection cups available. Analyze it just as you do the unknowns. Follow the procedures below and record your findings on the laboratory report.

Assignment

1. Swirl the specimen in the container to resuspend any sediments. Record the color and turbidity of the specimen.
2. Fill a clean test tube about three-fourths full with each urine sample to be analyzed. Use a glass-marking pen to label each tube.
3. Remove a 10SG-Multistix reagent strip from the jar while being careful not to touch the colored reagent squares. Compare the color of the reagent squares with the color chart provided, and note the color of negative and positive tests. Note that the resulting color is to be read only after a minimum number of seconds to obtain accurate results. *Be sure that you know how to read the tests before proceeding.*
4. Dip a 10SG-Multistix reagent strip into the urine sample so that all reagent squares are immersed. Remove the strip, tap off the excess urine on the side of the test tube, and lay it on a paper towel with the reagent squares up. Record the time.
5. Observing the minimum reading times, determine and record the results for each reagent square.
6. Discard the urine as directed by your instructor. Place the reagent strips, urine collecting cup, and paper towels in the biohazard bag.
7. Place the test tubes in the Amphyl solution provided. Wash your workstation with Amphyl.
8. Complete the laboratory report.

Exercise 25

THE REPRODUCTIVE SYSTEM

Objectives

After completing this exercise, you should be able to

1. Identify the parts of the male and female reproductive systems on charts or models and describe their functions.
2. Describe spermatogenesis and oogenesis.
3. Identify histological structures in prepared slides of ovary and testis.

Materials
Models of human reproductive organs
Prepared slides of human testis, ovary, uterine wall, and uterine tube

This study of the human reproductive systems includes (1) labeling anatomical illustrations, (2) locating the reproductive organs on classroom models, and (3) microscopic study of spermatogenesis and the tissues of selected organs.

The Male Organs

Figure 25.1 illustrates a sagittal section of the male reproductive system. The primary sex organs (gonads) are the **testes** (testicles), which are located in a pouch of skin called the **scrotum.** They produce the sperm and the male sex hormones. In a male fetus, the testes develop within the abdominal cavity and descend into the scrotum 1 to 2 months before birth.

Partially surrounding each testis is a downward extension of the peritoneum, the **tunica vaginalis** (label 27), that is carried into the scrotum during the descent of the testes. Between a testis and the scrotum is connective tissue and the **cremaster muscle** (label 21). Only a small portion of this muscle is shown in the illustration of testicular detail. The cremaster muscle contracts and relaxes to raise or lower the testis to maintain a testicular temperature of 94°F to 95°F. Higher temperatures prevent the production of viable sperm.

A fibrous capsule forms the outer wall of each testis. **Testicular septa,** formed of connective tissue, divide the testis into several lobules. Each lobule contains **seminiferous tubules** that produce **sperm.** All seminiferous tubules merge to form the **rete testis,** a complex network of tubules. Cilia within the tubules of the rete testis move the sperm through several **vasa efferentia** (label 23) and on into the highly coiled **epididymis** (label 6) where the sperm mature. Note that the epididymis wraps around the top and back of the testis. From the epididymis, sperm pass into the **vas (ductus) deferens,** the central canal within the **spermatic cord** (label 22). Note the histology of the vas deferens, x.s., especially the smooth muscle layers, in Figure 25.3.

Trace a vas deferens from the epididymis and note that it passes above the pubic bone and urinary bladder into the pelvic cavity. The distal end of each vas deferens is enlarged to form an **ampulla.** The ampulla merges with the duct from a **seminal vesicle** (label 2) to form the **ejaculatory duct.** The ejaculatory duct on each side empties into the **prostatic urethra** (label 11), the portion of the urethra that is within the **prostate.** The small **bulbourethral glands** (label 4) empty their secretions into the urethra at the base of the penis.

Semen is formed of sperm and the alkaline secretions of the accessory glands. Two-thirds of the semen is derived from secretions of the seminal vesicles, and about one-third comes from prostatic fluid. Secretions of the bulbourethral glands and sperm account for very little of the semen volume. The alkaline secretions protect the sperm from acidic environments, provide nutrients for the sperm, and activate their swimming movements.

Erection of the penis occurs when its three cylinders of erectile tissue fill with blood in response to sexual stimulation. The **corpus spongiosum** is the cylinder of tissue that surrounds the **penile urethra.** Its distal end is enlarged, to form the **glans penis,** and its proximal end is enlarged, forming the **bulb of the penis.** The two **corpora cavernosa** are located in the dorsal part of the penis and are separated by a medial

Unit 7: Reproduction

Figure 25.1 Male reproductive organs.

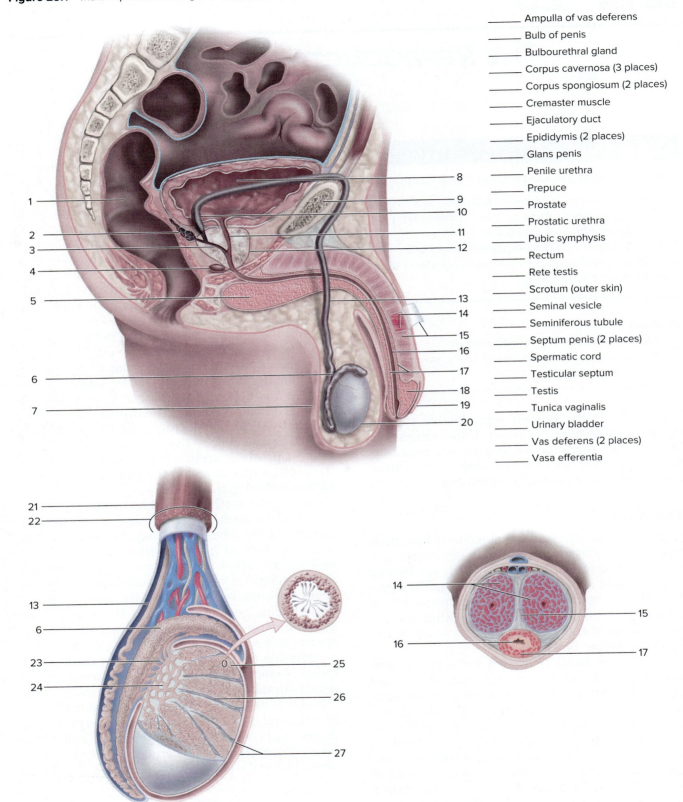

_____ Ampulla of vas deferens
_____ Bulb of penis
_____ Bulbourethral gland
_____ Corpus cavernosa (3 places)
_____ Corpus spongiosum (2 places)
_____ Cremaster muscle
_____ Ejaculatory duct
_____ Epididymis (2 places)
_____ Glans penis
_____ Penile urethra
_____ Prepuce
_____ Prostate
_____ Prostatic urethra
_____ Pubic symphysis
_____ Rectum
_____ Rete testis
_____ Scrotum (outer skin)
_____ Seminal vesicle
_____ Seminiferous tubule
_____ Septum penis (2 places)
_____ Spermatic cord
_____ Testicular septum
_____ Testis
_____ Tunica vaginalis
_____ Urinary bladder
_____ Vas deferens (2 places)
_____ Vasa efferentia

202 Reproduction

septum of connective tissue called the **septum penis** (label 15). Note in the cross section of the penis how the cylinders of erectile tissue are separately enveloped by connective tissue and bound together by additional connective tissue. Observe the histology of the penile urethra and erectile tissue in Figure 25.3.

A sheath of skin, the **prepuce,** begins just behind the glans and extends to cover it. Circumcision is a surgical procedure that removes the prepuce to facilitate sanitation.

During sexual stimulation, the secretion from the bulbourethral glands provides an alkaline environment within the urethra prior to the passage of the sperm. Peristaltic contractions move the sperm from the epididymis into the urethra. The secretions of the seminal vesicles are mixed with sperm in the ejaculatory ducts, and prostatic fluids are added in the proximal part of the urethra. During ejaculation, rhythmic contractions of the bulbospongiosus muscle at the base of the penis propel the semen through the urethra and out of the body.

Sperm motility requires an alkaline environment and nutrients as an energy source. The alkaline prostatic fluid increases the pH of the female vagina from about 4.0 to around 7.5 and activates the sperm. Fructose and other carbohydrates in seminal vesicle fluid serve as an energy source for sperm. Prostaglandins in seminal vesicle fluid stimulate wavelike contractions of the uterus and uterine tubes, helping to move sperm throughout the female reproductive tract.

Spermatogenesis

The process by which sperm are produced in the seminiferous tubules is called **spermatogenesis.** It begins at puberty and continues throughout life. Figure 25.2 illustrates a section of a seminiferous tubule and a diagram of the developmental stages of sperm formation. Both mitosis and meiosis are involved.

All sperm originate from **spermatogonia** that are located at the periphery of a seminiferous tubule. These cells contain 23 pairs of chromosomes, a total of 46 chromosomes, the same as all body cells. Since they contain both members of each chromosome pair, they are said to be diploid. The mitotic division of a spermatogonium forms one replacement spermatogonium and one **primary spermatocyte.**

Each diploid primary spermatocyte divides by **meiotic cell division,** a type of cell division that consists of one chromosome replication and two successive cell divisions. Meiotic division produces four haploid spermatids from the single diploid primary spermatocyte.

The first meiotic division forms two haploid **secondary spermatocytes.** Each secondary spermatocyte contains only 23 chromosomes, one member of each chromosome pair. These chromosomes are already replicated, so each chromosome consists of two chromatids joined at a centromere.

The second meiotic division occurs as each secondary spermatocyte divides to yield two haploid **spermatids,** each with 23 chromosomes. In this division, the chromatids separate to provide a haploid set of chromosomes for each spermatid. The spermatids subsequently mature to become sperm.

Assignment

1. Label Figure 25.1.
2. Locate the male reproductive organs on the models available.

Figure 25.2 Spermatogenesis.

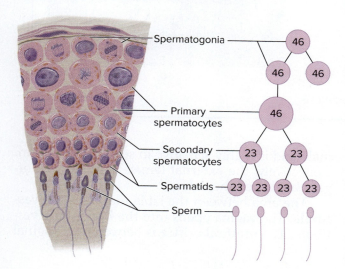

The Female Organs

Refer to Figures 25.4, 25.5, and 25.6 as you study this section.

External Genitalia

Two folds of skin lie on each side of the **vaginal orifice,** the labia majora and labia minora as shown in Figure 25.4. The larger and more lateral folds are the **labia majora,** which consist of rounded folds of adipose tissue covered by skin. They merge with the **mons pubis,** an elevation of fatty tissue over the pubic symphysis. The outer surfaces of the labia majora possess hair; their inner surfaces are smooth and moist. The smaller medial folds,

Figure 25.3 Male reproductive tissue. (A–D): Courtesy of Harold Benson

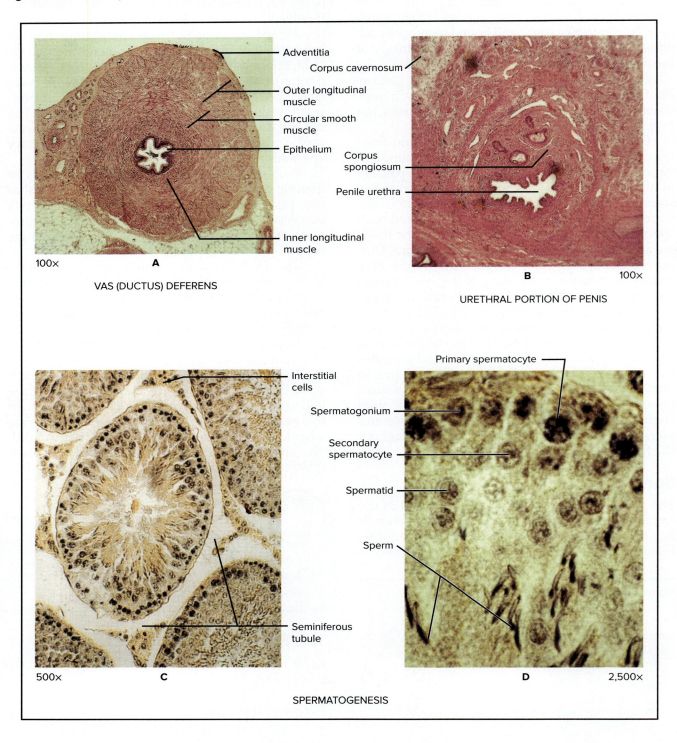

the **labia minora,** lack hair and merge in the back with the labia majora. In the front the labia minora join to form a hoodlike covering, the **prepuce** of the clitoris. The **clitoris** is a small protuberance of erectile tissue located at the junction of the labia minora. It is homologous to the penis in the male and is highly sensitive to sexual stimulation. Collectively, the external female reproductive organs are called the **vulva.**

The area between the labia minora is the **vestibule.** The vagina opens into the back of the vestibule. The **urethral orifice** is between the vaginal

Figure 25.4 Female genitalia.

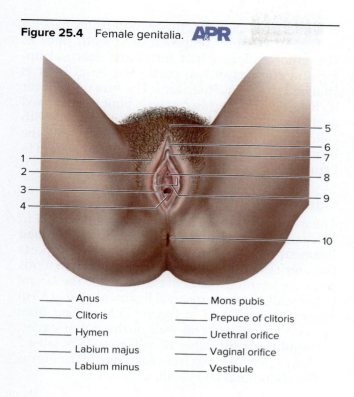

_____ Anus
_____ Clitoris
_____ Hymen
_____ Labium majus
_____ Labium minus
_____ Mons pubis
_____ Prepuce of clitoris
_____ Urethral orifice
_____ Vaginal orifice
_____ Vestibule

opening and the clitoris. On either side of the vaginal orifice are the openings of the vestibular glands, which provide a mucous secretion for vaginal lubrication. The **hymen** (label 4) is a thin mucous membrane that partially covers the vaginal opening. Its condition or absence is not a determiner of virginity.

Internal Organs

The internal female reproductive organs are the ovaries, uterine tubes, uterus, and vagina.

The Ovaries

The primary sex organs (gonads) of the female reproductive system are the **ovaries** (label 5 in Figure 25.5). They produce the **ova** (egg cells) and the female sex hormones. The ovaries are ovoid organs located against either side of the lateral walls of the pelvic cavity.

The Uterine Tubes

The **uterine tubes** (oviducts or fallopian tubes) extend from the ovaries to the uterus. Near the ovary, each tube is expanded to form a funnel-shaped

Figure 25.5 Posterior view of female internal reproductive organs.

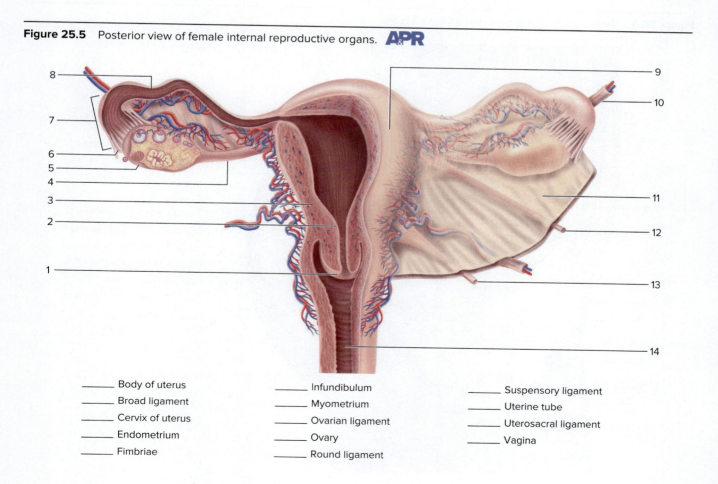

_____ Body of uterus
_____ Broad ligament
_____ Cervix of uterus
_____ Endometrium
_____ Fimbriae
_____ Infundibulum
_____ Myometrium
_____ Ovarian ligament
_____ Ovary
_____ Round ligament
_____ Suspensory ligament
_____ Uterine tube
_____ Uterosacral ligament
_____ Vagina

The Reproductive System **205**

infundibulum (label 7 in Figure 25.5) that bears a number of fingerlike extensions, the **fimbriae.** The fimbriae and infundibula receive the oocytes released from the ovaries. The oocytes are carried toward the uterus by peristaltic contractions of the uterine tubes and the beating cilia of the ciliated columnar cells that line the tubes. Fertilization usually occurs within the upper third of a uterine tube. Observe the histology of a uterine tube in Figure 25.8.

The Uterus
The **uterus** is a hollow, pear-shaped, thick-walled organ within which fetal development takes place. It is located above the vagina and between the rectum and the urinary bladder. The **fundus** of the uterus is a broad curvature that is above and in front of the junction with the uterine tubes. The **body** is the middle portion that forms most of the uterus, and the **cervix** is the lower third that extends into the upper portion of the vagina. The **cervical orifice** is the uterine opening into the vagina.

The uterine wall is composed of three layers. The outer **perimetrium** is a layer of the peritoneum. The thick middle layer, the **myometrium** (label 3 in Figure 25.5), is composed of smooth muscle. The inner **endometrium** forms the mucosal lining and consists of two parts. The basal layer is attached to the myometrium. The functional layer is closest to the uterine cavity, is built up and shed during each menstrual cycle, and is covered with columnar epithelium on its free surface. Note the histology of the myometrium and endometrium in Figure 25.8.

Ligaments

The ovaries, uterine tubes, and uterus are held in place and supported by several ligaments. The **broad ligament** is a fold of the peritoneum that supports the uterus, uterine tubes, and ovaries. It extends from the lateral surfaces of the uterus to the lateral pelvic walls. Two **uterosacral ligaments** (label 13 in Figure 25.5; label 13 in Figure 25.6) extend from the cervix to the sacral wall of the pelvic cavity. A **round ligament** (label 12 in Figure 25.5; label 7 in Figure 25.6) extends from each side of the uterus to the abdominal body wall.

Figure 25.6 Median section of female reproductive organs. **A&P R**

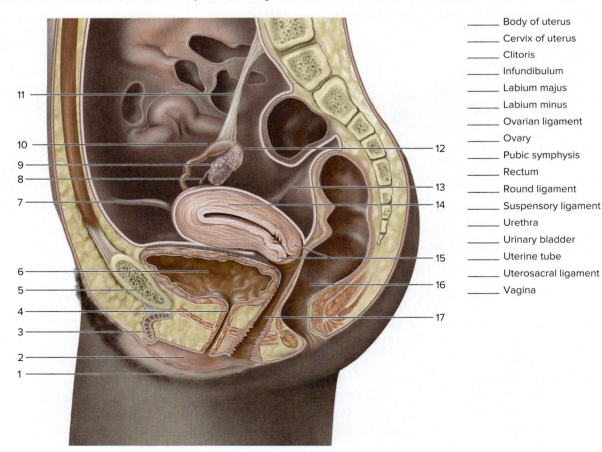

_____ Body of uterus
_____ Cervix of uterus
_____ Clitoris
_____ Infundibulum
_____ Labium majus
_____ Labium minus
_____ Ovarian ligament
_____ Ovary
_____ Pubic symphysis
_____ Rectum
_____ Round ligament
_____ Suspensory ligament
_____ Urethra
_____ Urinary bladder
_____ Uterine tube
_____ Uterosacral ligament
_____ Vagina

Figure 25.7 Oogenesis.

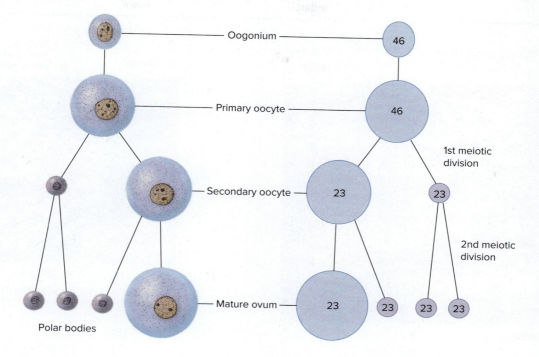

In addition to the broad ligament, each ovary is held in place by two ligaments. An **ovarian ligament** extends from the uterus to the medial surface of the ovary. The **suspensory ligament** (label 10 in Figure 25.5; label 11 in Figure 25.6) extends from the lateral surface of the ovary to the pelvic wall.

Vagina

The **vagina** is a fibromuscular canal extending from the uterus to the vestibule. It receives the penis during sexual intercourse and serves as the birth canal.

Oogenesis

The development of the egg cells is called **oogenesis,** and it involves both mitotic and meiotic divisions. See Figure 25.7. The **germinal epithelium** is formed early in the prenatal development of the ovaries. It occurs on the outer surface of the ovaries and consists of as many as 400,000 **oogonia** that form by mitotic division. By the end of the third month of development, mitotic division has ceased and some oogonia have migrated inward to become **primary oocytes.** Each is surrounded by a sphere of cells forming an **ovarian follicle.** Each primary oocyte is diploid because it contains 23 pairs of chromosomes, or a total of 46 chromosomes.

Starting at puberty, several primary oocytes are stimulated to develop further by FSH. Usually, only one of them will undergo meiotic division in each ovarian cycle. The first meiotic division forms one large **secondary oocyte** and one small nonfunctional **polar body.** Both of these cells are haploid because they contain only 23 chromosomes, one member of each chromosome pair. The chromosomes are already replicated and consist of two chromatids joined at the centromere. Each secondary oocyte is located within a **secondary ovarian follicle** that enlarges and fills with fluid to become a **mature ovarian follicle.** Ovulation occurs with the rupture of the mature ovarian follicle. The extruded secondary oocyte and first polar body enter the infundibulum and are carried toward the uterus by the uterine tube.

If a secondary oocyte is penetrated by a sperm (activation), it undergoes the second meiotic division, which forms the ovum and a polar body. The first polar body may also divide to form two polar bodies. Thus, the meiotic division of a diploid primary oocyte forms four haploid cells: one ovum and three polar bodies. Note that the bulk of the cytoplasm passes first to the secondary oocyte and then to the ovum. After the formation of the ovum, the egg nucleus and the sperm nucleus unite (fertilization) to form a diploid zygote. The polar bodies disintegrate.

Note that the orderly process of gametogenesis (spermatogenesis and oogenesis) results in the zygote receiving one member of each chromosome pair from each parent.

Figure 25.8 The uterus and uterine tube. (A–D): Courtesy of Harold Benson

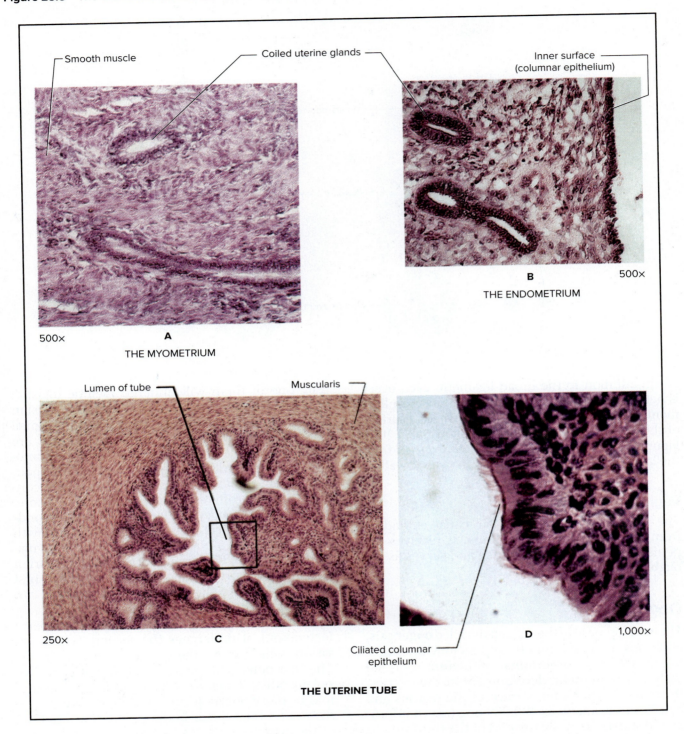

Assignment

1. Label Figures 25.4, 25.5, and 25.6.
2. Complete Sections A through E of the laboratory report.
3. Locate the female reproductive organs on the models available.

Microscopic Study

Examine the prepared slides noted below and compare your observations with Figures 25.3 and 25.8. Make labeled drawings of all your observations in Section F of the laboratory report.

Assignment

1. Examine a prepared slide of human testis with the low-power and high-dry objectives. Compare your observations with Figures 25.2 and 25.3.
   a. Note the connective tissue forming the walls of the seminiferous tubules. Locate a few interstitial cells. What is their function?
   b. Locate maturing sperm and the stages of spermatogenesis.
2. Examine a prepared slide of human ovary with the low-power and high-dry objectives. Compare your observations with Figure 17.10.
   a. Locate the cuboidal cells of the germinal epithelium at the outer edge of the ovary. Are these cells haploid or diploid?
   b. Locate the different developmental stages of ovarian follicles and find a secondary oocyte within a mature ovarian follicle.
3. Examine prepared slides of the uterine wall and compare your observations with Figure 25.8, A and B. In what tissue layer are uterine glands most abundant? What type of epithelium forms the surface lining of the endometrium?
4. Examine a prepared slide of uterine tube, x.s., and compare your observations with Figure 25.8, C and D. Note the irregularly shaped lumen. What type of epithelium lines the lumen?
5. Complete the laboratory report.

Notes

Laboratory Report 1

Student _____

Lab Section _____

INTRODUCTION TO HUMAN ANATOMY AND BODY ORGANIZATION

A. *Figures*

Write the labels for Figures 1.2 through 1.7 in the spaces provided.

Figure 1.2

Planes

1. _____
2. _____
3. _____

Body Surfaces

4. _____
5. _____
6. _____
7. _____
8. _____
9. _____
10. _____
11. _____
12. _____
13. _____
14. _____
15. _____
16. _____
17. _____
18. _____

Figure 1.3

Anterior View

1. _____
2. _____
3. _____
4. _____
5. _____
6. _____
7. _____
8. _____
9. _____
10. _____
11. _____
12. _____
13. _____
14. _____
15. _____
16. _____
17. _____
18. _____
19. _____
20. _____
21. _____
22. _____

Posterior View

1. _____
2. _____
3. _____
4. _____
5. _____

6. _____
7. _____
8. _____
9. _____
10. _____
11. _____
12. _____
13. _____

Figure 1.4

Abdominal Regions

1. _____
2. _____
3. _____
4. _____
5. _____
6. _____
7. _____
8. _____
9. _____

Abdominal Quadrants

1. _____
2. _____
3. _____
4. _____

Figure 1.5

Sagittal Section

1. _____
2. _____
3. _____
4. _____
5. _____
6. _____
7. _____
8. _____
9. _____
10. _____
11. _____

Frontal Section

1. _____
2. _____
3. _____
4. _____
5. _____
6. _____
7. _____
8. _____

Figure 1.6

Pleural Membranes

1. _____
2. _____
3. _____

Pericardial Membranes

4. _____
5. _____
6. _____

Figure 1.7

1. _____
2. _____
3. _____
4. _____
5. _____
6. _____

B. Directional Terms, Sections, and Surfaces

Write in the answer column the term from the list below that corresponds to the statements that follow.

Directional Terms

anterior   medial
distal   posterior
inferior   proximal
lateral   superior

Sections

frontal
median
paramedian
transverse

Directional Terms

1. The hand is _____ to the elbow.
2. The knee is _____ to the ankle.
3. The mouth is _____ to the nose.
4. The ear is _____ to the nose.
5. The teeth are _____ to the cheeks.

Sections

6. Divides the body into equal left and right halves.
7. Divides the body into top and bottom parts.
8. Divides the body into front and back parts.
9. Divides the body into unequal left and right parts.

Surfaces

Write the body surface on which the following are located.

10. Kneecap
11. Ear
12. Umbilicus (navel)
13. Buttocks
14. Palm of hand
15. Nose

1. _____
2. _____
3. _____
4. _____
5. _____
6. _____
7. _____
8. _____
9. _____
10. _____
11. _____
12. _____
13. _____
14. _____
15. _____

C. Body Regions and Surface Features

Write the term from the list below that corresponds to the following statements.

antebrachial
antecubital
axillary
brachial
calcaneal
carpal
cubital
digital
epigastric
femoral
gluteal
hypochondriac
hypogastric
inguinal
lumbar
pectoral
popliteal
tarsal

1. The ankle.
2. The forearm.
3. Heel of the foot.
4. Lateral chest region.
5. The back of the knee.
6. Depression at junction of thigh with abdomen.
7. The armpit.
8. Abdominal region overlying the urinary bladder.
9. Abdominal region above the lumbar region.
10. Abdominal region directly above the umbilical region.
11. The lower back lateral to the vertebral column.
12. The fingers and toes.
13. The buttocks.
14. The proximal portion of the upper limb.
15. The front of the elbow joint.

1. _____
2. _____
3. _____
4. _____
5. _____
6. _____
7. _____
8. _____
9. _____
10. _____
11. _____
12. _____
13. _____
14. _____
15. _____

D. Membranes

Write the names of the membranes that match the following statements.

Membranes

mesentery
parietal pericardium
parietal peritoneum
parietal pleura
visceral pericardium
visceral peritoneum
visceral pleura

1. Outer membrane around the heart.
2. Serous membrane attached to surface of lungs.
3. Serous membrane attached to surface of stomach.
4. Double-layered membrane supporting intestines.
5. Serous membrane lining wall of abdominal cavity.
6. Serous membrane tightly attached to the outer surface of the heart.
7. Serous membrane lining the inside of the pleural cavity.

1. _____
2. _____
3. _____
4. _____
5. _____
6. _____
7. _____

E. Complete the Chart

Organ System	Functions	Components
Integumentary		
Skeletal		
Muscular		
Nervous		
Endocrine		

Organ System	Functions	Components
Cardiovascular		
Lymphoid		
Respiratory		
Digestive		
Urinary		
Reproductive		

F. Terms
Define these terms and give examples of each.

Organ _____
Organ system _____
_____
_____

G. Matching
Write the name of the organ system(s) corresponding to the words or phrases that follow.

cardiovascular   lymphoid   respiratory
digestive   muscular   skeletal
endocrine   nervous   urinary
integumentary   reproductive

Functions
1. Protective covering of the body.
2. Transports materials throughout the body.
3. Enables gas exchange between blood and atmosphere.
4. Secretes hormones.
5. Digests food to form absorbable nutrients.
6. Provides supporting framework for the body.
7. Contractions enable body movements.
8. Removes wastes from the blood to form urine.
9. Enables rapid responses to environmental changes.
10. Collects tissue fluid from intercellular spaces.
11. Provides chemical control of body functions.
12. Perpetuates the species.
13. Produces blood cells.
14. Shields body from ultraviolet radiation.
15. Cleanses lymph and returns it to the blood.

Components
1. Stomach and intestines.
2. Pituitary, adrenal, and thyroid glands.
3. Trachea, bronchi, and lungs.
4. Arteries, veins, and capillaries.
5. Cartilages and ligaments.
6. Kidneys, ureters, and urethra.
7. Spleen, tonsils, and adenoids.
8. Epidermis and dermis.
9. Uterus, oviducts, and vagina.
10. Liver and pancreas.
11. Nasal cavity, pharynx, and larynx.
12. Brain and spinal cord.
13. Sensory receptors and nerves.
14. Pancreas, ovaries, and testes.
15. Vasa deferentia, urethra, and penis.

Introduction to Human Anatomy and Body Organization

Notes

Laboratory Report 2

Student _____

Lab Section _____

MICROSCOPES

A. Parts of the Microscope

Record in the answer column the microscope part described by the statements below.

arm	mechanical stage
base	objective
coarse-adjustment knob	ocular
condenser	revolving nosepiece
diaphragm	stage
fine-adjustment knob	stage aperture
lamp	

1. Platform on which slides are placed for viewing.
2. Lens that is closest to the microscope slide.
3. Controls amount of light entering the condenser.
4. Adjustment knob used with low-power objective only.
5. Concentrates light on object being observed.
6. Light source.
7. Serves as a "handle" in carrying a microscope.
8. Lens you look through to view the image.
9. Part to which objectives are attached.
10. Opening in center of the stage.
11. Enables precise movement of the slide.
12. Bottom of the microscope.
13. Adjustment knob used with all objectives.
14. Adjustment knob with the larger diameter.
15. May be rotated to bring different objectives into viewing position.

1. _____
2. _____
3. _____
4. _____
5. _____
6. _____
7. _____
8. _____
9. _____
10. _____
11. _____
12. _____
13. _____
14. _____
15. _____

B. Magnification

Record the magnification of the ocular(s) and objectives on your microscope and calculate the total magnification obtained when using each objective.

Ocular	×	Objective	=	Total Magnification
____ ×		____ ×		____ ×
____ ×		____ ×		____ ×
____ ×		____ ×		____ ×

C. True–False

Record your answer to these statements as true or false.

1. A microscope should be carried with two hands.
2. You should start your observations with the scanning objective.
3. Blue light gives better resolution than white light.
4. Malfunctions should be reported to your instructor.
5. If necessary, you should disassemble the ocular to clean it.

1. _____
2. _____
3. _____
4. _____
5. _____

D. Observations

1. Diagram the appearance of the letter *e* as observed with the following:

   unaided eye   low-power objective   high-dry objective

   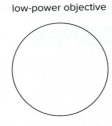 

2. Draw the appearance of the colored thread with the following:

   low-power objective   high-dry objective

3. Are human sperm larger or smaller than cheek epithelial cells? _____ Draw a few sperm as they appeared under the high-dry objective, and label the head and the flagellum.

4. Draw the appearance of the different types of blood cells that you observed as they appeared under the high-dry objective.

5. Draw a few cheek epithelial cells as they appeared under the high-dry objective.

220   Laboratory Report 2

E. Completion

Write your responses to the following statements in the answer column.

1. Tissue used to clean the lenses.
2. Objective with the greatest working distance.
3. Objective with the least working distance.
4. Usually, the condenser should be kept at its _____ position.
5. What effect (increase, no change, or decrease) does an increase in magnification of an objective have on the following?
   a. diameter of field
   b. working distance
   c. light intensity
6. The contrast of the image is _____ (increased, unchanged, or decreased) when light intensity is reduced.
7. Controls light intensity in microscopes with a constant light source.
8. Circle of light seen while looking into the ocular.
9. When the slide is moved to the left, the image moves to the _____.
10. Maximum resolution with a light microscope.
11. Two liquids that may be used to clean lenses.
12. When an object is in focus with one objective and is still in focus when another objective is rotated into position, the microscope is said to be _____.
13. Only adjustment knob used with high-dry and oil immersion objectives.

1. _____
2. _____
3. _____
4. _____
5a. _____
5b. _____
5c. _____
6. _____
7. _____
8. _____
9. _____
10. _____
11a. _____
11b. _____
12. _____
13. _____

Notes

Laboratory Report 3

Student _____

Lab Section _____

CELLULAR ANATOMY AND MITOSIS

A. Figure
List the labels for Figure 3.1.

1. _____
2. _____
3. _____
4. _____
5. _____
6. _____
7. _____
8. _____
9. _____
10. _____
11. _____
12. _____
13. _____
14. _____
15. _____
16. _____

B. Cell Structure and Organelle Function

1. Select the structure(s) from the following list that are described by the statements below.

centrioles   mitochondria
centrosome   nuclear membrane
chromatin   nucleolus
cilia   nucleus
cytoplasm   plasma membrane
flagellum   RER
Golgi complex   ribosomes
lysosomes   secretory vesicles
microvilli   SER

1. Short, hairlike projections on a cell surface.
2. Short bundles of microtubules oriented at right angles to each other.
3. Increase absorptive surface area of certain cells.
4. Provides motility of sperm.
5. A stack of membranous sacs near the ER.
6. Packages materials for extracellular transport.
7. Sites of ATP production via aerobic respiration.
8. Structures in nucleus composed of DNA and protein.
9. A double membrane surrounding the nucleus.
10. A spherical structure in the nucleus composed of RNA and protein.
11. ER studded with ribosomes.
12. Organelle with an inner membrane having many folds.
13. Controls passage of materials into and out of cells.
14. ER without ribosomes.
15. Sacs of digestive enzymes.
16. Sites of protein synthesis.
17. Integration center of the cell.
18. Assembles RNA and protein that will form ribosomes.
19. Sacs containing materials destined for export.
20. Contains the chromosomes of the cell.
21. Parts of chromosomes visible in nondividing cells.
22. Tiny cytoplasmic organelles composed of RNA and protein.

1. _____
2. _____
3. _____
4. _____
5. _____
6. _____
7. _____
8. _____
9. _____
10. _____
11. _____
12. _____
13. _____
14. _____
15. _____
16. _____
17. _____
18. _____
19. _____
20. _____
21. _____
22. _____

2. Write the terms that complete the sentences in the paragraphs below.

The plasma membrane consists of a bilayer of __1__ molecules in which __2__ molecules are embedded. Its __3__ permeability regulates the passage of materials into and out of the cell. In a similar manner, the nuclear envelope regulates the passage of materials between the nucleus and the __4__.

The nucleus is the integration center of the cell because the genetic information encoded in the __5__ molecules of the __6__ controls cellular functions. The nucleolus assembles __7__ and __8__, which will move into the cytoplasm to become ribosomes.

The __9__ provides passageways for the movement of materials within the cytoplasm. Proteins synthesized on __10__ of the RER are carried to the __11__, which packages them in condensing vesicles that carry the proteins toward the cell surface. Near the plasma membrane, condensing vesicles become __12__ that release their contents outside the cell.

Energy for cellular work is made available by aerobic respiration occurring in the __13__, which contain their own RNA and __14__ and are able to __15__ themselves.

1. _____
2. _____
3. _____
4. _____
5. _____
6. _____
7. _____
8. _____
9. _____
10. _____
11. _____
12. _____
13. _____
14. _____
15. _____

3. Summarize your understanding of cell structure by completing the following table.

Cell Component	Structure/Composition	Function
Plasma Membrane		
Nucleus		
Nucleolus		
RER		
SER		

Cell Component	Structure/Composition	Function
Ribosome		
Mitochondrion		
Golgi Apparatus		
Secretory Vesicle		
Lysosome		
Microtubule		
Microfilament		
Centriole		
Cilia		

C. **Matching**

Write the name of the correct response in the answer column.

anaphase   interphase   prophase
cytokinesis   metaphase   telophase

1. Chromosomes line up at equator of spindle.
2. Chromosomes become visible as rodlike structures.
3. Nuclear envelope is intact.
4. Chromatin is present.
5. Each pair of centrioles moves to opposite ends of cell.
6. Cleavage furrow forms.
7. Centrioles replicate.
8. Nuclear envelope disappears.
9. Starts with formation of cleavage furrow.
10. Chromosomes are replicated.
11. Sister chromatids separate and move to opposite poles of the spindle.
12. New nuclear envelopes are formed.
13. Daughter chromosomes uncoil and become less distinct.
14. Spindle is formed of microtubules.
15. Forms daughter cells.

1. _____
2. _____
3. _____
4. _____
5. _____
6. _____
7. _____
8. _____
9. _____
10. _____
11. _____
12. _____
13. _____
14. _____
15. _____

D. **Completion**

1. Briefly describe the process of mitosis. _____

2. There are 23 pairs of chromosomes in human body cells. How many chromosomes are in a parent cell? _____ In each daughter cell? _____
3. How many daughter chromosomes are in a human cell at anaphase? _____
4. Compare the genetic composition of parent cell and daughter cells. _____

Why is this relationship important? _____

226 Laboratory Report 3

E. Microscopic Study

Diagram the appearance of the whitefish cells (or onion root tip cells) in each phase of the cell cycle and label the aster (centrosome), astral fiber, spindle fiber, chromatin, nuclear envelope, replicated chromosome, sister chromatids, and daughter nuclei. List the key features of each stage below your drawings.

Interphase

Prophase

Metaphase

Anaphase

Telophase

Daughter Cells

Notes

Laboratory Report 4

Student _____

Lab Section _____

MEMBRANE TRANSPORT

A. Concepts

1. Define diffusion. _____
   _____
   _____

2. Define osmosis. _____
   _____
   _____

3. Describe the relationship between solute concentration and the osmotic pressure of an aqueous solution. _____
   _____
   _____

4. Consider two sodium chloride (NaCl) solutions separated by a semipermeable membrane. Solution A is a 5% solution; solution B is a 10% solution. Both water and NaCl can readily pass through the membrane. Indicate the following:

   The direction of net water movement. _____

   The direction of net NaCl movement. _____

   When this movement will stop. _____

   Is one of the solutions hypotonic? Why or why not? _____

5. If a 1.5% NaCl solution is separated by a semipermeable membrane from another NaCl solution that is isotonic, indicate the following:

   Salt concentration of the isotonic solution. _____

   Direction of net water movement. _____

B. Diffusion and Temperature

1. For each beaker, record the temperature of the water and the time required for molecules of potassium permanganate to diffuse throughout the water.

   Beaker A: Temp. _____   Diffusion Time _____

   Beaker B: Temp. _____   Diffusion Time _____

   Beaker C: Temp. _____   Diffusion Time _____

2. Describe the relationship between the rate of diffusion and temperature. _____
_____

C. Diffusion and Molecular Weight

1. Record the diameter of the colored area around each granule.

   Potassium permanganate: _____ mm   Methylene blue: _____ mm

2. Describe the relationship between the rate of diffusion and molecular weight. _____
_____

D. Osmosis

1. Record the height (in mm) of the water-sucrose columns during a 30-minute interval.

   A: At start: _____ mm   B: At start: _____ mm

       At end: _____ mm   At end: _____ mm

       Change: _____ mm   Change: _____ mm

2. Which osmometer has the higher concentration of sucrose? _____

   Explain your response. _____
_____
_____

E. Plasma Membrane Integrity

1. Record the degree of transparency and the effect on the red blood cells (lysis, crenation, not affected) of the solutions.
2. Indicate the number of the solution that is

   isotonic _____   hypertonic _____   hypotonic _____

Solution	Transparency	Cell Shape
1. 2.0% glucose		
2. 5.0% glucose		
3. 0.3% NaCl		
4. 0.9% NaCl		
5. 2.0% NaCl		

Laboratory Report 4

F. *Carrier-Mediated Active Transport*

1. Record the number of each item before the activity.
2. Record the number of each item after the activity.

Item	Before	After
Na⁺ outside of the cell		
Na⁺ inside of the cell		
K⁺ outside of the cell		
K⁺ inside of the cell		
ATP molecules inside the cell		
ATP molecules used (torn)		

3. In what direction is the Na⁺ gradient after the activity? _____
_____

4. How did the number positive charges outside the cell as compared to inside the cell change from before to after the activity? _____
_____

5. Can you come up with your own definition for an electrical gradient? _____
_____

6. Why couldn't students with Na⁺ but no ATP trade their Na⁺ for K⁺? _____
_____
_____

Notes

Laboratory Report 5

Student _____
Lab Section _____

HISTOLOGY

A. Epithelial Tissues
Select from the list of tissues those that are described by the following statements.

ciliated pseudostratified columnar
ciliated simple columnar
nonciliated simple columnar
nonciliated pseudostratified columnar
simple cuboidal
simple squamous
stratified columnar
stratified cuboidal
stratified squamous
transitional

1. Epidermis of the skin.
2. Lining of the intestine.
3. Lining of the blood vessels.
4. Lining of the urinary bladder.
5. Lining of anal canal.
6. Lining of uterine tubes.
7. Secretory epithelium in the pancreas.
8. Lining of large ducts in salivary glands.
9. Lining of nasal cavity and trachea.
10. Single layer of elongate cells without cilia.
11. Single layer of cubelike cells.
12. Multilayered; outer layer of thin, flat cells.
13. Ciliated cells appear multilayered but are not.
14. Multilayered; cells flatten when stretched.
15. Multilayered; outer layer of elongate cells.

1. _____
2. _____
3. _____
4. _____
5. _____
6. _____
7. _____
8. _____
9. _____
10. _____
11. _____
12. _____
13. _____
14. _____
15. _____

B. Connective Tissues
Select from the list of tissues those that are described by the statements that follow.

adipose tissue
areolar connective tissue
bone tissue
dense connective tissue
elastic cartilage
fibrocartilage
hyaline cartilage
reticular tissue

1. Superficial fascia.
2. Framework of lymph nodes.
3. Tendons and ligaments.
4. Supporting rings in trachea.
5. Supports pinna of external ear.
6. Occurs on ends of long bones.
7. Intervertebral discs.
8. Fat storage.
9. Most flexible supporting tissue.
10. Solid matrix of calcium salts.
11. Most rigid supporting tissues.
12. Cartilage tissue with many collagenous fibers.
13. White, glassy appearance.
14. Makes up most of the dermis.
15. Preforms most bones in fetal skeleton.

1. _____
2. _____
3. _____
4. _____
5. _____
6. _____
7. _____
8. _____
9. _____
10. _____
11. _____
12. _____
13. _____
14. _____
15. _____

C. Terminology
Write the term defined below in the answer column.

1. Protein in tendons.
2. Fiber-producing cell.
3. Mucus-secreting cell.
4. Lines medullary cavity.
5. Intercellular substance.
6. Produces bone matrix.
7. Cartilage cell.
8. Plates in spongy bone.
9. Rings of compact bone.
10. Space around osteocyte or chondrocyte.

1. _____
2. _____
3. _____
4. _____
5. _____
6. _____
7. _____
8. _____
9. _____
10. _____

D. Figure
Record the labels for Figure 5.22 in the answer column.

Figure 5.22

1. _____
2. _____
3. _____
4. _____
5. _____
6. _____
7. _____
8. _____
9. _____
10. _____
11. _____
12. _____
13. _____

E. Muscle Tissue Characteristics
Select the type of muscle tissue described by the statements below.

cardiac   skeletal   smooth

1. Attached to bones.
2. Rhythmic contractions.
3. Voluntary in function.
4. Cells tapered to a point at each end.
5. Cells without striations.
6. Fibers with striations and many nuclei.
7. Occurs in walls of blood vessels.
8. Cells separated by intercalated discs.
9. Striated cells with single nucleus.
10. Spontaneous contractions without neural activation.
11. Cells often branched to form network.
12. Forms muscle of heart.

1. _____
2. _____
3. _____
4. _____
5. _____
6. _____
7. _____
8. _____
9. _____
10. _____
11. _____
12. _____

F. Figures
List the labels for Figures 5.28 and 5.30.

Figure 5.28

1. _____
2. _____
3. _____
4. _____
5. _____
6. _____
7. _____
8. _____
9. _____
10. _____
11. _____
12. _____
13. _____
14. _____
15. _____
16. _____
17. _____

Figure 5.30

1. _____
2. _____
3. _____
4. _____
5. _____
6. _____
7. _____
8. _____
9. _____
10. _____
11. _____

G. Neuron Structure
Select the structure from the list that is described in the statements that follow.

axon
cell body
chromatophilic substance
collateral
dendrite
myelin sheath gaps
myelin sheath
neurilemma
neurofibrils
neurolemmocytes

1. Contains neuron nucleus.
2. Side branch of an axon.
3. Myelinated process of a motor neuron.
4. Myelinated process of a sensory neuron.
5. Process receiving action potentials.
6. Process transmitting action potentials to an effector.
7. Thin filaments in the cell body and processes.
8. Clumps of RER in the cell body.
9. Gaps between neurolemmocytes.
10. Formed by inner wrappings of neurolemmocytes.
11. Outer layer of a neurolemmocyte.
12. Cells covering all peripheral axons.

1. _____
2. _____
3. _____
4. _____
5. _____
6. _____
7. _____
8. _____
9. _____
10. _____
11. _____
12. _____

H. Neuron Types

Select the structural type of neuron described in the statements that follow.

bipolar   multipolar   pseudounipolar

1. Possesses a single process that forms a T-shaped branch.
2. Possesses a single dendrite and an axon.
3. Most common structural type.
4. Possesses multiple dendrites and an axon.
5. Spinal sensory neurons are examples.

1. _____
2. _____
3. _____
4. _____
5. _____

Select the functional type of neuron described in the statements that follow.

interneuron   motor   sensory

1. An afferent neuron.
2. An efferent neuron.
3. Totally within central nervous system.
4. Carries action potentials to a muscle or gland.
5. Carries action potentials to the central nervous system.

1. _____
2. _____
3. _____
4. _____
5. _____

I. Microscopic Study

Use the space below and additional sheets of paper as necessary to draw the appearance of each tissue observed. Add labels and make notations that will help you recognize these tissues in a lab practicum.

Laboratory Report 6

The Integumentary System

A. Matching

Select the structure from the list below that is described by the following statements and record it in the answer column.

apocrine sweat gland
arrector pili muscle
dermal papilla
dermis
eccrine sweat gland
epidermis
lamellar corpuscle
sebaceous gland
stratum basale
stratum corneum
stratum granulosum
stratum spinosum
subcutaneous layer
tactile corpuscle

1. Layer of epidermis containing melanocytes.
2. Contains blood vessels that nourish hair follicle.
3. Touch receptor in dermal papillae.
4. Sweat gland emptying secretion into hair follicle.
5. Inner layer of skin formed of connective tissue.
6. Outermost layer of epidermis.
7. Epidermal layer beneath the stratum granulosum.
8. Most abundant type of sweat gland.
9. Pressure receptor located deep in dermis.
10. Secretes oily lubricant into hair follicle.
11. Superficial fascia.
12. Activation produces goose bumps.
13. Formed of areolar connective and adipose tissues.
14. Four or five layers of stratified squamous epithelium.
15. Has numerous collagen fibers that give strength to skin.
16. Forms watery secretion containing waste materials.
17. Epidermal layer in which cell division occurs.
18. Secretion is responsible for body odor.
19. Epidermal layer with granule precursors of keratin.
20. Responsible for the pattern of fingerprints.
21. Sweat gland activated primarily by thermal stimuli.
22. Epidermal layer producing cells that form a hair.
23. Contains insulating adipose tissue.
24. Epidermal layer preventing evaporative water loss.
25. Skin layer containing nerves and blood vessels.
26. Epidermal layer of flat, dead, keratinized cells.
27. Contains numerous sensory receptors.
28. Larger, but less numerous, sweat glands.
29. Secretes sebum.
30. Contains coiled portion of eccrine sweat glands.

1. _____
2. _____
3. _____
4. _____
5. _____
6. _____
7. _____
8. _____
9. _____
10. _____
11. _____
12. _____
13. _____
14. _____
15. _____
16. _____
17. _____
18. _____
19. _____
20. _____
21. _____
22. _____
23. _____
24. _____
25. _____
26. _____
27. _____
28. _____
29. _____
30. _____

B. **Figures**

Record the labels for Figures 6.1 and 6.2 in the answer column.

C. **Completion**

1. Describe the function of melanin. _____
   _____
   _____
   _____
   _____

2. Explain why a "tan" is temporary. _____
   _____
   _____
   _____
   _____
   _____
   _____

D. **Microscopic Study**

Use the space below and additional sheets of paper to make drawings and notations that will help you understand skin structure and recognize skin and its components on a lab practicum.

Figure 6.1

1. _____
2. _____
3. _____
4. _____
5. _____
6. _____
7. _____
8. _____
9. _____
10. _____
11. _____
12. _____
13. _____
14. _____
15. _____
16. _____
17. _____
18. _____
19. _____
20. _____
21. _____
22. _____

Figure 6.2

1. _____
2. _____
3. _____
4. _____
5. _____
6. _____
7. _____
8. _____
9. _____
10. _____
11. _____
12. _____

E. Density of Touch Receptors

Indicate the minimum distance between the tips of the calipers that produced a two-point sensation.

Back of neck _____ mm   Inner surface of forearm _____ mm

Palm of hand _____ mm   Tip of index finger _____ mm

Is there a correlation between the density of touch receptors and the importance of touch sensations for the body parts tested? _____

Notes

Laboratory Report 7

Student _____
Lab Section _____

Skeletal Overview

A. *Figures*
List the labels for Figures 7.1, 7.2, and 7.3.

Figure 7.1

1. _____
2. _____
3. _____
4. _____
5. _____
6. _____
7. _____
8. _____
9. _____
10. _____
11. _____
12. _____

Figure 7.2

1. _____
2. _____
3. _____
4. _____
5. _____
6. _____
7. _____
8. _____
9. _____
10. _____
11. _____
12. _____
13. _____
14. _____
15. _____
16. _____
17. _____
18. _____
19. _____
20. _____
21. _____
22. _____
23. _____
24. _____

Figure 7.3

A. _____
B. _____
C. _____
D. _____
E. _____
F. _____
G. _____
H. _____
I. _____
J. _____

B. **Long Bone Structure**

Select the response that matches the statements that follow and write it in the answer column.

articular cartilage	epiphysis
compact bone	medullary cavity
diaphysis	periosteum
endosteum	red bone marrow
epiphysial line	spongy bone
epiphysial plate	yellow bone marrow

1. Shaft portion of a long bone.
2. Chamber in the bone shaft.
3. Type of bone marrow in the medullary cavity.
4. Enlarged end of a long bone.
5. Lining of the medullary cavity.
6. Type of bone tissue forming most of the diaphysis.
7. Connective tissue membrane covering bone.
8. Site of longitudinal growth.
9. Protective surface on ends of a long bone.
10. Type of bone tissue forming inside of epiphyses.
11. Reduces friction in joints formed by long bones.
12. Type of bone marrow in spongy bone of the femur.
13. Formed by fusion of epiphysis and diaphysis.

1. _____
2. _____
3. _____
4. _____
5. _____
6. _____
7. _____
8. _____
9. _____
10. _____
11. _____
12. _____
13. _____

Provide these answers on the basis of your study of the split long bones.

1. Describe the distribution of compact and spongy bone. _____

   Compact bone _____

   Spongy bone _____

2. Is the periosteum elastic or nonelastic? _____ Weak or strong? _____

   Is the periosteum firmly or loosely attached to the bone? _____

3. How is a tendon or ligament attached to the bone? _____

4. Feel the surface of an articular cartilage. Describe how it feels. _____

5. How do articular cartilages aid movement at a joint formed by two long bones? _____

6. Had the beef bone attained full growth? _____ Explain. _____

C. Parts of the Skeleton
Select the response that matches the statements that follow and write it in the answer column.

clavicle   patella   skull
coxal bone pubic symphysis sternum
femur   radius   thoracic cage
fibula   rib   tibia
humerus   sacrum   ulna
hyoid   scapula   vertebrae

1. Shoulder blade.
2. Collarbone.
3. Breastbone.
4. Shinbone.
5. Kneecap.
6. Arm bone.
7. Bones of vertebral column.
8. Thighbone.
9. Lateral bone of forearm.
10. Joint between coxal bones.
11. Horseshoe-shaped bone under lower jaw.
12. Hipbone.
13. Medial bone of forearm.
14. Thin bone lateral to tibia.
15. Bony framework of the head.
16. Two bones forming the shoulder girdle.
17. Three bones forming the elbow joint.
18. Two bones forming the knee joint.
19. Bones the rib connect to in the back.
20. Limb bone articulating with a scapula.
21. Limb bone articulating with a coxal bone.
22. Forms protective framework for thorax.
23. Part of vertebral column attached to coxal bones.
24. Articulates with humerus and clavicle.

1. _____
2. _____
3. _____
4. _____
5. _____
6. _____
7. _____
8. _____
9. _____
10. _____
11. _____
12. _____
13. _____
14. _____
15. _____
16a. _____
16b. _____
17a. _____
17b. _____
17c. _____
18a. _____
18b. _____
19. _____
20. _____
21. _____
22. _____
23. _____
24. _____

Skeletal Overview

D. **Surface Features**
Select the term that matches the descriptions that follow.

condyle   sinus
crest   spine
fissure   sulcus
foramen   trochanter
fossa   tubercle
head   tuberosity
meatus

1. An enlarged end of a bone that is supported by a narrow neck.
2. A passageway for nerves or blood vessels.
3. An air-filled cavity.
4. A shallow depression.
5. A sharp, slender process.
6. A very large process.
7. A groove or furrow.
8. A knucklelike process that articulates with another bone.
9. A narrow slit.
10. A large canal or passageway.
11. A narrow ridge.
12. A low, roughened process to which a muscle attaches.
13. A small, rounded process.

E. **Bone Fractures**
Select the type of fracture that matches the statements that follow and write it in the answer column.

comminuted   oblique
compacted   segmental
fissured   spiral
greenstick   transverse

1. Results from twisting of the bone.
2. Complete fracture at 90° to bone axis.
3. Break on one side only due to bending of the bone.
4. One part of a bone is forced into another part of the same bone.
5. A linear, incomplete splitting of the bone.
6. Complete fracture *not* at 90° to bone axis.
7. Bone broken into several pieces.
8. A single piece is broken out of the bone.

Write in the answer column the type of fracture that is described by the statements below.

9. Fractured bone breaks through the skin.
10. Fractured bone is not broken clear through.
11. Fracture caused by excessive vertical force.
12. Fractured bone does not break through the skin.

1. _____
2. _____
3. _____
4. _____
5. _____
6. _____
7. _____
8. _____
9. _____
10. _____
11. _____
12. _____
13. _____

1. _____
2. _____
3. _____
4. _____
5. _____
6. _____
7. _____
8. _____
9. _____
10. _____
11. _____
12. _____

Laboratory Report 8

Student _____

Lab Section _____

THE AXIAL SKELETON

A. *Figures*

List the labels for Figures 8.1 through 8.7 and Figure 8.9 in the answer columns.

Figure 8.1

1. _____
2. _____
3. _____
4. _____
5. _____
6. _____
7. _____
8. _____
9. _____
10. _____
11. _____
12. _____
13. _____
14. _____
15. _____
16. _____
17. _____
18. _____
19. _____
20. _____
21. _____
22. _____

Figure 8.2

1. _____
2. _____
3. _____
4. _____
5. _____
6. _____
7. _____
8. _____
9. _____
10. _____
11. _____
12. _____
13. _____
14. _____
15. _____
16. _____
17. _____
18. _____
19. _____

Figure 8.3

1. _____
2. _____
3. _____
4. _____
5. _____
6. _____
7. _____
8. _____
9. _____
10. _____
11. _____
12. _____
13. _____
14. _____
15. _____

Figure 8.4

1. _____
2. _____
3. _____
4. _____
5. _____
6. _____
7. _____
8. _____
9. _____
10. _____
11. _____
12. _____
13. _____
14. _____
15. _____
16. _____
17. _____

Figure 8.5

1. _____
2. _____
3. _____
4. _____
5. _____
6. _____
7. _____
8. _____
9. _____
10. _____

Figure 8.6

1. _____
2. _____
3. _____
4. _____

Figure 8.7

1. _____
2. _____
3. _____
4. _____
5. _____
6. _____
7. _____
8. _____
9. _____
10. _____
11. _____
12. _____
13. _____
14. _____
15. _____

Figure 8.9

1. _____
2. _____
3. _____
4. _____
5. _____
6. _____
7. _____
8. _____
9. _____
10. _____
11. _____
12. _____

B. Bones

Select the bones that are described or that have the structures noted in the statements below.

ethmoid   nasal   sphenoid
frontal   nasal conchae   temporal
lacrimal occipital   vomer
mandible palatine   zygomatic
maxilla   parietal

1. Upper bone of nasal septum
2. Foramen magnum
3. Lower jaw
4. Cheekbone
5. Mental foramen
6. Scroll-like bones in nasal cavity
7. Coronoid process
8. Zygomatic process
9. Internal acoustic meatus
10. Carotid canal
11. Mandibular fossa
12. Mastoid process
13. Ramus
14. Sella turcica
15. Styloid process
16. Supraorbital foramen
17. Upper jaw
18. Cribriform plates

1. _____
2. _____
3. _____
4. _____
5. _____
6. _____
7. _____
8. _____
9. _____
10. _____
11. _____
12. _____
13. _____
14. _____
15. _____
16. _____
17. _____
18. _____

C. Sutures

Select the sutures described below.

coronal   median palatine   squamous
lambdoid   sagittal

1. Joins parietal bones.
2. Joins frontal and parietal bones.
3. Joins palatine processes of maxillae.
4. Joins parietal and occipital bones.
5. Joins parietal and temporal bones.

1. _____
2. _____
3. _____
4. _____
5. _____

D. **Completion**
Write your response in the answer column.

1. Two bones forming the hard palate.
2. Two bones forming the zygomatic arch.
3. Allow compression of skull during birth.
4. Four bones that contain sinuses.
5. Point of attachment of skull with first vertebra.
6. Allow sound waves to enter skull.
7. Form bridge of nose.
8. Contain groove for tear duct.
9. Forms lateral wall of orbital cavity.
10. Forms back of the orbital cavity.
11. Forms roof of the orbital cavity.
12. Depression into which the pituitary gland fits.
13. Two bones forming cheekbones.
14. Two bones forming nasal septum.
15. Membranous areas at junctions of cranial bones in a fetal skull.

1. _____
2. _____
3. _____
4. _____
5. _____
6. _____
7. _____
8. _____
9. _____
10. _____
11. _____
12. _____
13. _____
14. _____
15. _____

E. **Figures**
List the labels for Figures 8.10, 8.11, and 8.12.

Figure 8.10

1. _____
2. _____
3. _____
4. _____
5. _____
6. _____
7. _____
8. _____
9. _____
10. _____
11. _____
12. _____
13. _____

14. _____
15. _____
16. _____
17. _____
18. _____
19. _____
20. _____
21. _____
22. _____
23. _____
24. _____
25. _____
26. _____
27. _____
28. _____
29. _____

Figure 8.11

1. _____
2. _____
3. _____
4. _____
5. _____
6. _____
7. _____

Figure 8.12

1. _____
2. _____
3. _____
4. _____
5. _____
6. _____
7. _____
8. _____
9. _____
10. _____
11. _____
12. _____
13. _____
14. _____
15. _____

F. **Parts of the Vertebral Column**
Select the response from the list at the right that matches the statement. Answers may be used more than once.

1. Facet on transverse processes.
2. Transverse foramina.
3. First cervical vertebra.
4. Rib without anterior attachment.
5. Lower part of breastbone.
6. Middle part of breastbone.
7. Second cervical vertebra.
8. Upper part of breastbone.
9. Vertebrae of the neck.
10. Rib attaching to sternum by costal cartilage.
11. Rib with fused costal cartilage.
12. Breastbone.
13. Tailbone.
14. Possesses dens.
15. Has largest vertebral foramen.

atlas
axis
cervical vertebrae
coccyx
false rib
floating rib
lumbar vertebrae
manubrium
sacrum
sternum
thoracic vertebrae
true rib
vertebral body
xiphoid process

1. _____
2. _____
3. _____
4. _____
5. _____
6. _____
7. _____
8. _____
9. _____
10. _____
11. _____
12. _____
13. _____
14. _____
15. _____

G. **Numbers**
Indicate the number of components of the following.

1. Cervical vertebrae.
2. Fused vertebrae composing coccyx.
3. Fused vertebrae composing sacrum.
4. Lumbar vertebrae.
5. Pairs of false ribs.
6. Pairs of floating ribs.
7. Pairs of true ribs.
8. Total of all movable vertebrae.
9. Spinal curvatures.
10. Thoracic vertebrae.

1. _____
2. _____
3. _____
4. _____
5. _____
6. _____
7. _____
8. _____
9. _____
10. _____

H. **Completion**
Provide the terms or phrases that match the statements.

1. Cushion between vertebral bodies.
2. Articulates with inferior articulating facets.
3. Bony framework around vertebral foramen.
4. Part of vertebra bearing most load.
5. Vertebral type with largest bodies.
6. Forms back of pelvis.
7. Vertebrae with ribs.
8. Attach ribs to sternum.
9. Articulate with facets on transverse processes.
10. Openings through which spinal nerves exit.

1. _____
2. _____
3. _____
4. _____
5. _____
6. _____
7. _____
8. _____
9. _____
10. _____

Laboratory Report 9

Student _____

Lab Section _____

The Appendicular Skeleton

A. Figures

List the labels for Figures 9.1 through 9.4.

Figure 9.1

1. _____
2. _____
3. _____
4. _____
5. _____
6. _____
7. _____
8. _____

Figure 9.2

1. _____
2. _____
3. _____
4. _____
5. _____
6. _____
7. _____
8. _____
9. _____
10. _____
11. _____
12. _____
13. _____
14. _____
15. _____
16. _____
17. _____
18. _____
19. _____
20. _____
21. _____
22. _____
23. _____
24. _____
25. _____
26. _____
27. _____
28. _____

Figure 9.3

1. _____
2. _____
3. _____
4. _____
5. _____
6. _____
7. _____
8. _____
9. _____
10. _____
11. _____
12. _____
13. _____
14. _____
15. _____

Figure 9.4

1. _____
2. _____
3. _____
4. _____
5. _____
6. _____
7. _____
8. _____
9. _____
10. _____
11. _____
12. _____
13. _____
14. _____
15. _____
16. _____
17. _____
18. _____
19. _____
20. _____
21. _____
22. _____
23. _____
24. _____
25. _____

B. **True–False**

Compare the male and female pelves, side by side, to determine if the following statements are true or false.

1. The opening of the female pelvis is larger and more oval than that of the male pelvis.
2. The acetabula of the female pelvis face forward more than those of the male pelvis.
3. The iliac crests of the female pelvis protrude laterally more than those of the male pelvis.
4. The distance between the ischial spines is greater in the female pelvis.
5. The angle of the pubic arch below the pubic symphysis is wider in the female pelvis.
6. The sacrum is more concave in the female pelvis than in the male pelvis.

1. _____
2. _____
3. _____
4. _____
5. _____
6. _____

C. Completion

Provide the bones or structures described by the statements.

1. Fossa on scapula that articulates with humerus.
2. Condyle of humerus that articulates with the radius.
3. Scapular process that articulates with clavicle.
4. Large process lateral to head of humerus.
5. Process forming point of elbow.
6. Fossa of ulna that articulates with the humerus.
7. Wrist bones.
8. Bones of palm of the hand.
9. Scapular process projecting forward under clavicle.
10. Fossa on distal front of humerus.
11. Process on distal lateral margin of radius.
12. Bones of the fingers and toes.
13. Tuberosity just below neck of radius.
14. Fossa on lower back portion of humerus.
15. Part of radius that articulates with the humerus.
16. Upper bone of a coxal bone.
17. Process at angle of ischium.
18. Heelbone.
19. Large foramen in a coxal bone formed by pubis and ischium.
20. Joint between the pubic bones.
21. Large process lateral to neck of femur.
22. Fossa of a coxal bone that articulates with head of femur.
23. Joint between a coxal bone and sacrum.
24. Uppermost process on back of ischium.
25. Upper edge of ilium.
26. Carpal bone that articulates with tibia and fibula.
27. Process below neck on medial surface of femur.
28. Proximal process on tibia below kneecap.
29. Process on medial surface at distal end of tibia.
30. Forms the arch of the foot.

1. _____
2. _____
3. _____
4. _____
5. _____
6. _____
7. _____
8. _____
9. _____
10. _____
11. _____
12. _____
13. _____
14. _____
15. _____
16. _____
17. _____
18. _____
19. _____
20. _____
21. _____
22. _____
23. _____
24. _____
25. _____
26. _____
27. _____
28. _____
29. _____
30. _____

Notes

Laboratory Report 10

Student _____

Lab Section _____

ARTICULATIONS AND BODY MOVEMENTS

A. Figures

List the labels for Figures 10.1 through 10.4.

Figure 10.1

1. _____
2. _____
3. _____
4. _____
5. _____
6. _____
7. _____
8. _____
9. _____
10. _____

Figure 10.2

1. _____
2. _____
3. _____
4. _____
5. _____
6. _____
7. _____
8. _____
9. _____
10. _____
11. _____

Figure 10.3

1. _____
2. _____
3. _____
4. _____
5. _____
6. _____
7. _____
8. _____
9. _____
10. _____
11. _____
12. _____
13. _____
14. _____

Figure 10.4

1. _____
2. _____
3. _____
4. _____
5. _____
6. _____
7. _____
8. _____
9. _____

B. Types and Subtypes

Identify the basic articulation type to which the subtypes listed belong.

amphiarthrosis   diarthrosis   synarthrosis

1. Ball-and-socket
2. Condyloid
3. Gliding
4. Hinge
5. Pivot
6. Saddle
7. Suture
8. Symphysis
9. Synchondrosis
10. Syndesmosis

1. _____
2. _____
3. _____
4. _____
5. _____
6. _____
7. _____
8. _____
9. _____
10. _____

C. Joint Characteristics

Select the joints that have the characteristics described by the statements.

ball-and-socket   saddle
condyloid   suture
gliding   symphysis
hinge   synchondrosis
pivot   syndesmosis

1. Bone ends are flat or slightly convex.
2. Slightly movable; fibrocartilaginous pad.
3. Slightly movable; bones joined by interosseous ligament.
4. Immovable; bones bonded by cartilaginous disc.
5. Rotational movement only.
6. Freely movable in one direction only.
7. Freely movable in two planes.
8. Angular movement in all directions.
9. Immovable; bonded by dense connective tissue.
10. Bone ends are convex in one direction and concave in the other.

1. _____
2. _____
3. _____
4. _____
5. _____
6. _____
7. _____
8. _____
9. _____
10. _____

D. Joint Classification

By examining and manipulating the bones of an articulated skeleton, classify the following joints according to the list in Section C.

1. Femur–tibia.
2. Humerus–ulna.
3. Vertebra–vertebra.
4. Phalanges–phalanges.
5. Phalanges–metacarpals.
6. Humerus–scapula.
7. Atlas–axis.
8. Carpals–carpals.
9. Tibia–fibula, distal ends.
10. Left pubis–right pubis.
11. Parietal–frontal.
12. Trapezium–metacarpal of thumb.
13. Diaphysis–epiphysis of femur in child.
14. Radius–ulna, distal ends.
15. Femur–coxal bone.

1. _____
2. _____
3. _____
4. _____
5. _____
6. _____
7. _____
8. _____
9. _____
10. _____
11. _____
12. _____
13. _____
14. _____
15. _____

LABORATORY REPORT 10

E. Muscle Arrangements

Match the terms in the list below with the statements that follow.

agonists   fixators
antagonists   synergists

1. Muscles that assist prime movers.
2. Prime movers.
3. Muscles opposing prime movers.
4. Muscles holding structures steady to allow prime movers to act smoothly.

1. _____
2. _____
3. _____
4. _____

F. Movements

Match the terms in the list below with the statements that follow.

abduction   hyperextension
adduction   inversion
circumduction   plantar flexion
dorsiflexion   pronation
eversion   rotation
extension   supination
flexion

1. Movement of hand from palm down to palm up.
2. Movement of limbs away from midline of the body.
3. Movement of 200° around a central point.
4. Flexion of foot upward (specific term).
5. Decrease in the angle of bones at a joint.
6. Movement of hand from palm up to palm down.
7. Movement of limb tracing a circle.
8. Turning the sole of the foot inward.
9. Increase in the angle of bones at a joint.
10. Backward movement of head.
11. Movement of a limb toward midline of the body.
12. Moving sole of foot downward (specific term).
13. Turning the sole of the foot outward.

1. _____
2. _____
3. _____
4. _____
5. _____
6. _____
7. _____
8. _____
9. _____
10. _____
11. _____
12. _____
13. _____

Notes

Laboratory Report 11

Student _____

Lab Section _____

MUSCLE ORGANIZATION AND CONTRACTIONS

A. Figures
List the labels for Figure 11.1.

Figure 11.1

1. _____
2. _____
3. _____
4. _____
5. _____
6. _____
7. _____
8. _____
9. _____
10. _____
11. _____
12. _____
13. _____

B. Muscle Structure
Match the terms below with the statements that follow.

aponeurosis   fasciculus
deep fascia   ligament
distal attachment   perimysium
endomysium   proximal attachment
epimysium   tendon

1. Bundle of muscle fibers.
2. Connective tissue around each fasciculus.
3. End of a muscle farther from trunk.
4. Thin layer of connective tissue around each muscle fiber.
5. Broad sheet of connective tissue attaching a muscle to another muscle or bone.
6. Connective tissue that surrounds a group of fasciculi.
7. End of a muscle nearest the trunk.
8. Outer layer of connective tissue that envelops entire muscle.
9. Narrow band of connective tissue attaching muscle to bone or other muscle.
10. Not part of a muscle.

1. _____
2. _____
3. _____
4. _____
5. _____
6. _____
7. _____
8. _____
9. _____
10. _____

C. Figure
List the labels for Figure 11.2.

1. _____
2. _____
3. _____
4. _____
5. _____
6. _____
7. _____
8. _____
9. _____
10. _____

D. Neuromuscular Junction
Select the terms that match the statements below.

acetylcholine   synaptic vesicle
sarcolemma   terminal boutons
synaptic cleft

1. Space between terminal bouton and sarcolemma.
2. Embedded in invaginations of the sarcolemma.
3. Secreted into synaptic cleft when action potential reaches the terminal bouton.
4. Membrane covering muscle fiber.
5. Membranous container of acetylcholine.
6. Reacts with receptors on sarcolemma to initiate an action potential in the muscle fiber.

1. _____
2. _____
3. _____
4. _____
5. _____
6. _____

List the sequence of events in the transmission of an action potential from an axon terminal branch to a muscle fiber.

_____
_____
_____
_____

E. Myofibril Structure
Select the terms that match the statements below.

A band   thick myofilament
I band   thin myofilament
sarcomere   Z disc

1. Part of myofibril between two Z discs.
2. Myofilament with cross-bridges.
3. Myofilament attached to Z discs.
4. Main myofilament within the A band.
5. Lies between two I bands.
6. Myofilament within I bands.

1. _____
2. _____
3. _____
4. _____
5. _____
6. _____

Laboratory Report 11

F. **Mechanics of Contraction**

Select the terms that match the statements below.

ATP sarcomere
calcium ions sarcoplasmic reticulum
myosin heads

1. Releases calcium ions when depolarized by action potential.
2. Expose the active sites on the thin myofilament.
3. Shortens during contraction.
4. Provides energy for contraction.
5. Pull thin myofilaments toward the center of the A band.

1. _____
2. _____
3. _____
4. _____
5. _____

List the sequence of events in the contraction of a muscle fiber after it has been activated by an action potential from a motor neuron. _____
_____
_____
_____
_____
_____
_____
_____

G. **Experimental Muscle Contraction**

1. Indicate the length of the muscle strands before and after contraction and calculate the percentage of contraction.

   Before contraction: _____ mm

   After contraction: _____ mm

   Percentage of contraction: _____

2. Draw the pattern of the striations as they appear in relaxed and contracted muscle fibers.

Relaxed

Contracted

Muscle Organization and Contractions

Notes

Laboratory Report 12

Student _____

Lab Section _____

AXIAL MUSCLES

A. Figures

List the labels for Figures 12.1 through 12.2.

Figure 12.1

1. _____
2. _____
3. _____
4. _____
5. _____
6. _____
7. _____
8. _____
9. _____
10. _____
11. _____

Figure 12.2

1. _____
2. _____
3. _____
4. _____
5. _____
6. _____
7. _____
8. _____
9. _____

B. **Head and Neck Muscles**
Select the muscle that matches the following statements. Muscles can be used more than once.

buccinator	orbicularis oris
depressor anguli oris	platysma
frontal belly	sternocleidomastoid
masseter	temporalis
orbicularis oculi	zygomaticus

1. Primary muscle of chewing.
2. Frowning.
3. Wrinkles forehead and raises eyebrows.
4. Smiling.
5. Compresses cheeks.
6. Contraction on one side produces a "wink."
7. Draws head downward toward chest.
8. Acts synergistically with the masseter.
9. Closes eyelids.
10. Closes and puckers lips.
11. Pulls scalp backward.
12. Assists in opening the mouth.
13. Holds food between teeth during chewing.
14. Contraction of muscle on right side turns head to the left.

1. _____
2. _____
3. _____
4. _____
5. _____
6. _____
7. _____
8. _____
9. _____
10. _____
11. _____
12. _____
13. _____
14. _____

C. **Respiratory, Abdominal, and Vertebral Muscles**
Select the muscle described in the following statements. Muscles can be used more than once.

erector spinae	internal oblique
external intercostals	rectus abdominis
external oblique	transversospinalis group
internal intercostals	transversus abdominis

1. Innermost abdominal muscle.
2. Superficial muscles of the vertebral column.
3. Elevates thoracic cage in breathing.
4. Depresses thoracic cage during breathing.
5. Segmented muscle extending from ribs to pubis.
6. Outermost abdominal muscle.
7. Lies just under the external oblique.
8. Deepest muscles of the vertebral column.

1. _____
2. _____
3. _____
4. _____
5. _____
6. _____
7. _____
8. _____

Laboratory Report 13

Student _____

Lab Section _____

APPENDICULAR MUSCLES

A. Figure
List the labels for Figure 13.1.

Figure 13.1

1. _____
2. _____
3. _____
4. _____
5. _____
6. _____
7. _____
8. _____
9. _____

B. Muscles Moving the Pectoral Girdle
Select the muscle described in the following statements. Muscles can be used more than once.

coracobrachialis   serratus anterior
deltoid   supraspinatus
infraspinatus   teres major
latissimus dorsi   trapezius
pectoralis major

1. Flexes and adducts the arm.
2. Rotates arm medially only.
3. Adducts and elevates scapula.
4. Rotates arm laterally only.
5. Extends, adducts, and rotates arm medially.
6. Draws scapula downward and forward.
7. Adducts and flexes arm medially.
8. Abducts, flexes, and extends arm.
9. Draws head backward.
10. Primary muscle of the shoulder.
11. Large surface muscle of the upper chest.
12. Triangular surface muscle of upper back.
13. Assists deltoid in abducting arm.
14. Hyperextends head.
15. Superficial muscle covering lower back.

1. _____
2. _____
3. _____
4. _____
5. _____
6. _____
7. _____
8. _____
9. _____
10. _____
11. _____
12. _____
13. _____
14. _____
15. _____

C. **Figures**
List the labels for Figures 13.2 and 13.3.

Figure 13.2

A. _____
B. _____
C. _____
D. _____
E. _____

Figure 13.3

1. _____
2. _____
3. _____
4. _____
5. _____
6. _____
7. _____
8. _____
9. _____
10. _____
11. _____
12. _____
13. _____

D. *Muscles Moving the Forearm*
Select the muscle described in the statements below. Answers may be used more than once.

biceps brachii   pronator teres
brachialis   supinator
brachioradialis   triceps brachii

1. Extends the forearm.
2. Supinates forearm.
3. Three muscles that flex the forearm.
4. Antagonist of the brachialis.
5. Flexes and supinates the forearm.
6. Superficial muscle on front of arm.
7. Covers the back of humerus.
8. Superficial muscle on lateral surface of forearm.
9. Located just under biceps brachii.
10. Primary muscle pronating the forearm.

1. _____
2. _____
3. _____
4. _____
5a. _____
5b. _____
5c. _____
6. _____
7. _____
8. _____
9. _____
10. _____

E. Muscles Moving the Hand
Select the muscle described in the statements below.

extensor carpi radialis   flexor carpi radialis
extensor carpi ulnaris   flexor carpi ulnaris
extensor digitorum   flexor digitorum

1. Flexes and adducts the hand.
2. Flexes and abducts the hand.
3. Extends fingers 2–5.
4. Flexes fingers 2–5.
5. Extends and abducts the hand.
6. Extends and adducts the hand.

1. _____
2. _____
3. _____
4. _____
5. _____
6. _____

F. Figures
List the labels for Figures 13.4 through 13.6.

Figure 13.4

1. _____
2. _____
3. _____
4. _____
5. _____
6. _____
7. _____
8. _____
9. _____
10. _____

Figure 13.5

1. _____
2. _____
3. _____
4. _____
5. _____
6. _____
7. _____
8. _____
9. _____
10. _____

Figure 13.6

1. _____
2. _____
3. _____
4. _____
5. _____
6. _____

G. **Muscles Moving the Thigh**
Select the muscle(s) from the list that is (are) described by the statements below.

adductor longus
adductor magnus
gluteus maximus
gluteus medius
gluteus minimus
iliacus
pectineus
psoas major

1. Extends and rotates thigh laterally.
2. Proximal attachment is the iliac fossa.
3. Three muscles that adduct, flex, and rotate thigh laterally.
4. Two muscles that abduct and rotate thigh medially.
5. Strongest adductor muscle.
6. Superficial buttocks muscle.
7. Innermost gluteus muscle.
8. Longest adductor muscle.
9. Proximal attachment on lumbar vertebrae.

1. _____
2. _____
3a. _____
3b. _____
3c. _____
4a. _____
4b. _____
5. _____
6. _____
7. _____
8. _____
9. _____

H. **Muscles Moving the Leg**
Select the muscle(s) that apply to the following statements.

biceps femoris
gracilis
rectus femoris
sartorius
semimembranosus
semitendinosus
tensor fasciae latae
vastus intermedius
vastus lateralis
vastus medialis

1. Four muscles extending the leg.
2. Adducts thigh and flexes leg.
3. Flexes the leg and thigh.
4. Three hamstrings flexing the leg.
5. Deepest quadriceps muscle.
6. Superficial middle quadriceps muscle.
7. Lateral hamstring muscle.
8. Medial hamstring muscle.

1a. _____
1b. _____
1c. _____
1d. _____
2. _____
3. _____
4. _____
5a. _____
5b. _____
5c. _____
6. _____
7. _____
8. _____

I. Muscles Moving the Foot

Select the muscle(s) that apply to the following statements. Muscles may be used more than once.

extensor digitorum longus   soleus
gastrocnemius   tibialis anterior
fibularis longus

1. Two muscles with a common calcaneal tendon.
2. Flexes the leg and plantar flexes the foot.
3. Muscle that plantar flexes and evert the foot.
4. Extends toes 2–5.
5. Dorsiflexes and inverts the foot.
6. Two calf muscles that plantar flex the foot.
7. Supports the transverse arch of the foot.

1a. _____
1b. _____
2. _____
3. _____
4. _____
5. _____
6. _____
7. _____

J. Major Surface Muscles

List the labels for Figure 13.7.

Anterior

1. _____
2. _____
3. _____
4. _____
5. _____
6. _____
7. _____
8. _____
9. _____
10. _____
11. _____
12. _____
13. _____
14. _____
15. _____
16. _____
17. _____
18. _____
19. _____
20. _____

Posterior

1. _____
2. _____
3. _____
4. _____
5. _____
6. _____
7. _____
8. _____
9. _____
10. _____
11. _____
12. _____
13. _____
14. _____
15. _____

Notes

Laboratory Report 14

Student _____
Lab Section _____

THE SPINAL CORD AND REFLEX ARCS

A. Figures
List the labels for Figures 14.1, 14.2, and 14.5

Figure 14.1

1. _____
2. _____
3. _____
4. _____
5. _____
6. _____
7. _____
8. _____
9. _____
10. _____
11. _____
12. _____
13. _____
14. _____
15. _____
16. _____
17. _____
18. _____
19. _____
20. _____
21. _____

Figure 14.2

1. _____
2. _____
3. _____
4. _____
5. _____
6. _____
7. _____
8. _____
9. _____
10. _____
11. _____
12. _____
13. _____
14. _____
15. _____
16. _____

Figure 14.5

1. _____
2. _____
3. _____
4. _____
5. _____
6. _____

1. _____
2. _____
3. _____
4. _____
5. _____
6. _____
7. _____

B. Spinal Nerves and Plexuses
Select the terms that match the statements below.

brachial plexus
cervical plexus   5 pairs
lumbar plexus   8 pairs
sacral plexus   12 pairs

1. The number of pairs of cervical nerves.
2. The number of pairs of thoracic nerves.
3. The number of pairs of lumbar nerves.
4. Formed of upper cervical nerves.
5. Formed of lower lumbar and sacral nerves.
6. Formed of lower cervical nerves plus T1.
7. Formed of most lumbar nerves.

271

C. **The Spinal Cord**
Select the terms that match the statements below.

arachnoid mater
central canal
dura mater
epidural space
gray matter
pia mater
subarachnoid space
subdural space
vertebral canal
white matter

1. Meninx adhered to surface of spinal cord.
2. Outermost meninx.
3. Meninx between other meninges.
4. Channel continuous with ventricles of the brain.
5. Outer portion of spinal cord.
6. Inner portion of spinal cord.
7. Space between outer meninx and vertebrae.
8. Space around spinal cord containing cerebrospinal fluid.
9. Part of spinal cord containing unmyelinated neuron processes.
10. Part of dorsal aspect containing spinal cord.

1. _____
2. _____
3. _____
4. _____
5. _____
6. _____
7. _____
8. _____
9. _____
10. _____

D. **Reflex Arcs**
Select the terms that match the statements below.

effector
gray matter
interneuron
motor neuron
postganglionic efferent neuron
preganglionic efferent neuron
receptor
sensory neuron
spinal ganglion
white matter

1. A muscle or gland.
2. Carries an action potential from receptor to spinal cord.
3. Carries an action potential from spinal cord to an effector.
4. Receives stimuli and forms action potentials.
5. Part of spinal cord with many myelinated neuron processes.
6. Contains cell bodies of sensory neurons.
7. Carries an action potential from spinal cord to autonomic ganglia.
8. Neuron within the gray matter.
9. Carries action potentials from autonomic ganglia to viscera.
10. Where synapses occur in spinal cord.

1. _____
2. _____
3. _____
4. _____
5. _____
6. _____
7. _____
8. _____
9. _____
10. _____

E. **Diagnostic Reflexes**
Record the results of your responses.

Biceps reflex: left arm _____   right arm _____

Triceps reflex: left arm _____   right arm _____

Patellar reflex: left leg _____   right leg _____

Calcaneal reflex: left foot _____   right foot _____

Explain why reinforcement techniques improve responses. _____

Laboratory Report 15

Student _____

Lab Section _____

THE BRAIN

A. Figures

List the labels for Figures 15.1, 15.2, 15.3, 15.4, and 15.6.

Figure 15.1

1. _____
2. _____
3. _____
4. _____
5. _____
6. _____
7. _____
8. _____
9. _____
10. _____

Figure 15.2

1. _____
2. _____
3. _____
4. _____
5. _____
6. _____
7. _____
8. _____
9. _____
10. _____
11. _____
12. _____

Figure 15.3

1. _____
2. _____
3. _____
4. _____
5. _____
6. _____
7. _____
8. _____
9. _____
10. _____
11. _____
12. _____

Figure 15.4

1. _____
2. _____
3. _____
4. _____
5. _____
6. _____
7. _____
8. _____
9. _____
10. _____
11. _____
12. _____
13. _____
14. _____
15. _____
16. _____
17. _____
18. _____
19. _____
20. _____

Figure 15.6

1. _____
2. _____
3. _____
4. _____
5. _____
6. _____
7. _____
8. _____
9. _____
10. _____
11. _____
12. _____
13. _____
14. _____
15. _____
16. _____
17. _____
18. _____
19. _____
20. _____
21. _____

B. The Meninges
Select the structures described by the statements below. Words may be used more than once or not used at all.

arachnoid mater   epidural space   subarachnoid space
dura mater   pia mater   subdural space

1. The fibrous meninx attached to cranial bones.
2. Meninx attached to the brain surface.
3. The middle meninx.
4. Space containing cerebrospinal fluid.
5. Meninx containing superior sagittal sinus.

1. _____
2. _____
3. _____
4. _____
5. _____

C. The Ventricles and Cerebrospinal Fluid
Provide the words to correctly complete the paragraph. Cerebrospinal fluid is released into the ventricles from the __1__ in each ventricle. It passes from the lateral ventricles into the third ventricle via an opening, the __2__ foramen, and then into the fourth ventricle through the __3__. From the fourth ventricle, the fluid passes via the __4__ into the __5__ below the cerebellum. From here, some of the fluid moves upward around the cerebellum and cerebrum and diffuses into the superior sagittal sinus. Most of the fluid flows downward in the subarachnoid space along the __6__ surface of the spinal cord, upward along the __7__ surface of the spinal cord, around the cerebrum, and finally diffuses from the __8__ into the blood in the __9__.

1. _____
2. _____
3. _____
4. _____
5. _____
6. _____
7. _____
8. _____
9. _____

D. Fissures and Sulci
Select the structures described by the statements below.

central sulcus   longitudinal cerebral fissure
lateral sulcus   parieto-occipital sulcus

1. Between frontal and parietal lobes.
2. Separates cerebral hemispheres.
3. Between temporal and parietal lobes.
4. Between occipital and parietal lobes.

1. _____
2. _____
3. _____
4. _____

E. *Brain Regions and Functions*

Select the parts of the brain described by the statements below.

cerebellum   hypothalamus   midbrain
cerebrum   medulla oblongata   pons

1. Controls heart and breathing rates.
2. Functions in will, intelligence, and memory.
3. Controls body temperature and water balance.
4. Controls muscular coordination.
5. Pathway for axons between the cerebrum and the pons, cerebellum, and medulla oblongata.

1. _____
2. _____
3. _____
4. _____
5. _____

F. *Functional Areas of the Cerebrum*

1. List the labels for Figure 15.7.

1. _____
2. _____
3. _____
4. _____
5. _____
6. _____
7. _____
8. _____
9. _____
10. _____
11. _____
12. _____
13. _____

2. Select the functional areas described by the statements below.

auditory association   primary motor
common interpretive   primary somatosensory
motor speech   primary taste
prefrontal   primary visual
premotor   sensory speech
primary auditory   somatosensory association
primary olfactory   visual association

1. Precentral gyrus of frontal lobe.
2. Postcentral gyrus of parietal lobe.
3. Voluntary control of body movements.
4. Occipital lobe.
5. Controls speaking.
6. Visual images.
7. Upper temporal lobe.
8. Medial surface of temporal lobe.
9. Touch sensations of the skin.
10. Coordinates learned motor skills.
11. Odor sensations.
12. Sound sensations.
13. Provides assessment of sensory input.
14. Interprets meaning of written and spoken words.
15. Site of judgment, critical thinking, and personality.

1. _____
2. _____
3. _____
4. _____
5. _____
6. _____
7. _____
8. _____
9. _____
10. _____
11. _____
12. _____
13. _____
14. _____
15. _____

G. Summary of Structures and Functions

1. Record the name of the brain structure described below.

   1. The second largest portion of the brain.
   2. Composed of the midbrain, pons, and medulla oblongata.
   3. Fluid within the subarachnoid space.
   4. Secretes the hormone melatonin.
   5. Provides an uncritical sensory awareness.
   6. Generates action potentials to keep the cerebrum alert.
   7. Lowest portion of the brainstem.
   8. The largest part of the brain.
   9. Fissure separating the cerebral hemispheres.
   10. Controls heart and breathing rates.

1. _____
2. _____
3. _____
4. _____
5. _____
6. _____
7. _____
8. _____
9. _____
10. _____

2. Select the structures that are described in the statements below.

cerebral peduncles   hypothalamus
corpora quadrigemina   infundibulum
corpus callosum   optic chiasm
fornix   thalamus

   1. Receives sensory action potentials and relays them to the cerebrum.
   2. Contains tracts associated with the sense of smell.
   3. Regulates body temperature.
   4. Consists of axons connecting cerebral hemispheres.
   5. Stalk supporting the pituitary gland.
   6. Enables cerebrospinal fluid to diffuse into the blood.
   7. Allows cerebrospinal fluid to pass from the lateral ventricles into the third ventricle.
   8. Crossing of optic axons.
   9. Main pathway of motor axons from the cerebrum.
   10. Reflex center for movements in response to auditory and visual stimuli.
   11. Connects third and fourth ventricles.
   12. Secretes cerebrospinal fluid into the ventricles.

1. _____
2. _____
3. _____
4. _____
5. _____
6. _____
7. _____
8. _____
9. _____
10. _____
11. _____
12. _____

H. Cranial Nerves

Select the cranial nerves described by the statements below.

I. Olfactory
II. Optic
III. Oculomotor
IV. Trochlear
V. Trigeminal
VI. Abducens
VII. Facial
VIII. Vestibulocochlear
IX. Glossopharyngeal
X. Vagus
XI. Accessory
XII. Hypoglossal

1. Innervates abdominal viscera.
2. Carries action potentials from the internal ear.
3. Carries motor action potentials for swallowing reflexes.
4. Innervates teeth and gums.
5. Carries motor action potentials to four muscles moving eyeball.
6. Two nerves each innervating one extrinsic eye muscle.
7. Innervates muscles of the tongue.
8. Carries action potentials from light receptors in the eye.
9. Innervates facial muscles and salivary glands.
10. Innervates the trapezius and muscles of the pharynx and larynx.
11. Carries action potentials that are interpreted as odors.

1. _____
2. _____
3. _____
4. _____
5. _____
6. _____
7. _____
8. _____
9. _____
10. _____
11. _____

Notes

Laboratory Report 16

Student _____

Lab Section _____

SENSES

A. Figures
List the labels for Figures 16.1, 16.2, and 16.3.

Figure 16.1

1. _____
2. _____
3. _____
4. _____
5. _____
6. _____
7. _____
8. _____
9. _____

Figure 16.2

1. _____
2. _____
3. _____
4. _____
5. _____
6. _____
7. _____

Figure 16.3

1. _____
2. _____
3. _____
4. _____
5. _____
6. _____
7. _____
8. _____
9. _____
10. _____
11. _____
12. _____
13. _____
14. _____
15. _____

B. Lacrimal Apparatus
Provide the words that correctly complete the paragraph. The __1__ lines the inner surface of the eyelids and covers the front of the eye. Fluid, secreted by the __2__, keeps the eye surface and __3__ moist. At the medial corner of the eye, the fluid flows through two tiny openings, the __4__, into the superior and inferior __5__. Then, the fluid flows into the __6__ and on into the __7__, which empties the fluid into the __8__.

1. _____
2. _____
3. _____
4. _____
5. _____
6. _____
7. _____
8. _____

C. Extrinsic Muscles
Select the muscles described by statements below.

inferior oblique   superior levator palpebrae
inferior rectus   superior oblique
lateral rectus   superior rectus
medial rectus

1. Rotates eye downward.
2. Raises the upper eyelid.
3. Attached to the medial surface of the eye.
4. Passes through a cartilaginous loop above the eye.
5. Rotates eye outward.

1. _____
2. _____
3. _____
4. _____
5. _____

D. Structures
Select the structures described by the statements below.

aqueous humor   macula lutea
choroid   optic disc
ciliary body   pupil
cornea   retina
fovea centralis   sclera
iris   suspensory ligaments
lens   vitreous body

1. Contains light receptors: cones and rods.
2. Clear window at front of eye that bends light rays.
3. Fills cavity in front of the lens.
4. Outer layer of the eye.
5. Opening in center of iris.
6. Attaches lens to ciliary body.
7. Fills cavity behind the lens.
8. Precisely focuses light on retina.
9. Area of critical vision.
10. Controls shape of the lens.
11. Controls amount of light entering the eye.
12. Yellow spot containing only cones.
13. Black middle coat of the eyeball.
14. Provides strength to the eyeball wall.
15. Lacks rods and cones.
16. Layer of eyeball containing blood vessels.
17. Place where axons of retina leave eyeball.
18. Place where blood vessels enter eyeball.
19. Jellylike substance within eyeball.
20. Colored ring in front of the lens.

1. _____
2. _____
3. _____
4. _____
5. _____
6. _____
7. _____
8. _____
9. _____
10. _____
11. _____
12. _____
13. _____
14. _____
15. _____
16. _____
17. _____
18. _____
19. _____
20. _____

E. Beef Eye Dissection

1. Which layer of the eyeball is the toughest? _____
2. What is the shape of the pupil? _____
3. When you hold up the lens, look at a distant object. What is unusual about the image? _____
4. When placed on a printed page, does the lens make print appear larger or smaller? _____
5. Is the lens denser near the center or periphery? _____
6. What is the clear jellylike substance? _____
7. What holds the retina in place in an intact eyeball? _____

F. Physiology of Vision

1. Name the two types of photosensitive cells in the retina. _____
2. Which type of photosensitive cell is most abundant? _____
3. Which type of photosensitive cell provides color vision? _____
4. Which type of photosensitive cell occurs in the fovea centralis? _____
5. Name the photosensitive pigment in rods. _____
6. What are the two compounds formed by the photochemical breakdown of the photosensitive pigment in rods? _____
_____
7. What vitamin is necessary for the re-formation of rhodopsin? _____
8. The three types of cones are sensitive to three different colors of light. Name the colors. _____
_____
9. Explain the cause of temporary blindness when you enter a dark room from a brightly lighted area. _____

    Explain the brief time lapse before dim-light vision is restored. _____
    _____

G. Visual Tests

Record the results obtained in the following tests.

Test	Right Eye	Left Eye
Blind Spot (inches)		
Near Point (inches)		
Visual Acuity (20/x)		
Astigmatism (present or absent)		

H. Light Intensity and Pupil Size

1. Did the pupil size become larger or smaller in the

    exposed left eye? _____   Unexposed right eye? _____

2. Explain these responses. _____
_____
_____

3. Suggest a possible benefit resulting from this reflex pattern. _____
_____

I. Color Blindness Test

Have your laboratory partner record your responses to each test plate. The mark *x* indicates inability to read correctly.

Plate Number	Subject's Response	Correct Response	Response If Red-Green Deficiency	Response If Totally Color-blind
1		12	12	12
2		8	3	x
3		5	2	x
4		29	70	x
5		74	21	x
6		7	x	x
7		45	x	x
8		2	x	x
9		x	2	x
10		16	x	x
11		traceable	x	x
			red cones absent	green cones absent	
12		35	5	3	x
13		96	6	9	x
14		can trace two lines	purple	red	x

J. Figure

List the labels for Figure 16.11.

1. _____
2. _____
3. _____
4. _____
5. _____
6. _____
7. _____
8. _____
9. _____
10. _____
11. _____
12. _____

13. _____
14. _____
15. _____
16. _____
17. _____
18. _____
19. _____
20. _____
21. _____
22. _____
23. _____

K. Hearing (Structures)
Select the structures described in the statements below.

basilar membrane
cochlear duct
endolymph
hair cells
incus
malleus
middle ear
perilymph
round window
scala tympani
scala vestibuli
stapes
tectorial membrane
tympanic membrane
vestibular membrane

1. Membrane receiving sound waves via air.
2. Ossicle attached to the tympanic membrane.
3. Ossicle inserted into oval window.
4. Cochlear chamber into which oval window opens.
5. Cochlear chamber with the round window.
6. Fluid within the cochlear duct.
7. Fluid between the bony and membranous labyrinths.
8. Receptors for sound stimuli.
9. Contains 20,000 transverse fibers.
10. Cochlear chamber containing the spiral organ.
11. Membrane in which hair tips are embedded.
12. Membrane allowing movement of perilymph.
13. Transfers vibrations from fluid to hair cells.
14. Cavity traversed by the ear ossicles.
15. Activated by the movement of the perilymph.
16. Set in motion by the movement of the stapes in the oval window.

L. Hearing (True–False)

1. Sound waves may reach the cochlea through the skull bones as well as via the tympanic membrane.
2. The more hair cells sending action potentials, the louder will be the sound sensation.
3. Pitch is determined by the part of the basilar membrane that is activated.
4. Bending of the hair cells causes the formation of action potentials.
5. There is no cure for conduction deafness, but nerve deafness may often be corrected.

M. Equilibrium
Select the structures described by the statements below.

ampulla
crista ampullaris
macula
otoliths
saccule
utricle

1. Two portions of the membranous labyrinth containing sensory receptors for static equilibrium.
2. Sensory hair cells in ampullae of semicircular canals.
3. Hair cells stimulated by the movement of the otoliths.
4. Hair cells stimulated by the force of endolymph.
5. Sensory receptors for dynamic equilibrium.
6. Enlarged base of a semicircular canal.
7. Calcium carbonate crystals.

N. **Hearing Tests**
Record your results of the hearing tests.
Rinne Test

Left ear _____

Right ear _____

State a conclusion from this test. _____

Weber Test

Left ear _____

Right ear _____

State a conclusion from this test. _____

O. **Static Balance Test**
Record the degree of swaying as small, moderate, or great.

Standing on both feet with eyes open _____

Standing on both feet with eyes closed _____

Standing on one foot with eyes closed _____

Rate the importance of visual input in static balance. _____

_____

Laboratory Report 17

Student _____
Lab Section _____

THE ENDOCRINE SYSTEM

A. Figures
List the labels for Figures 17.1 and 17.4.

Figure 17.1
1. _____
2. _____
3. _____
4. _____
5. _____
6. _____
7. _____
8. _____
9. _____
10. _____

Figure 17.4
1. _____
2. _____
3. _____
4. _____
5. _____
6. _____

B. Hormone Sources
Select the glandular tissue that produces the hormones listed.

adrenal cortex
adrenal medulla
anterior lobe of the pituitary gland
hypothalamus
ovary: corpus luteum
ovary: ovarian follicles
pancreas
parathyroid gland
pineal gland
posterior lobe of the pituitary gland
testis
thymus
thyroid gland

1. ACTH.
2. ADH.
3. Aldosterone.
4. Calcitonin.
5. Cortisol.
6. Epinephrine.
7. Estrogens.
8. FSH.
9. Glucagon.
10. Growth hormone.
11. Insulin.
12. LH.
13. Melatonin.
14. Oxytocin.
15. Parathyroid hormone.
16. Prolactin.
17. Progesterone.
18. Testosterone.

1. _____
2. _____
3. _____
4. _____
5. _____
6. _____
7. _____
8. _____
9. _____
10. _____
11. _____
12. _____
13. _____
14. _____
15. _____
16. _____
17. _____
18. _____

19. Thymosin.
20. TSH.
21. Thyroid hormone.

19. _____
20. _____
21. _____

C. **Hormone Functions**
Select the hormones that perform the functions listed below.

Group 1

aldosterone   epinephrine
cortisol   melatonin

1. Increases resistance to stress.
2. Promotes excretion of K⁺.
3. Increases reabsorption of Na⁺.
4. Increases blood glucose level (2).
5. Prepares body for response to emergencies.
6. Increases heart and respiration rate.

Group 1
1. _____
2. _____
3. _____
4a. _____
4b. _____
5. _____
6. _____

Group 2

calcitonin   parathyroid hormone
glucagon   somatostatin
insulin   thyroid hormone

1. Promotes bone tissue deposition.
2. Promotes bone tissue reabsorption.
3. Increases blood glucose level.
4. Decreases blood glucose level.
5. Stimulates metabolism.
6. Raises blood calcium level.
7. Lowers blood calcium level.
8. Aids entrance of glucose into cells.

Group 2
1. _____
2. _____
3. _____
4. _____
5. _____
6. _____
7. _____
8. _____

Group 3

ACTH   LH
ADH   oxytocin
estrogens   progesterone
FSH   testosterone
growth hormone   TSH

1. Stimulates descent of testes.
2. Promotes breast development.
3. Promotes release of milk.
4. Stimulates thyroxin production.
5. Stimulates ovarian follicular development.
6. Stimulates glucocorticoid production.
7. Stimulates growth of body cells.
8. Prepares uterus for implantation of embryo.
9. Rapid increase causes ovulation.
10. Stimulates spermatogenesis.
11. Promotes water reabsorption by kidneys.
12. Strengthens uterine contractions.

Group 3
1. _____
2. _____
3. _____
4. _____
5. _____
6. _____
7. _____
8. _____
9. _____
10. _____
11. _____
12. _____

D. **Hormonal Imbalance**
Select the hormones involved in the maladies listed and indicate if the condition is due to an excess (+) or a deficiency (−).

glucagon
glucocorticoids
growth hormone
insulin
mineralocorticoids
parathyroid hormone
thyroid hormone

1. Acromegaly.
2. Addison disease.
3. Cretinism.
4. Cushing syndrome.
5. Diabetes mellitus.
6. Dwarfism.
7. Gigantism.
8. Hypoglycemia.
9. Myxedema.
10. Tetany.

1. _____
2. _____
3. _____
4. _____
5. _____
6. _____
7. _____
8. _____
9. _____
10. _____

E. *Figure*
List the labels for Figure 17.11.

F. *Microscopic Study*
Make labeled drawings of your observations of the prepared microscope slides of the endocrine glands. Use additional sheets of paper as necessary.

Figure 17.11

1. _____
2. _____
3. _____
4. _____
5. _____
6. _____
7. _____
8. _____
9. _____
10. _____
11. _____
12. _____
13. _____
14. _____

G. Endocrine Hormones and Actions

Summarize your understanding of endocrine glands by completing the chart below.

Hormone	Endocrine Gland	Action
Adrenocorticotropic hormone (ACTH)		
Aldosterone		
Antidiuretic hormone (ADH)		
Calcitonin		
Cortisol		
Epinephrine		
Estrogen		
Follicle-stimulating hormone (FSH)		
Glucagon		
Growth hormone (GH)		
Insulin		
Interstitial-cell-stimulating hormone (ICSH)		
Luteinizing hormone (LH)		
Melatonin		
Oxytocin		
Parathyroid hormone (PH)		
Progesterone		
Prolactin (PRL)		
Somatostatin (SST)		
Thymosin		
Thyroid hormone		
Thyroid-stimulating hormone (TSH)		

Laboratory Report 18

Student _____

Lab Section _____

BLOOD TESTS

A. Formed Elements

Select the blood cell type described in the statements below.

basophils   lymphocytes   neutrophils
eosinophils   monocytes   platelets
erythrocytes

1. Agranulocytes with spherical nucleus and little cytoplasm.
2. Biconcave, enucleate cells.
3. Cells with bilobed to five-lobed nucleus and tiny lavender cytoplasmic granules.
4. Smallest of the formed elements.
5. Cells with red-orange cytoplasmic granules and bilobed or U-shaped nucleus.
6. Large agranulocytes with a lobed or U-shaped nucleus.
7. Cell with bilobed or U-shaped nucleus and bluish purple cytoplasmic granules.
8. 4,000,000 to 6,000,000 per µl.
9. Initiate clotting process.
10. Produce antibodies.
11. Transport oxygen and carbon dioxide.
12. Counteract products of allergy reactions.
13. Release histamine in allergy reactions.
14. 50% to 70% of the leukocytes.
15. 20% to 30% of the leukocytes.
16. Two types of cells that phagocytize pathogens.
17. 150,000 to 400,000 per µl.
18. Contain hemoglobin.
19. 2% to 6% of the leukocytes.

1. _____
2. _____
3. _____
4. _____
5. _____
6. _____
7. _____
8. _____
9. _____
10. _____
11. _____
12. _____
13. _____
14. _____
15. _____
16a. _____
16b. _____
17. _____
18. _____
19. _____

B. Differential White Blood Cell Count

Tabulate your identification of the various types of leukocytes in the table below. Count a total of 100 white cells. Calculate the percentage of each type of leukocyte by dividing the total count for each type by 100.

Neutrophils	Lymphocytes	Monocytes	Eosinophils	Basophils
Totals				
Percentage				

C. **Blood Test Results**

Record your test results in the table below.

Test	Healthy Range (varies by testing source)	Test Results	Evaluation (over, under, average)
Differential WBC Count	Neutrophils: 50%–70%		
	Lymphocytes: 20%–30%		
	Monocytes: 2%–6%		
	Eosinophils: 1%–3%		
	Basophils: 0.5%–1%		
Hematocrit	Males: 40%–54% (Av. 47%)		
	Females: 37%–47% (Av. 42%)		

D. **Blood Typing**

1. Indicate your blood type.   ABO group _____   Rh _____
2. Complete the table below to indicate compatibility (o) and incompatibility (+) of possible transfusions. A plus (+) indicates agglutination of the erythrocytes.

Blood Type/Antigen of Donor	Blood Type and Antibodies of Recipient			
	O Anti-A Anti-B	A Anti-B	B Anti-A	AB none
O				
A				
B				
AB				

Laboratory Report 19

Student _____

Lab Section _____

THE HEART

A. Figures

List the labels for Figures 19.1 and 19.2.

Figure 19.1

1. _____
2. _____
3. _____
4. _____
5. _____
6. _____
7. _____
8. _____
9. _____
10. _____
11. _____
12. _____
13. _____
14. _____
15. _____
16. _____
17. _____
18. _____
19. _____
20. _____

Figure 19.2

1. _____
2. _____
3. _____
4. _____
5. _____
6. _____
7. _____
8. _____
9. _____
10. _____
11. _____
12. _____
13. _____
14. _____
15. _____
16. _____
17. _____
18. _____
19. _____
20. _____
21. _____
22. _____
23. _____
24. _____
25. _____
26. _____
27. _____
28. _____
29. _____
30. _____
31. _____
32. _____
33. _____

B. **Structures**

Select the structures described by the statements below.

aortic semilunar valve
endocardium
epicardium
interventricular septum
left atrium
left ventricle
mitral valve
myocardium
papillary muscles
pulmonary semilunar valve
right atrium
right ventricle
tendinous cords
tricuspid valve

1. Partition between right and left ventricles.
2. Muscular portion of cardiac wall.
3. Lines inside of the heart.
4. Receives deoxygenated blood from the venae cavae.
5. Receives oxygenated blood from the pulmonary veins.
6. Prevents backflow of blood into right atrium.
7. Prevents backflow of blood into left atrium.
8. Prevents backflow of blood into right ventricle.
9. Prevents backflow of blood into left ventricle.
10. Pumps blood into the pulmonary artery.
11. Pumps blood into the aorta.
12. Chamber with the thickest walls.
13. Portion of ventricular wall to which tendinous cords are attached.
14. Strands of fibrous tissue attaching valve cusps to the inner wall of the ventricles.
15. Receives blood from the coronary sinus.

1. _____
2. _____
3. _____
4. _____
5. _____
6. _____
7. _____
8. _____
9. _____
10. _____
11. _____
12. _____
13. _____
14. _____
15. _____

C. **Circulation**

Provide the words to correctly complete the paragraph. During diastole, deoxygenated blood returns to the __1__ via the __2__ and __3__ and enters the right ventricle. Simultaneously, oxygenated blood returns to the __4__ via the __5__ and enters the left ventricle. During systole, the right ventricle pumps blood into the __6__, which carries blood to the lungs. Simultaneously, the left ventricle pumps blood into the __7__, which carries blood to all parts of the body except the lungs.

1. _____
2. _____
3. _____
4. _____
5. _____
6. _____
7. _____

D. **Sheep Heart Dissection**

1. Indicate the number of cusps composing the heart valves. _____

   Mitral _____   Tricuspid _____   Aortic _____   Pulmonary _____

2. Compare the thickness of the valve cusps. _____

   Atrioventricular valves _____

   Semilunar valves _____

3. Which ventricle has the thicker wall? _____

4. Are the tendinous cords fragile or strong? _____

   Describe their function. _____

   _____

5. Is the ventricular septum a thin sheet of connective tissue or a thick muscular wall? _____

_____

E. **Heart Sounds**

Select the heart sound described by the statements below.

first sound   third sound
second sound   murmur

1. Vibration of ventricular walls and valve cusps.
2. Closure of the semilunar valves.
3. Closure of the atrioventricular valves.
4. Damaged heart valves.
5. Detected best in supine subject.

1. _____
2. _____
3. _____
4. _____
5. _____

F. *Auscultatory Areas*

Select the auscultatory area described by the statements below.

aortic semilunar area   pulmonary semilunar area
bicuspid area   tricuspid area

1. About 2 in left of sternum at fifth rib.
2. On left edge of sternum just below second rib.
3. On left edge of sternum at the fifth rib.
4. On right edge of sternum at the second rib.
5. Site used to detect the third heart sound.

1. _____
2. _____
3. _____
4. _____
5. _____

Notes

Laboratory Report 20

Student _____

Lab Section _____

BLOOD VESSELS

A. The Circulatory Plan
Select the blood vessels described by the statements below.

aorta
hepatic artery
hepatic portal vein
hepatic veins
inferior vena cava
pulmonary artery
pulmonary vein
renal artery
renal vein
superior mesenteric artery
superior vena cava

1. Carries oxygenated blood to the body, except the lungs.
2. Carries blood from a kidney to the inferior vena cava.
3. Carries deoxygenated blood to the lungs.
4. Carries blood from the head and neck to the right atrium.
5. Carries oxygenated blood to the left atrium.
6. Carries blood from the aorta to the intestines.
7. Carries blood from regions below the heart to the right atrium.
8. Carries blood from the intestines to the liver.
9. Carries blood from the liver to the inferior vena cava.
10. Carries blood from the aorta to the liver.

1. _____
2. _____
3. _____
4. _____
5. _____
6. _____
7. _____
8. _____
9. _____
10. _____

B. Completion: Arteries and Veins
Provide the name of the artery or vein described by the statements.

1. Vein commonly used for venipuncture.
2. Branches to form right subclavian artery and right common carotid artery.
3. Supplies most of small intestine and part of large intestine.
4. Arterial segment between the subclavian artery and the brachial artery.
5. Veins of the neck emptying into the subclavian veins.
6. Large superficial vein draining the lower limb.
7. Supplies blood to the stomach.
8. Merge to form the inferior vena cava.
9. Branches to form the anterior and posterior tibial arteries.
10. Merge to form the superior vena cava.
11. Carries blood to the kidney.
12. Carries blood to the large intestine and rectum.
13. Branch from the distal end of the aorta.
14. Carries blood from the aortic arch to the left side of the head.
15. Return blood from the brain to brachiocephalic veins.

1. _____
2. _____
3. _____
4. _____
5. _____
6. _____
7. _____
8. _____
9. _____
10. _____
11. _____
12. _____
13. _____
14. _____
15. _____

C. *Figures*

List the labels for Figures 20.2, 20.3, and 20.5.

Figure 20.2

1. _____
2. _____
3. _____
4. _____
5. _____
6. _____
7. _____
8. _____
9. _____
10. _____
11. _____
12. _____
13. _____
14. _____
15. _____
16. _____
17. _____
18. _____
19. _____
20. _____
21. _____
22. _____
23. _____
24. _____
25. _____
26. _____

Figure 20.3

1. _____
2. _____
3. _____
4. _____
5. _____
6. _____
7. _____
8. _____
9. _____
10. _____
11. _____
12. _____
13. _____
14. _____
15. _____
16. _____
17. _____
18. _____
19. _____
20. _____
21. _____
22. _____
23. _____
24. _____

Figure 20.5

1. _____
2. _____
3. _____
4. _____
5. _____
6. _____
7. _____
8. _____
9. _____
10. _____
11. _____
12. _____
13. _____
14. _____
15. _____
16. _____
17. _____
18. _____

D. *Microscopic Study*

Make a drawing of the slides.

Healthy Artery   Atherosclerotic Artery   Healthy Vein

E. *Fetal Circulation*

Select the structures described by the statements below.

ductus arteriosus   umbilical arteries
ductus venosus   umbilical vein
foramen ovale

1. Opening between the atria.
2. Extensions of the hypogastric arteries.
3. Brings oxygenated blood to the fetus.
4. Connects the pulmonary artery with the aorta.
5. Carry fetal blood to the placenta.
6. Becomes the ligamentum venosum after birth.
7. Becomes the round ligament after birth.
8. Becomes the ligamentum arteriosum after birth.

1. _____
2. _____
3. _____
4. _____
5. _____
6. _____
7. _____
8. _____

F. *Blood Pressure*

1. Complete the following statement.

   The reading on the pressure gauge at the first Korotkoff sound is the _____ pressure; the reading when the Korotkoff sounds cease is the _____ pressure.

2. Indicate your blood pressure before and after exercise.

   Systolic: before exercise _____   after exercise _____
   Diastolic: before exercise _____   after exercise _____

Blood Vessels   297

3. Explain the cause of a higher blood pressure after exercise.
   _____
   _____

4. Explain how chronic high blood pressure (hypertension) may cause the rupture of an artery in the brain. _____
   _____
   _____

G. The Pulse

1. Record your pulse rates (beats/min) in the chart below.

Supine	Sitting	Standing	After Exercise

2. How long did it take for your pulse rate to return to the sitting value after exercise?
   _____

3. The strength of the pulse decreases in the right radial artery when you raise your right hand above your head. Why?
   _____
   _____

4. What is the value of a nurse's knowing the pressure points where the pulse may be taken? _____
   _____
   _____
   _____

5. Calculate your pulse pressure.

   Systolic pressure _____ − Diastolic pressure _____ = Pulse pressure _____

Laboratory Report 21

Student _____

Lab Section _____

THE LYMPHOID SYSTEM

A. Lymphoid System
Select the terms described by the statements below.

cisterna chyli   lymph capillaries   lymphatic vessels
lymph   lymph nodes   right lymphatic duct
                                     thoracic duct

1. Collects extracellular fluid from tissues.
2. Fluid in lymphatic vessels.
3. Empties lymph into right subclavian vein.
4. Saclike structure at abdominal end of thoracic duct.
5. Cleanses lymph as it passes through.
6. Carries lymph to left subclavian vein.

1. _____
2. _____
3. _____
4. _____
5. _____
6. _____

B. Lymphocytes and Immunity
Select the terms described by the statements below.

antibodies   helper T cells
antibody-mediated immunity lymphoid tissue
antigen-presenting cell   memory B and T cells
B cells   phagocytosis
blood   plasma cells
cell-mediated immunity   stem cells in red bone
cytokines   marrow
cytotoxic T cells   T cells

1. Source of all lymphocytes.
2. Lymphocytes maturing in thymus.
3. Lymphocytes maturing in red bone marrow.
4. T cells that kill diseased cells with chemicals.
5. Lymphocytes that produce antibodies.
6. Immunity enabled by T cells.
7. Where most mature lymphocytes reside.
8. Chemicals that bind to antigens.
9. Chemicals that activate B cells.
10. Lymphocytes that start a secondary immune response.
11. Chemicals secreted by helper T cells.
12. Immunity enabled by B cells.
13. T cells that start a primary immune response.
14. Immunity that targets extracellular pathogens.
15. Destroys antigens bound to antibodies.

1. _____
2. _____
3. _____
4. _____
5. _____
6. _____
7. _____
8. _____
9. _____
10. _____
11. _____
12. _____
13. _____
14. _____
15. _____

Notes

Laboratory Report 22

Student _____
Lab Section _____

THE RESPIRATORY SYSTEM

A. Figures
List the labels for Figures 22.1 and 22.2.

Figure 22.1

1. _____
2. _____
3. _____
4. _____
5. _____
6. _____
7. _____
8. _____
9. _____
10. _____
11. _____
12. _____
13. _____
14. _____
15. _____
16. _____
17. _____
18. _____
19. _____
20. _____
21. _____
22. _____

Figure 22.2

1. _____
2. _____
3. _____
4. _____
5. _____
6. _____
7. _____
8. _____
9. _____
10. _____
11. _____
12. _____
13. _____
14. _____
15. _____

B. Air Pathway
Complete the following paragraph.

During inspiration, air passes through the __1__ into the nasal cavity, where it is warmed. The surface area of the nasal cavity is increased by the three __2__. From the nasal cavity, air passes through the __3__ and __4__ and into the larynx, which contains the __5__ folds. From the larynx, air passes down the __6__, which divides into the primary __7__ that enter each lung. The tubes continue to branch and ultimately form small tubules called __8__, which lead to the pulmonary alveoli. Gas exchange occurs between __9__ in the pulmonary alveoli and __10__ in the capillaries surrounding the pulmonary alveoli. During __11__, air flow is reversed through this same pathway.

1. _____
2. _____
3. _____
4. _____
5. _____
6. _____
7. _____
8. _____
9. _____
10. _____
11. _____

C. **Respiratory Structures**
Select the structures described by the statements below.

bronchi
bronchioles
epiglottis
hard palate
larynx
lingual tonsils
nasal conchae
nasopharynx
oral cavity
oral vestibule
oropharynx
palatine tonsils
pharyngeal tonsils
pleural cavity
pulmonary alveoli
soft palate
trachea

1. Cavity between lips and teeth.
2. Tonsils in oropharynx.
3. Increase surface area of nasal cavity.
4. Sites of gas exchange.
5. Tubes branching from lower end of trachea.
6. Prevents food from entering larynx.
7. Space between visceral and parietal pleurae.
8. Small tubes leading to pulmonary alveoli.
9. Tonsils in nasopharynx.
10. Separated from nasal cavity by hard palate.
11. Tiny air sacs in lung.
12. Portion of palate supported by bone.
13. Contains the vocal folds.
14. Where swallowing reflex is initiated.
15. Tonsils at base of the tongue.

1. _____
2. _____
3. _____
4. _____
5. _____
6. _____
7. _____
8. _____
9. _____
10. _____
11. _____
12. _____
13. _____
14. _____
15. _____

D. **Sheep Pluck Dissection**

1. Describe the shape of the cartilaginous rings of the trachea.
   _____

2. What contributes to the smooth and slimy nature of the inside of the trachea? _____
   _____

3. Indicate the number of lobes in the

   left lung: _____   right lung: _____

4. Lung tissue seems to be full of small spaces. What pulmonary structures contribute most to this appearance?
   _____

E. **Microscopic Study**

1. Diagram a section of tracheal wall at 100× magnification and label the ciliated epithelial lining and hyaline cartilage.
2. Diagram the epithelial lining at 400×.
3. Diagram healthy lung tissue at 100× magnification. Label the pulmonary alveoli.

Tracheal Wall
100×

Epithelial Lining
400×

Healthy Lung Tissue
100×

F. **Control of Breathing**

Complete the following paragraph.

Breathing is controlled by the __1__ rhythmicity center located in the __2__ of the brain. The most important stimulus for breathing is the direct stimulation of this center by an __3__ concentration of __4__ in the blood and cerebrospinal fluid. __5__ also acts directly on the rhythmicity center. In contrast, __6__ acts indirectly on the rhythmicity center. Its concentration is detected by __7__ in the carotid arteries and aorta, which send action potentials to the rhythmicity center. Usually, __8__ concentration is of little importance in stimulating inspiration.

1. _____
2. _____
3. _____
4. _____
5. _____
6. _____
7. _____
8. _____

G. **Breathing Experiments**

Hyperventilation

1. Describe how you felt after hyperventilating. _____
_____
_____

2. Explain why it was easier to breathe into the paper bag. _____
_____
_____
_____

3. Indicate how long you were able to hold your breath. _____

   After resting breathing _____ sec   After hyperventilating _____ sec

   Explain your results. _____
   _____
   _____

Deglutition Apnea

1. Indicate whether the urge to breathe decreased or increased when you sipped the water through the straw.
_____

2. Is this phenomenon a reflex or a voluntary response? _____

The Respiratory System

H. *Spirometry and Lung Volumes and Capacities*

Record your lung volumes determined by spirometry in the table below. Circle those values that are 22% or more below average.

Lung Capacities	Average Healthy Value (ml)	Your Capacities
Tidal Volume (TV)	500	
Minute Ventilation (MV) (MV = TV × Resp. Rate)	6,000	
Expiratory Reserve Volume (ERV)	1,100	
Vital Capacity (VC) (VC = TV + ERV + IRV)	See Appendix A	
Inspiratory Capacity (IC) (IC = VC − ERV)	3,000	
Inspiratory Reserve Volume (IRV) (IRV = IC − TV)	2,500	

If you did the FEV_T test, provide your results below.

FEV_1 _____ % FEV_3 _____ %

Attach your spirograms to the laboratory report.

I. Clinical Applications

1. Emphysema is characterized by the rupture of numerous pulmonary alveoli and a decrease in lung elasticity.

   Explain why emphysematous patients have a below average expiratory reserve volume. _____
   _____
   _____

   How would this affect the vital capacity? _____
   _____

2. Asthma causes inflammation of the bronchial tree and is classified as an obstructive disorder. On an FEV_1 test, would you expect an asthmatic patient to have a below average, average, or above average result? _____

   Explain. _____
   _____
   _____

Laboratory Report 23

Student _____

Lab Section _____

The Digestive System

A. Figures

List the labels for Figures 23.1 through 23.5.

Figure 23.1

1. _____
2. _____
3. _____
4. _____
5. _____
6. _____
7. _____
8. _____
9. _____
10. _____
11. _____

Figure 23.2

1. _____
2. _____
3. _____
4. _____
5. _____
6. _____
7. _____
8. _____
9. _____
10. _____
11. _____
12. _____

Figure 23.3

1. _____
2. _____
3. _____
4. _____
5. _____
6. _____
7. _____
8. _____
9. _____

Figure 23.4

1. _____
2. _____
3. _____
4. _____
5. _____
6. _____
7. _____
8. _____

Figure 23.5

1. _____
2. _____
3. _____
4. _____
5. _____
6. _____
7. _____
8. _____
9. _____
10. _____
11. _____
12. _____
13. _____
14. _____
15. _____

16. _____
17. _____
18. _____
19. _____
20. _____
21. _____
22. _____
23. _____
24. _____
25. _____
26. _____
27. _____
28. _____
29. _____

B. Oral Cavity

Select the structures described by the statements below.

cementum
crown of tooth
dentin
enamel
filiform papillae
fungiform papillae
gingiva
lingual tonsils
oropharynx
palatine tonsils
parotid gland
pulp cavity
root of tooth
sublingual gland
submandibular gland
vallate papillae

1. Portion of a tooth covered with enamel.
2. Papillae giving rough texture to the tongue.
3. Mucous membrane covering the neck of each tooth.
4. Salivary gland located in front of the ear.
5. Substance forming most of a tooth.
6. Bonelike substance between root and periodontal ligament.
7. Eight to twelve large papillae located at back of tongue.
8. Salivary gland located in the floor of oral cavity.
9. Contains nerves and blood vessels of a tooth.
10. Where swallowing reflex is initiated.
11. Papillae with taste buds scattered over surface of tongue.
12. Tonsils located in oropharynx.
13. Salivary gland secreting salivary amylase.
14. Salivary glands secreting mucin.

1. _____
2. _____
3. _____
4. _____
5. _____
6. _____
7. _____
8. _____
9. _____
10. _____
11. _____
12. _____
13. _____
14a. _____
14b. _____

C. Esophagus to Anus
Select the structures described by the statements below.

cecum
colon
common bile duct
cystic duct
duodenum
esophagus
hepatic duct
ileocecal valve
ileum
jejunum
lower esophageal sphincter
pancreatic duct
pyloric sphincter
rectum
stomach
villi

1. Valve that opens to let food into the stomach.
2. Valve that opens to let chyme out of stomach.
3. Valve between the small and large intestines.
4. Pouch from which appendix extends.
5. Tube that carries food to the stomach.
6. The first 10 to 12 in of the small intestine.
7. Duct that carries bile from the liver.
8. Duct that carries bile to and from gallbladder.
9. The terminal 5 to 6 in of the large intestine.
10. Duct that carries bile and pancreatic juice to the duodenum.
11. Microscopic projections through which absorption of nutrients occurs.
12. Site of bacterial decomposition of food residue.
13. Middle portion of the small intestine.
14. Where food is mixed with gastric juice.

1. _____
2. _____
3. _____
4. _____
5. _____
6. _____
7. _____
8. _____
9. _____
10. _____
11. _____
12. _____
13. _____
14. _____

D. Microscopic Study
Make drawings from these slides. Label pertinent parts.

**Vallate Papilla
with Taste Buds**

Stomach Wall

The Digestive System 307

Ileum Wall   **Jejunum Wall**

Liver Lobules

E. Completion

1. Explain the meaning of enzymatic hydrolysis. _____

2. What determines the three-dimensional shape of an enzyme? _____

3. Explain the importance of the shape of an enzyme. _____

4. Explain how temperature and pH changes may alter the shape of an enzyme. _____

F. Digestive Enzymes

Select the enzyme described by the statements below.

amylase
carboxypeptidase
chymotrypsin
lactase
lipase
maltase
pepsin
peptidase
sucrase
trypsin

1. Enzyme present in saliva and pancreatic juice.
2. Converts lipids to fatty acids and glycerol.
3. Intestinal enzyme converting peptides to amino acids.
4. Converts starch to maltose.
5. Converts maltose to glucose.
6. Converts sucrose to glucose and fructose.
7. Converts lactose to glucose and galactose.
8. Gastric enzyme converting proteins to peptides.
9. Two pancreatic enzymes converting proteins to peptides only.
10. Pancreatic enzyme converting peptides to peptide residues and amino acids.
11. Enzyme found in gastric, pancreatic, and intestinal juices.

1. _____
2. _____
3. _____
4. _____
5. _____
6. _____
7. _____
8. _____
9a. _____
9b. _____
10. _____
11. _____

G. Completion

Indicate the end products of digestion for the following:

Carbohydrates _____

Lipids _____

Proteins _____

The Digestive System 309

H. Starch Digestion Experiments

Record your data from the experiments in the tables below. In the starch digestion row, use a "0" to indicate no digestion, a "+" to indicate some digestion, and a "C" to indicate complete digestion.

Table I Amylase activity at 20°C

Tube No.	1	2	3	4	5	6	7	8	9	10
Color										
Starch Digestion										

Table II Amylase activity at 37°C

Tube No.	1	2	3	4	5	6	7	8	9	10
Color										
Starch Digestion										

Table III Amylase activity when boiled

Tube No.	1	2	3
Color			
Starch Digestion			

Table IV pH and amylase activity

Time	2 min			4 min			6 min		
Tube No.	1	2	3	4	5	6	7	8	9
pH	5	7	9	5	7	9	5	7	9
Color									
Starch Digestion									

I. Conclusions

1. How long did it take for starch digestion to be

    evident?   20°C _____ min   37°C _____ min

    complete? 20°C _____ min   37°C _____ min

2. Did you expect this pattern of results? _____ Why? _____

3. What effect did boiling have on amylase? _____

4. At which pH was digestion most rapid? _____ How does this compare with the pH of saliva? _____

5. Gastric juice has a pH of 2 or less. What will its effect be on salivary amylase? _____

    Do you expect any starch digestion in the stomach? _____

Laboratory Report 24

Student _____

Lab Section _____

THE URINARY SYSTEM

A. **Figures**

List the labels for Figures 24.1, 24.2, and 24.3.

Figure 24.1

1. _____
2. _____
3. _____
4. _____
5. _____
6. _____
7. _____
8. _____
9. _____
10. _____
11. _____
12. _____
13. _____
14. _____
15. _____
16. _____

Figure 24.2

1. _____
2. _____
3. _____
4. _____
5. _____
6. _____
7. _____
8. _____
9. _____

Figure 24.3

1. _____
2. _____
3. _____
4. _____
5. _____
6. _____
7. _____
8. _____
9. _____
10. _____
11. _____
12. _____

B. **Microscopic Study**

Diagram a glomerulus and glomerular capsule as observed on the prepared slide of kidney tissue.

C. Anatomy

Select the structures described by the statements below.

calyces nephron renal pelvis
collecting duct renal column renal pyramids
fibrous capsule renal cortex ureters
glomerular capsule renal medulla urethra
glomerulus renal papilla

1. Tubes that drain the kidneys.
2. Tube that drains the urinary bladder.
3. Portion of kidney that contains renal corpuscles.
4. Cone-shaped portions of the renal medulla.
5. Part of kidney containing collecting ducts.
6. Functional unit of the kidney.
7. Capillary tuft within renal corpuscle.
8. Tip of renal pyramid.
9. Tube receiving urine from several nephrons.
10. Thin, fibrous covering of the kidney.
11. Cortical tissue between renal pyramids.
12. Short tubes receiving urine from renal pyramids.
13. Part of kidney composed of renal pyramids and renal columns.
14. Funnel-like structure receiving urine from calyces.
15. Cuplike structure enveloping a glomerulus.

1. _____
2. _____
3. _____
4. _____
5. _____
6. _____
7. _____
8. _____
9. _____
10. _____
11. _____
12. _____
13. _____
14. _____
15. _____

D. Table 24.1

Calculate the concentrations (grams/liter) of the following substances in Table 24.1 and record them below.

Substance	Blood	Glomerular Filtrate	Urine
Proteins	____	____	____
Chloride ions	____	____	____
Potassium ions	____	____	____
Sodium ions	____	____	____
Glucose	____	____	____
Creatinine	____	____	____
Urea	____	____	____
Uric acid	____	____	____

1. Why do so few proteins enter the glomerular filtrate? _____

2. What substances have different concentrations in the plasma and in the glomerular filtrate? _____

3. What substances are completely reabsorbed? _____

4. What substances are more concentrated in urine than in the glomerular filtrate? _____

5. Does urine formation remove all nitrogenous wastes or only reduce their concentrations in the plasma? _____

6. What percentage of the glomerular filtrate is reabsorbed? _____

    The reabsorption of what substance accounts for most of the volume reduction and the concentration of urine solutes? _____

7. List the three processes of urine formation. _____

E. Terminology
Write the term defined below in the answer column.

1. Inflammation of the kidney (general).
2. Albumin in the urine.
3. A measure of the concentration of solutes in urine.
4. Erythrocytes in the urine.
5. Inflammation of the urinary bladder.
6. Most abundant inorganic compound in urine.
7. Leukocytes or pus components in the urine.
8. pH range of healthy urine.
9. Hemoglobin in the urine.
10. More than a trace of glucose in the urine.
11. Ketones in the urine.
12. Inflammation of the kidney involving glomeruli.
13. Accumulations of materials hardened in tubules.
14. Most abundant nitrogenous waste.
15. Excessive urine production.
16. Bile pigment in the urine.
17. Inflammation of the urethra.
18. Kidney stones.
19. Little or no urine production.
20. Most abundant inorganic solute in urine.

F. Clinical Significance
Select the name of the possible clinical condition from the list below that is indicated by the urinalysis results. Write your answer in the answer column.

Clinical Conditions

calculi
diabetes insipidus
diabetes mellitus
glomerulonephritis
hemolytic anemia
liver pathology
renal failure
urinary infection

Urinalysis Results

1. Anuria.
2. Pyuria, hematuria, kidney/ureter pain.
3. Hemoglobinuria.
4. Polyuria, very low specific gravity.
5. Urobilinogenuria and/or bilirubinuria.
6. Polyuria, glycosuria, ketonuria, high specific gravity.
7. Pyuria, bacteriuria.
8. Low pH, albuminuria, pyuria, casts.

G. *Urinalysis*

Record on the chart below the results of the analysis of each urine specimen. All of the following tests except color and turbidity are easily done with a 10SG-Multistix.

Test	Typical Healthy Values	Unknowns No. ___	Unknowns No. ___	Student's Urine
Leukocytes	None			
Nitrite	None			
Urobilinogen	<2 mg/dl			
Protein	None			
pH	4.8 to 7.5			
Blood	None			
Sp. Gravity	1.002–1.030			
Ketone	None			
Bilirubin	None			
Glucose	None or trace			
Color	Pale yellow to amber			
Turbidity	Clear			

Indicate the possible clinical conditions associated with the unknowns.

No. _____ _____

_____

No. _____ _____

_____

Laboratory Report 25

Student _____
Lab Section _____

THE REPRODUCTIVE SYSTEM

A. *Figures*
List the labels for Figures 25.1, 25.4, 25.5, and 25.6.

Figure 25.1

1. _____
2. _____
3. _____
4. _____
5. _____
6. _____
7. _____
8. _____
9. _____
10. _____
11. _____
12. _____
13. _____
14. _____
15. _____
16. _____
17. _____
18. _____
19. _____
20. _____
21. _____
22. _____
23. _____
24. _____
25. _____
26. _____
27. _____

Figure 25.4

1. _____
2. _____
3. _____
4. _____
5. _____
6. _____
7. _____
8. _____
9. _____
10. _____

Figure 25.5

1. _____
2. _____
3. _____
4. _____
5. _____
6. _____
7. _____
8. _____
9. _____
10. _____
11. _____
12. _____
13. _____
14. _____

Figure 25.6

1. _____
2. _____
3. _____
4. _____
5. _____
6. _____
7. _____
8. _____
9. _____
10. _____
11. _____
12. _____
13. _____
14. _____
15. _____
16. _____
17. _____

B. Gametogenesis
Select the cells described by the statements below.

Spermatogenesis	Oogenesis
primary spermatocytes	oogonia
secondary spermatocytes	ovum
spermatids	polar bodies
spermatogonia	primary oocyte
sperm	secondary oocyte

1. Haploid male gametes.
2. Haploid female gamete.
3. Divide mitotically to form primary spermatocytes.
4. Cells composing germinal epithelium in ovary.
5. Large haploid cell released in ovulation.
6. Male cells formed by first meiotic division.
7. Large female cell formed by first meiotic division.
8. Male cells in which first meiotic division occurs.
9. Female cells in which first meiotic division occurs.
10. Large female cell formed by second meiotic division.

1. _____
2. _____
3. _____
4. _____
5. _____
6. _____
7. _____
8. _____
9. _____
10. _____

C. Male Organs
Select the structures described by the statements below.

bulbourethral gland   prostate
corpora cavernosa   prostatic urethra
corpus spongiosum   seminal vesicle
ejaculatory duct   seminiferous tubule
epididymis   testes
penis   vas (ductus) deferens
prepuce

1. Male copulatory organ.
2. Male gonads.
3. Sheath of skin over the glans penis.
4. Carries sperm from epididymis to ejaculatory duct.
5. Tube from urinary bladder into prostate.
6. Empties secretions into base of penile urethra.
7. Empties secretions into prostatic urethra.
8. Empties secretions into ejaculatory duct.
9. Erectile tissue forming glans penis.
10. Two dorsally located cylinders of erectile tissue.
11. Slender tube containing sperm-forming cells.
12. Site of sperm storage and maturation.

1. _____
2. _____
3. _____
4. _____
5. _____
6. _____
7. _____
8. _____
9. _____
10. _____
11. _____
12. _____

D. **Female Organs**

Select the structures described by the statements below.

broad ligament
cervix
clitoris
fimbriae
hymen
infundibulum
labia majora
labia minora
mons pubis
myometrium
ovarian ligament
ovaries
suspensory ligament
uterine tubes
uterosacral ligament
vagina
vestibular glands

1. Female copulatory organ.
2. Muscular wall of the uterus.
3. Narrow neck of the uterus.
4. Female gonads.
5. Fingerlike projections around infundibulum.
6. Source of ovum.
7. Provides vaginal lubrication during intercourse.
8. Duct carrying oocyte toward uterus.
9. Inner folds of vulva surrounding vestibule.
10. Nodule of erectile tissue.
11. Outer fatty folds of vulva merging with mons pubis.
12. Funnel-like opening of uterine tube.
13. Narrow ligament attaching ovary to uterus.
14. Sheetlike ligament supporting ovary, uterine tubes, and uterus.
15. Attaches uterus to rear pelvic wall.
16. Ligament from ovary to lateral pelvic wall.

1. _____
2. _____
3. _____
4. _____
5. _____
6. _____
7. _____
8. _____
9. _____
10. _____
11. _____
12. _____
13. _____
14. _____
15. _____
16. _____

E. **Completion**

1. The inner lining of the uterus is the _____.
2. The mixture of sperm and secretions of accessory glands is called _____.
3. Testes are located in a sac called the _____.
4. Fertilization usually occurs in the upper third of the _____.

1. _____
2. _____
3. _____
4. _____

F. **Microscopic Study**

Make labeled drawings of your observations below. Use additional sheets of paper as necessary.

Notes

Appendix A

VITAL CAPACITIES

TABLE A.1
Predicted Vital Capacities (in milliliters) for Females

Age	146	148	150	152	154	156	158	160	162	164	166	168	170	172	174	176	178	180	182	184	186	188	190	192	194
16	2950	2990	3030	3070	3110	3150	3190	3230	3270	3310	3350	3390	3430	3470	3510	3550	3590	3630	3670	3715	3755	3800	3840	3880	3920
18	2920	2960	3000	3040	3080	3120	3160	3200	3240	3280	3320	3360	3400	3440	3480	3520	3560	3600	3640	3680	3720	3760	3800	3840	3880
20	2890	2930	2970	3010	3050	3090	3130	3170	3210	3250	3290	3330	3370	3410	3450	3490	3525	3565	3605	3645	3695	3720	3760	3800	3840
22	2860	2900	2940	2980	3020	3060	3095	3135	3175	3215	3255	3290	3330	3370	3410	3450	3490	3530	3570	3610	3650	3685	3725	3765	3800
24	2830	2870	2910	2950	2985	3025	3065	3100	3140	3180	3220	3260	3300	3335	3375	3415	3455	3490	3530	3570	3610	3650	3685	3725	3765
26	2800	2840	2880	2920	2960	3000	3035	3070	3110	3150	3190	3230	3265	3300	3340	3380	3420	3455	3495	3530	3570	3610	3650	3685	3725
28	2775	2810	2850	2890	2930	2965	3000	3040	3070	3115	3155	3190	3230	3270	3305	3345	3380	3420	3460	3495	3535	3570	3610	3650	3685
30	2745	2780	2820	2860	2895	2935	2970	3010	3045	3085	3120	3160	3195	3235	3270	3310	3345	3385	3420	3460	3495	3535	3570	3610	3645
32	2715	2750	2790	2825	2865	2900	2940	2975	3015	3050	3090	3125	3160	3200	3235	3275	3310	3350	3385	3425	3460	3495	3535	3570	3610
34	2685	2725	2760	2795	2835	2870	2910	2945	2980	3020	3055	3090	3130	3165	3200	3240	3275	3310	3350	3385	3425	3460	3495	3535	3570
36	2655	2695	2730	2765	2805	2840	2875	2910	2950	2985	3020	3060	3095	3130	3165	3205	3240	3275	3310	3350	3385	3420	3460	3495	3530
38	2630	2665	2700	2735	2770	2810	2845	2880	2915	2950	2990	3025	3060	3095	3130	3170	3205	3240	3275	3310	3350	3385	3420	3455	3490
40	2600	2635	2670	2705	2740	2775	2810	2850	2885	2920	2955	2990	3025	3060	3095	3135	3170	3205	3240	3275	3310	3345	3380	3420	3455
42	2570	2605	2640	2675	2710	2745	2780	2815	2850	2885	2920	2955	2990	3025	3060	3100	3135	3170	3205	3240	3275	3310	3345	3380	3415
44	2540	2575	2610	2645	2680	2715	2750	2785	2820	2855	2890	2925	2960	2995	3030	3060	3095	3130	3165	3200	3235	3270	3305	3340	3375
46	2510	2545	2580	2615	2650	2685	2715	2750	2785	2820	2855	2890	2925	2960	2995	3030	3060	3095	3130	3165	3200	3235	3270	3305	3340
48	2480	2515	2550	2585	2620	2650	2685	2715	2750	2785	2820	2855	2890	2925	2960	2995	3030	3060	3095	3130	3160	3195	3230	3265	3300
50	2455	2485	2520	2555	2590	2625	2655	2690	2720	2755	2785	2820	2855	2890	2925	2955	2990	3025	3060	3090	3125	3155	3190	3225	3260
52	2425	2455	2490	2525	2555	2590	2625	2655	2690	2720	2755	2790	2820	2855	2890	2925	2955	2990	3020	3055	3090	3125	3155	3190	3220
54	2395	2425	2460	2495	2530	2560	2590	2625	2655	2690	2720	2755	2790	2820	2855	2885	2920	2950	2985	3020	3050	3085	3115	3150	3180
56	2365	2400	2430	2460	2495	2525	2560	2590	2625	2655	2690	2720	2755	2790	2820	2855	2885	2920	2950	2980	3015	3045	3080	3110	3145
58	2335	2370	2400	2430	2460	2495	2525	2560	2590	2625	2655	2690	2720	2750	2785	2815	2850	2880	2920	2945	2975	3010	3040	3075	3105
60	2305	2340	2370	2400	2430	2460	2495	2525	2560	2590	2625	2655	2685	2720	2750	2780	2810	2845	2875	2915	2940	2970	3000	3035	3065
62	2280	2310	2340	2370	2405	2435	2465	2495	2525	2560	2590	2620	2655	2685	2715	2745	2775	2810	2840	2870	2900	2935	2965	2995	3025
64	2250	2280	2310	2340	2370	2400	2430	2465	2495	2525	2555	2585	2620	2650	2680	2710	2740	2770	2805	2835	2865	2895	2920	2955	2990
66	2220	2250	2280	2310	2340	2370	2400	2430	2460	2495	2525	2555	2585	2615	2645	2675	2705	2735	2765	2800	2825	2860	2890	2920	2950
68	2190	2220	2250	2280	2310	2340	2370	2400	2430	2460	2490	2520	2550	2580	2610	2640	2670	2700	2730	2760	2795	2820	2850	2880	2910
70	2160	2190	2220	2250	2280	2310	2340	2370	2400	2425	2455	2485	2515	2545	2575	2605	2635	2665	2695	2725	2755	2780	2810	2840	2870
72	2130	2160	2190	2220	2250	2280	2310	2335	2365	2395	2425	2455	2480	2510	2540	2570	2600	2630	2660	2685	2715	2745	2775	2805	2830
74	2100	2130	2160	2190	2220	2245	2275	2305	2335	2360	2390	2420	2450	2475	2505	2535	2565	2590	2620	2650	2680	2710	2740	2765	2795

Source: E. DeF. Baldwin and E. W. Richards, Jr., "Pulmonary Insufficiency 1, Physiologic Classification, Clinical Methods of Analysis, Standard Values in Normal Subjects," *Medicine*, vol. 27, p. 243.

TABLE A.2
Predicted Vital Capacities (in milliliters) for Males

Age	146	148	150	152	154	156	158	160	162	164	166	168	170	172	174	176	178	180	182	184	186	188	190	192	194
16	3765	3820	3870	3920	3975	4025	4075	4130	4180	4230	4285	4335	4385	4440	4490	4540	4590	4645	4695	4745	4800	4850	4900	4955	5005
18	3740	3790	3840	3890	3940	3995	4045	4095	4145	4200	4250	4300	4350	4405	4455	4505	4555	4610	4660	4710	4760	4815	4865	4915	4965
20	3710	3760	3810	3860	3910	3960	4015	4065	4115	4165	4215	4265	4320	4370	4420	4470	4520	4570	4625	4675	4725	4775	4825	4875	4930
22	3680	3730	3780	3830	3880	3930	3980	4030	4080	4135	4185	4235	4285	4335	4385	4435	4485	4535	4585	4635	4685	4735	4790	4840	4890
24	3635	3685	3735	3785	3835	3885	3935	3985	4035	4085	4135	4185	4235	4285	4330	4380	4430	4480	4530	4580	4630	4680	4730	4780	4830
26	3605	3655	3705	3755	3805	3855	3905	3955	4000	4050	4100	4150	4200	4250	4300	4350	4395	4445	4495	4545	4595	4645	4695	4740	4790
28	3575	3625	3675	3725	3775	3820	3870	3920	3970	4020	4070	4115	4165	4215	4265	4310	4360	4410	4460	4510	4555	4605	4655	4705	4755
30	3550	3595	3645	3695	3740	3790	3840	3890	3935	3985	4035	4080	4130	4180	4230	4275	4325	4375	4425	4470	4520	4570	4615	4665	4715
32	3520	3565	3615	3665	3710	3760	3810	3855	3905	3950	4000	4050	4095	4145	4195	4240	4290	4340	4385	4435	4485	4530	4580	4625	4675
34	3475	3525	3570	3620	3665	3715	3760	3810	3855	3905	3950	4000	4045	4095	4140	4190	4225	4285	4330	4380	4425	4475	4520	4570	4615
36	3445	3495	3540	3585	3635	3680	3730	3775	3825	3870	3920	3965	4010	4060	4105	4155	4200	4250	4295	4340	4390	4435	4485	4530	4580
38	3415	3465	3510	3555	3605	3650	3695	3745	3790	3840	3885	3930	3980	4025	4070	4120	4165	4210	4260	4305	4350	4400	4445	4495	4540
40	3385	3435	3480	3525	3575	3620	3665	3710	3760	3805	3850	3900	3945	3990	4035	4085	4130	4175	4220	4270	4315	4360	4410	4455	4500
42	3360	3405	3450	3495	3540	3590	3635	3680	3725	3770	3820	3865	3910	3955	4000	4050	4095	4140	4185	4230	4280	4325	4370	4415	4460
44	3315	3360	3405	3450	3495	3540	3585	3630	3675	3725	3770	3815	3860	3905	3950	3995	4040	4085	4130	4175	4220	4270	4315	4360	4405
46	3285	3330	3375	3420	3465	3510	3555	3600	3645	3690	3735	3780	3825	3870	3915	3960	4005	4050	4095	4140	4185	4230	4275	4320	4365
48	3255	3300	3345	3390	3435	3480	3525	3570	3615	3655	3700	3745	3790	3835	3880	3925	3970	4015	4060	4105	4150	4190	4235	4280	4325
50	3210	3255	3300	3345	3390	3430	3475	3520	3565	3610	3650	3695	3740	3785	3830	3870	3915	3960	4005	4050	4090	4135	4180	4225	4270
52	3185	3225	3270	3315	3355	3400	3445	3490	3530	3575	3620	3660	3705	3750	3795	3835	3880	3925	3970	4010	4055	4100	4140	4185	4230
54	3155	3195	3240	3285	3325	3370	3415	3455	3500	3540	3585	3630	3670	3715	3760	3800	3845	3890	3930	3975	4020	4060	4105	4145	4190
56	3125	3165	3210	3255	3295	3340	3380	3425	3465	3510	3550	3595	3640	3680	3725	3765	3810	3850	3895	3940	3980	4025	4065	4110	4150
58	3080	3125	3165	3210	3250	3290	3335	3375	3420	3460	3500	3545	3585	3630	3670	3715	3755	3800	3840	3880	3925	3965	4010	4050	4095
60	3050	3095	3135	3175	3220	3260	3300	3345	3385	3430	3470	3500	3555	3595	3635	3680	3720	3760	3805	3845	3885	3930	3970	4015	4055
62	3020	3060	3110	3150	3190	3230	3270	3310	3350	3390	3440	3480	3520	3560	3600	3640	3680	3730	3770	3810	3850	3890	3930	3970	4020
64	2990	3030	3080	3120	3160	3200	3240	3280	3320	3360	3400	3440	3490	3530	3570	3610	3650	3690	3730	3770	3810	3850	3900	3940	3980
66	2950	2990	3030	3070	3110	3150	3190	3230	3270	3310	3350	3390	3430	3470	3510	3550	3600	3640	3680	3720	3760	3800	3840	3880	3920
68	2920	2960	3000	3040	3080	3120	3160	3200	3240	3280	3320	3360	3400	3440	3480	3520	3560	3600	3640	3680	3720	3760	3800	3840	3880
70	2890	2930	2970	3010	3050	3090	3130	3170	3210	3250	3290	3330	3370	3410	3450	3480	3520	3560	3600	3640	3680	3720	3760	3800	3840
72	2860	2900	2940	2980	3020	3060	3100	3140	3180	3210	3250	3290	3330	3370	3410	3450	3490	3530	3570	3610	3650	3680	3720	3760	3800
74	2820	2860	2900	2930	2970	3010	3050	3090	3130	3170	3200	3240	3280	3320	3360	3400	3440	3470	3510	3550	3590	3630	3670	3710	3740

Height in Centimeters

Source: E. DeF. Baldwin and E. W. Richards, Jr., "Pulmonary Insufficiency 1, Physiologic Classification, Clinical Methods of Analysis, Standard Values in Normal Subjects," *Medicine*, vol. 27, p. 243.

Appendix B

DIRECTIONAL TERMS IN ANATOMY

A description of the relative positions of anatomical parts requires the use of directional terms. These terms occur in pairs with members of a pair having opposite meanings. Common directional terms are described below, and most are illustrated in the accompanying figure.

To observe some anatomical structures, it is necessary to view them in sections that have been cut along specific planes through the body or a body part. The planes are described and shown in the figure below.

Plane	Definition
Sagittal Plane	Divides the body into left and right portions and is parallel to the longitudinal axis of the body.
Median Plane	A sagittal plane along the midline of the body dividing the body into equal left and right portions.
Frontal (Coronal) Plane	Divides the body into front and back sections and is parallel to the longitudinal axis to the body.
Transverse Plane	A horizontal plane that divides the body into top and bottom sections and is perpendicular to the longitudinal axis of the body.

Directional term	Definition
Left	To the left of the body (not *your* left, the subject's)
Right	To the right of the body or structure being studied
Lateral	Toward the side; away from the median plane
Medial	Toward the median plane; away from the side
Anterior	Toward the front of the body
Posterior	Toward the back (rear)
Superior	Toward the top of the body
Inferior	Toward the bottom of the body
Dorsal	Along (or toward) the vertebral surface of the body
Ventral	Along (toward) the belly surface of the body
Caudad (caudal)	Toward the tail
Cephalad	Toward the head
Proximal	Toward the trunk (describes relative position in a limb or other appendage)
Distal	Away from the trunk or point of attachment
Visceral	Toward an internal organ; away from the outer wall (describes positions inside a body cavity)
Parietal	Toward the wall; away from internal structures
Deep	Toward the inside of a part; away from the surface
Superficial	Toward the surface of a part; away from the inside
Medullary	Refers to an inner region, or *medulla*
Cortical	Refers to an outer region, or *cortex*

Appendix C

LABORATORY SAFETY PROCEDURES

Because certain experiments in this manual use a variety of laboratory equipment, hazardous chemicals, and new, unfamiliar procedures, it is apparent that you must be alerted to regulations that will promote safety and laboratory harmony. The following laboratory rules must be observed:

1. Avoid tardiness to class. It is during the first 15 minutes of the period that your instructor will outline procedures for the session and alert you to any problems that you may encounter.
2. Keep your work area uncluttered at all times. Unnecessary books, lunches, purses, etc., should be stored somewhere other than on your desktop.
3. Always be aware of the incompatibility of electricity and water when using electronic equipment.
4. Keep the open flame of Bunsen burners out of reach of clothing and hair (beards included).
5. Before attempting to use a piece of equipment, be sure you are completely familiar with its operation. When in doubt ask your instructor.
6. When transferring fluids by pipette use mechanical pipetting devices. Pipetting by mouth is prohibited in any laboratory.
7. When working with chemical agents and body fluids, especially body fluids of other persons, be certain to wear protective disposable gloves to avoid skin contact. Although biologicals that are extracts from human blood products are manufactured under strict controls, it is still prudent to avoid skin contact.
8. When dissecting fresh or preserved specimens, wear protective disposable gloves to protect your skin against pathogens or harsh preservative chemicals.
9. Keep your hands away from your mouth while in the laboratory. Remember to moisten labels with tap water, not your tongue.
10. Sponge down your desktop and wash all laboratory utensils at the end of the period.
11. Return all equipment and glassware to designated storage areas at the end of the period.
12. Before leaving the laboratory at the end of the period, wash your hands with soap and water. If a skin disinfectant is available, use it.